Pedestrian Facilities

ice

Institution of Civil Engineers

publishing

Pedestrian Facilities
Geometric design for safety and mobility
Second edition

John G Schoon
University of Southampton

Published by ICE Publishing, One Great George Street, Westminster, London SW1P 3AA

Full details of ICE Publishing representatives and distributors can be found at:
https://www.icebookshop.com/contact.aspx

Other titles by ICE Publishing:

Designing for Cycle Traffic, International Principles and Practice (2018)
John Parkin ISBN 978-0-7277-6349-5
Practical Road Safety Auditing, Third edition (2015)
Martin Belcher, Steve Proctor and Phil Cook ISBN 978-0-7277-6016-6
Highways, Fifth edition: The Location, Design, Construction and Maintenance of Road Pavements (2015)
Coleman O'Flaherty (ed.) with David Hughes ISBN 978-0-7277-5993-1

www.icebookshop.com

A catalogue record for this book is available from the British Library.

ISBN 978-0-7277-6309-9

Cover photo: People in motion blur, crossing the street near Big Ben and Westminster Palace in London

Commissioning editor: Michael Fenton
Production editor: Madhubanti Bhattacharyya
Marketing specialist: April Asta Brodie

Typeset by Academic + Technical, Bristol
Index created by Matthew Gale
Printed and bound in Great Britain by Bell and Bain, Glasgow

Dedication

This book is dedicated to George and Jenny, my parents, whose walks in the Pentland Hills brought them happy days, and to Marion my wife, for her limitless encouragement and cheerfulness.

Contents

Preface

Engineering the geometric design of facilities to improve the safety and mobility of pedestrians and other road users is the focus of this book. Based on practice in the UK, the emphasis is on numerical, functional aspects of designing footways, uncontrolled and controlled crossings along roads and at junctions, roundabouts and driveways where pedestrian and vehicular traffic interact. These are the places where most pedestrian fatalities and injuries due to collisions with motor vehicles occur. They are the places where people are most concerned about how safe walking is and, therefore, about whether to walk or use some other mode of transport for their desired mobility – or, significantly for many, whether to travel at all, with resulting adverse social, health and economic implications.

Also addressed are pedestrian safety and the comfort of facilities designed to achieve inclusive mobility – embodied in the Equality Act 2010 – and recent developments in computerised demand and design methods. The design of facilities for other non-motorised users is addressed in terms of joint use with pedestrian facilities, and the book complements other works on the needs of bicycle and equestrian travel.

Numerous books and research papers on the geometric design of highways focus on planning and design to ensure collision avoidance and reduction of travel time for

Outline of chapter arrangement

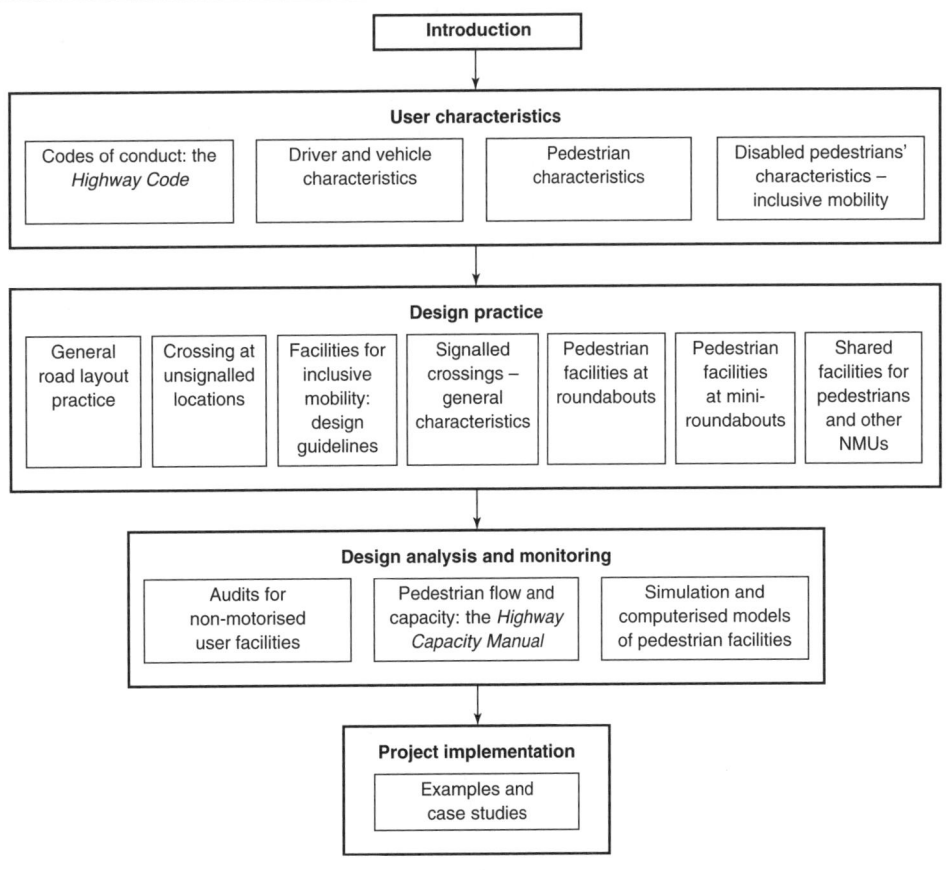

vehicular traffic. This book does the same for pedestrians. Approximately 5000 pedestrians are killed or seriously injured annually in the UK, with a further 20 000 'slightly' injured – and these are casualties which occurred to people who walked. Several million more are reluctant to walk at all because of the perceived, and actual, danger – with adverse effects on health, welfare and social, environmental and economic impacts.

Intended primarily for practitioners and students of civil engineering and design for highways and public spaces, the book is also intended to assist urban planners, architects and the wide range of people who plan and design the public realm in the UK. Inconsistencies in design procedures related to user characteristics, operational procedure and advice are addressed and, because evidence-based analysis and design of pedestrian facilities is often unavailable, it is hoped that the book will assist investigations into the interaction of pedestrians, infrastructure and vehicles. Hopefully, this will help to establish a body of engineering and design knowledge on pedestrian facilities that is comparable to that for vehicular transport.

Engineering and geometric design, which determine the layout and dimensions of the visible elements of the footways and crossings, are integral parts of the aesthetic, social and community aspects of providing for pedestrians, other non-motorised traffic and vehicular traffic. This book, therefore, complements and augments the many texts and guidance documents on these subjects and on the public realm environment and highway engineering. It draws together material from many, often overlapping, sources.

The concentration throughout is on the detailed, numerical, engineering analysis and design of individual elements of a pedestrian's route. This focus recognises that, no matter how well the overall route and its environs are planned and executed, or how attractive aesthetically the route may be, if only one junction is perceived by a pedestrian to be unsafe, the entire route can be rendered unacceptable. A basic premise, therefore, is that the design of pedestrian facilities and their environs should be based on the *pedestrian's* perspective of functionality and safety. Consequently, design procedures should, wherever possible, incorporate evidence-based physical and movement characteristics of people who use the facility, in accordance with established codes of conduct.

In this second edition, developments in cycle facilities and shared use are also addressed – albeit as they relate to pedestrian traffic and rather briefly – in recognition of their own extensive and often complex needs. Cyclist characteristics, flow and capacity, auditing and the combined use of facilities are included. A consideration of shared space and related concepts includes examples from the UK and continental Europe. This aspect of the pedestrian's environment is recognised as contributing to the desirability of walking and 'sojourning' as a part of the social experience. However, in most traditional inner suburbs and rural areas, the emphasis has historically been on design for vehicular traffic, often at the expense of pedestrians' safety and convenience. In such locations, as well as in new developments, improved geometric design will continue to be of value.

The geometry of roads (which include pedestrian ways) where pedestrians interact with vehicles is responsive to and determined by a wide variety of physical, operational and human factors. Physical factors include the carriageway and footway dimensions and horizontal and vertical alignments, the proximity of obstructions, and property lines. Operational features include vehicle speeds and stopping distances, road markings, signs and signals, substantially as described in the *Highway Code* and related publications. Human factors include cultural, physical and mental abilities, reaction times, walking speeds, visual ability and the extent of pedestrians' disabilities or encumbrances. Overlaying these direct variables on the design process are the policy, educational, enforcement and cost factors affecting all road users and responsible authorities – matters that must be left to works other than this one to address.

A wide variety of government agencies and transportation and urban design organisations act as a source of references for guidance during the design process. Often, the bases for such guidance are expressed in qualitative, rather than numerical, terms, and the associated features and dimensions are based on accumulated experience and assumptions about how reasonable road users might behave. Consequently, several of the chapters in this book repeat the main dimensional features and elements given in guidance documents, without further comment on the analytical aspects of how such features were derived. These documents may include dimensions, such as space requirements of pedestrians (especially disabled people and their space needs). They therefore provide a guide to pedestrian facilities design as practised by local, regional and national authorities.

Appropriate detailed geometric design of the facilities along a pedestrian's route is an essential element in a person's deciding whether a walking trip is acceptable or, for many disabled or older people or children (nearly 20% of the population, depending on the precise definition), even possible. Safety – the underlying requirement for acceptable design, as presented here – is regarded as an essential prerequisite for mobility. Of course, safety, mobility and accessibility must exist alongside other key factors, such as urban design, aesthetics, and environmental, community, policy and related issues. Too often, crossings where pedestrians cannot adequately see oncoming vehicles, footways that are too narrow to adequately pass, refuges that are too narrow to accommodate an adult with small children, crossings at junctions with large radii and signals that do not offer sufficient time to cross can intimidate pedestrians and reduce the number of walking trips. The contrast with the comfort and perceived safety of an equivalent trip by car could not be greater. The results of such a contrast are excessive dependence on car use, decreased health, decreased access to essential services and, often, a feeling of exclusion for the large portion of people who must depend on walking as a means of attaining independence and social interaction.

The approach adopted in this book also assumes that, because design of pedestrian facilities is usually inseparable from that of adjacent facilities for vehicular traffic, a basic knowledge of highway engineering is essential. Elements may also assist forensic engineering in that reference is sometimes made to underlying principles where assumptions about UK design practice are not founded on documented evidence. Furthermore, in the light of increasing concerns about the environment and attempts to reduce motor vehicle usage, such evidence may well underlie the reluctance of people to walk rather than drive for many short trips. The author hopes that this book will help to address many of these concerns, thereby improving the safety and attractiveness of walking and, in doing so, improve the sustainability of our increasingly vulnerable environment.

The main features of the approach adopted here are shown in the figure. First, an introduction section describes the current framework for most design of pedestrian facilities in the UK. It outlines features of the context in which geometric design of pedestrian facilities is undertaken; these include highway and street classification, information sources, the analysis and design process and the institutional and legal setting. Importantly, the introduction also indicates some areas where design may be made more responsive to pedestrians' needs. The main features of the book are outlined next.

Chapters 1 to 4: User characteristics

These chapters address the underlying theory and available parameters associated with driver and pedestrian behaviour and design rationale, including material documented in British publications on which the UK's design guidance is based. A brief description of the *Highway Code* is included as an important, although not often quoted, descriptor of required road users' conduct, which physical designs must recognise. The characteristics described in this part also mention shortcomings in current design analysis. This is intended to assist research focusing on improvements to pedestrian facilities as they affect all users' behaviour, mobility needs and associated design parameters.

Chapters 5 to 11: Design practice

This part of the book mostly covers current practice, as described in governmental guidance and advice. Chapters on general road layout practice (including home zones and shared spaces or surfaces), at-grade crossings at unsignalled crossings and an introduction to design practice to ensure inclusive mobility lead to chapters on the geometric design of the pedestrian elements of signalled junctions, roundabouts and facilities shared with other non-motorised users. Owing to extensive coverage in numerous governmental and institutional publications, the material presented is selective and is intended to address key aspects as a guide to more detailed engineering design.

Chapters 12 to 14: Design analysis and monitoring

Processes and methods of evaluating pedestrian schemes in order to improve and refine specific designs are described here. Auditing of pedestrian facilities, capacity estimation and simulation using computerised methods are the main interests.

Chapter 15: Project implementation

This final chapter briefly provides a selection of examples and projects – extensively based on pedestrian facilities – incorporating aspects of analysis that assist in key elements of design and evaluation.

Summary – the way ahead

Provision of adequate pedestrian facilities affects and is affected by many, often conflicting, factors, from the skill of workers repairing a broken paving slab to a national government's transport policy. The focus in this book is on one essential element of the spectrum of factors, i.e. detailed geometric design. This focus is deliberate, perhaps to the excessive exclusion of the other elements. It is hoped that readers will understand the intent of this approach.

About the author

John G Schoon MSCE PhD CEng MICE MASCE FITE PE is a visiting fellow with the Transportation Research Group, University of Southampton, and Emeritus Professor of Civil Engineering, Northeastern University, Boston. His consulting experience of multimodal transportation planning, highway engineering and community involvement brings a wide perspective on transportation engineering and planning. Other experience includes forensic engineering in traffic accident reconstruction, project management and consulting in several countries for governmental agencies and the United Nations. He is the author of numerous project reports and technical papers and several books.

Acknowledgements

As with the first edition of this book, many people have assisted in the preparation of this second edition. In the basic outline of essential topics, Professor Nick Hounsell, Kit Mitchell and Richard Hall were particularly helpful. In the overall approach to civil engineering, transport and their importance to the quality of everyday life, Harlan Lunn, Don Courten, Harold Michael and Herbert Levinson were early guides. Goff Jacobs and David Jeffery provided detailed comments, John Parkin provided a wider perspective on bicycle facilities, Oksana Koltsova assisted with descriptions of simulation methods and Andrew Baxter of BWB Consulting and Andrew Thompson of Buchanan and Associates helped with case studies. In preparing the book, Michael Fenton, commissioning editor, and the suggestions of Madhubanti Bhattacharyya, production editor, and Alison Lees, copy-editor, made the book eminently more readable. Members of the Transportation Research Group, at the University of Southampton, helped to make the presentations consistent with the needs of professionals, educators and people concerned with the far-reaching aspects of pedestrians' needs.

The subject of pedestrian safety and mobility varies widely; opinions between policy makers, users and the professional community often differ. Hopefully, the approach adopted in this book reflects an objective introduction to these differences, so enabling analysis, design and planning to respond effectively to specific cases. Of course, any errors or omissions in the book are mine and I would appreciate being told about them, together with suggestions for possible improvements.

Conversion of units

$1 \text{ in} = 25.4 \text{ mm}$ $1 \text{ mm} = 0.034 \text{ in}$

$1 \text{ ft} = 0.305 \text{ m}$ $1 \text{ m} = 3.281 \text{ ft}$

$1 \text{ ft}^2 = 0.093 \text{ m}^2$ $1 \text{ m}^2 = 10.76 \text{ ft}^2$

Cautionary note

This book, particularly the sections on design guidance and advice, includes edited excerpts from government and other documents. This edition focuses on the geometric design of pedestrian facilities, often embedded in other aspects of analysis and design. The descriptions in this book should be read, and professional experience and judgement employed, in the knowledge that relevant, updated and additional source documents, including those not mentioned in this book, must be consulted and appropriate advice sought in any application and actual design task and its context. In general, the design guidance and advice sources described apply to cases in England; there may be differences in those applicable in Northern Ireland, Scotland and Wales.

Schoon, John G
ISBN 978-0-7277-6309-9
https://doi.org/10.1680/pfse.63099.001

Introduction

The context of designing pedestrian facilities, in particular, the geometric design associated with highways and streets, is the focus of this introduction. Most collisions that significantly affect pedestrians occur where pedestrians interact with vehicles. An increasing number, but still a very small proportion, of these vehicles are bicycles and mobility scooters. Many pedestrian casualties also occur as the result of inadequate surfacing of footways and footpaths, where the geometry of the location is not a key issue.

The geometric layout of a location, affecting the ability of drivers and pedestrians to interact safely, is therefore one of the greatest concerns for a person undertaking a walking trip and, especially for older and disabled people, may affect the decision over whether to make the trip at all. The implications of this for encouraging walking instead of using a motor vehicle, and for the mobility of those unable to use a vehicle, are clear – safety and convenience for pedestrians are essential for fostering prevailing social, economic and environmental policies. This introduction addresses key points of the framework for the engineering and geometric design of pedestrian facilities – a necessary consideration in the successful application of design principles and practice.

Encouraging walking

Walking significantly improves personal health due to the physical activity involved. As well as enhancing environmental quality and sustainability, improvements in pedestrian facilities often make access to essential activities and services easier and safer.

Numerous government and other documents address the need for improved pedestrian facilities in order to increase the use of walking as a major transport mode. As reported in the *Manual for Streets* (*MfS*) (DfT, 2007) for local travel, walking is the most important means of transport and offers the greatest potential for replacing shorter car trips, particularly those under 2 km. It states that authorities should identify the network of routes and locations where the needs and safety of pedestrians will be given priority and describes measures that will support that objective. Further advice and information, with an emphasis on sustainability in planning decisions, is contained in *National Planning Policy Framework: An Overview* (DCLG, 2012).

Authorities should pay particular attention to the design of new developments to help promote walking as a prime means of access. The publications *Planning for Walking* (CIHT, 2015a) and *Designing for Walking* (CIHT, 2015b) elaborate on many of the key issues and provide helpful advice to local authorities and non-government organisations.

Measures to encourage walking include widening footways and reallocating road space. Road crossings should be provided that give pedestrians greater priority at traffic signals and avoid long detours and walking times, indirect footbridges or underpasses. The Department for Transport emphasise the key roles of walking and cycling as the main modes used for short trips, especially to local facilities, such as shops and schools, and in providing access to public transport for longer journeys.

The role of improving walking and cycling facilities in helping to create liveable towns and cities and to promote health and social inclusion has not always been recognised. Around 25% of all journeys, and 80% of journeys of less than one mile, are made on foot (DfT, 2004). Over the past decade, the average distance travelled both on foot and by cycle has remained fairly constant; more than 40% of trips are under 2 miles and a quarter of car journeys are shorter than this. Many of these shorter journeys could be made on foot or by cycle (DfT, 2007).

Pedestrian safety

Of key concern in any system is the matter of safety – in the present context, an essential component of mobility. One cannot expect a product or service that does not provide the necessary level of safety to be accepted as a normal part of human activity. Pedestrian facilities are no exception. Unless footways, crossings, platforms and the many varieties of areas used by pedestrians are safe (and perceived to be safe), it is unreasonable to expect them to be adequately used. In many cases, walking as a mode of transport competes with the perceived safety and comfort of cars. Therefore, the need for safety in the pedestrian environment is of paramount importance in encouraging walking.

Figure I.1 Reported fatalities for the four largest casualty groups, per billion miles travelled (DfT, 2014, Chart 7)
© Crown Copyright, 2014
This information is licensed under the Open Government Licence v3.0. To view this licence, visit http://www.nationalarchives.gov.uk/doc/open-government-licence/ **OGL**

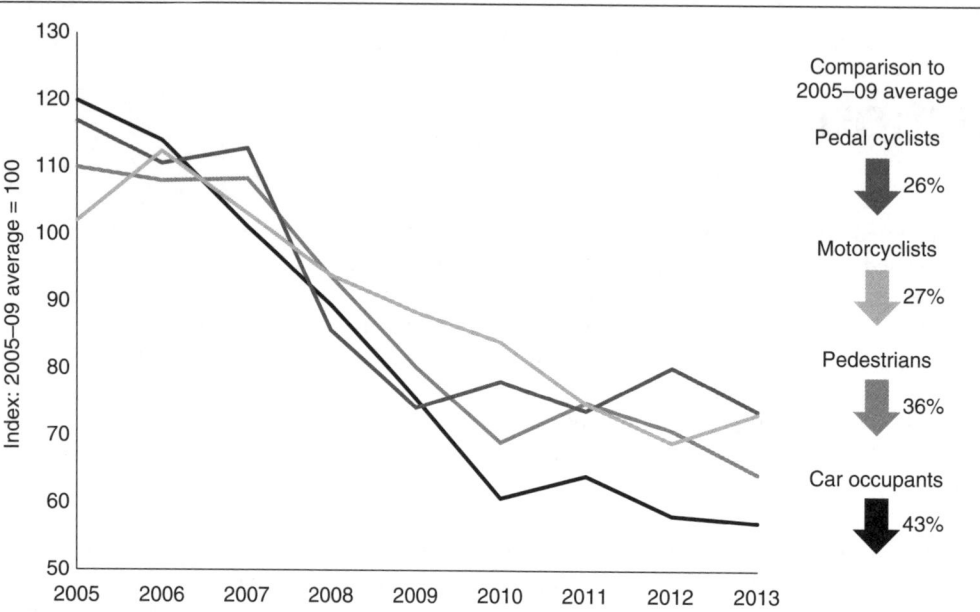

In general, there has been a reduction in the fatality rate for four main casualty groups over the past decade, as shown in Figure I.1, with a broadly continuous drop between 2005 and 2010 and subsequent slowing down or flattening of the rate. The biggest overall improvement in fatality rate has been seen for car occupants. Rates for pedal cyclists and motor-cyclists saw slower decreases and were around a quarter less than the 2005–09 average in 2013 (DfT, 2014). A later break-down of key pedestrian casualty data, shown in Table I.1, indicates an increase in fatalities of 10% from 2015 to 2016 – more than any other mode. Further details are shown in Figure I.2.

With regard to casualties by time of day (DfT, 2015), the following statistics are worth noting.

- Between 2009 and 2013, roughly 40% of pedestrian casualties (fatalities or serious injuries) occurred during Monday to Friday between 8 a.m. and 9 a.m. or between 3 p.m. and 8 p.m. During these hours, there is likely to be a large number of pedestrians walking to and from work or school.
- Pedestrian casualties (fatalities or serious injuries) (13%) also occur on Friday or Saturday night between 5 p.m. and midnight; many people go out to social events during

Table I.1 Fatalities by road user type in 2016 (DfT, 2017a)

	Fatalities in 2016	% share in 2016	% change since 2015	% change from 2010–2014 average
Car occupants	816	46	↑8	↓1
Pedestrians	448	25	↑10	↑6
Motorcyclists	319	18	↓13	↓10
Pedal cyclists	102	6	↑2	↓9
Other	107	6	↑4	↑19

© Crown Copyright, 2017
This information is licensed under the Open Government Licence v3.0. To view this licence, visit http://www.nationalarchives.gov.uk/doc/open-government-licence/ **OGL**

Figure I.2 Pedestrian casualty characteristics 2016 (DfT, 2017a)
© Crown Copyright, 2017
This information is licensed under the Open Government Licence v3.0. To view this licence, visit http://www.nationalarchives.gov.uk/doc/open-government-licence/ **OGL**

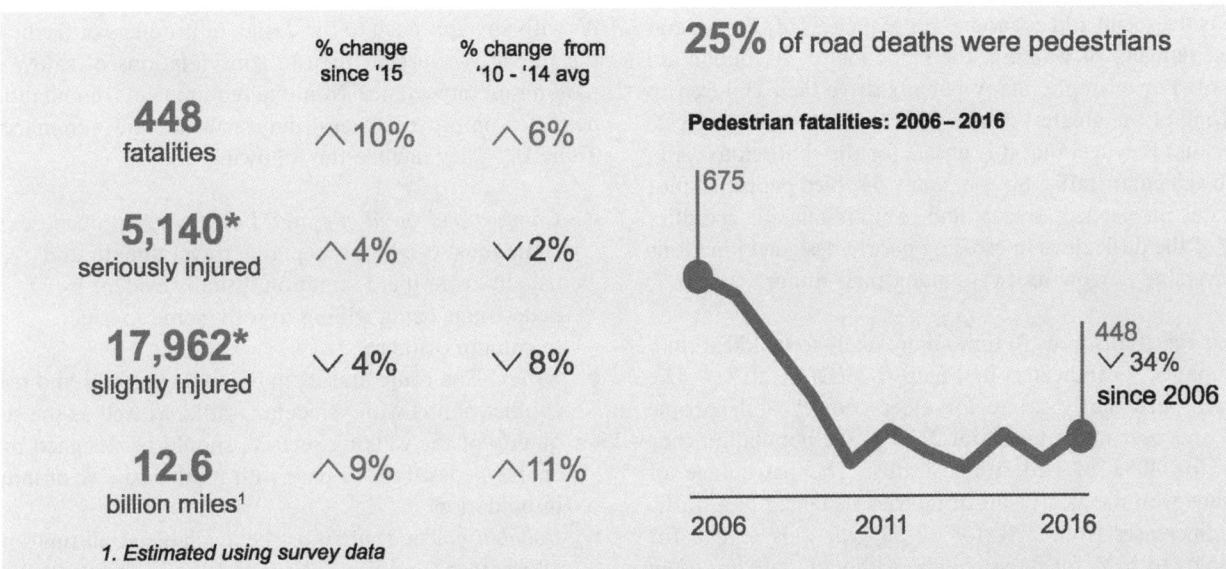

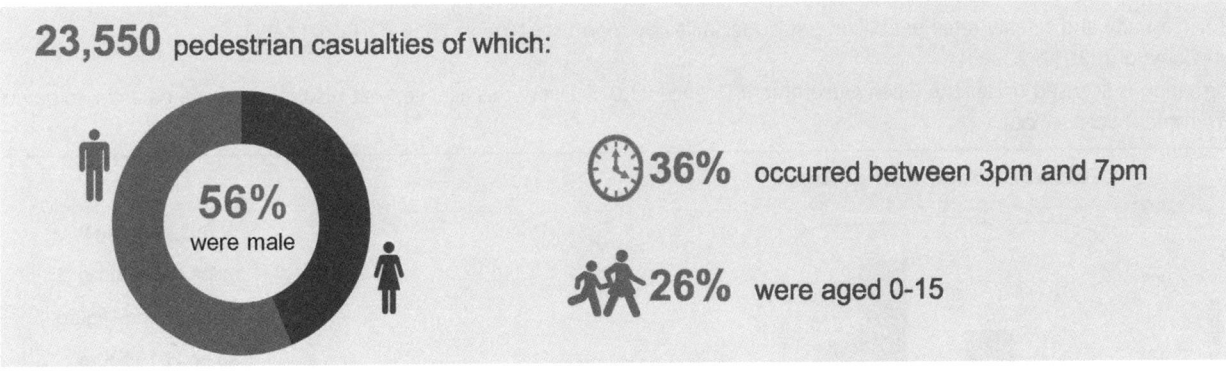

these hours. In particular, some of the destinations will be parties, bars, restaurants and pubs in urban areas. These destinations will often offer alcohol and will be more easily accessed on foot than by driving. This increases the number of pedestrians travelling and increases the number of them under the influence of alcohol.

■ A further 5% of pedestrian casualties (fatalities or serious injuries) occur in the early morning hours of Saturday and Sunday (from midnight to 4 a.m.).

In 2016, car occupants accounted for 46% of road deaths, pedestrians 25%, motorcyclists 18% and pedal cyclists 6%.

In 2016, pedestrian fatalities, after falling for some years, rose to 448. This has remained much the same since 2010. Another 5140 pedestrians were seriously injured, and 17 962 were slightly injured – this last number being the fewest recorded. It should be noted that pedestrian casualties decreased by 34% over this period yet the equivalent reduction for vehicle occupants was 49%.

Collisions in 2016 between pedestrians and bicycles resulted in 3 fatalities, 112 serious injuries and 312 slight injuries. Those involving motor scooters in 2016 resulted in 12 seriously injured and 32 slightly injured pedestrians (DfT, 2017a). Reporting methods for collisions involving motor scooters are under review, and the conditions under which a motor scooter user is considered a pedestrian and entitled to operate on the footway have yet to be resolved.

As well as the social and economic consequences of the real and perceived dangers of walking, the reactions of the public are significant. For example, many parents drive their children to school (one of the greatest causes of urban area traffic congestion) because they feel that it is unsafe for the children to walk, owing to vehicular traffic. So too, many disabled people cannot participate in needed social and welfare-related activities because of the difficulties in crossing poorly designed junctions or in traversing narrow footways along their routes.

In 2016, pedestrians were 16 times more likely to be killed than car occupants, as indicated in Figure I.3 (DfT, 2017a). The effects are particularly severe for older people; while people aged 60 and over make up about 21% of the population they account for 40% of pedestrian deaths. The percentage of pedestrians who die as a result of injuries sustained in a traffic accident increases from $2\frac{1}{2}$% for 30–50 year olds to 6% for 70 year olds to 10% for 80 year olds. Although data are often difficult to compare, the decline in pedestrians' deaths might have come about because people walk less.

In a European context, the share of pedestrians' fatalities in the UK is above average – for example, more than twice as high as for the Netherlands, as shown in Figure I.4 (European Commission, 2015).

As with any approach to the design of products or facilities that the public is expected to use, considerations of safety are of paramount importance. Minimal requirements, highlighting the need for improved pedestrian facilities, are summarised in Table I.2. They include the following.

- *Comfort and speed of travel.* For any given purpose, the route must offer an acceptable travel time to and usually from the destination, usually evident by pedestrians being willing to walk some specific maximum distance.
- *Safety.* The route and its individual crossings and other conflict points with vehicular traffic, as well as the surface quality of the walking surface, should be designed to enable pedestrians to pass with minimum risk of injury or intimidation.
- *Reliability.* The route must be available at all times and in all weather conditions if it is to be used frequently.
- *Security.* Freedom from crime or intimidation is a key requirement for a route to be used reliably.

Figure I.3 Casualty and fatality rates per billion passenger miles by road user type in 2016 (DfT, 2017a)
© Crown Copyright, 2017
This information is licensed under the Open Government Licence v3.0. To view this licence, visit http://www.nationalarchives.gov.uk/doc/open-government-licence/ **OGL**

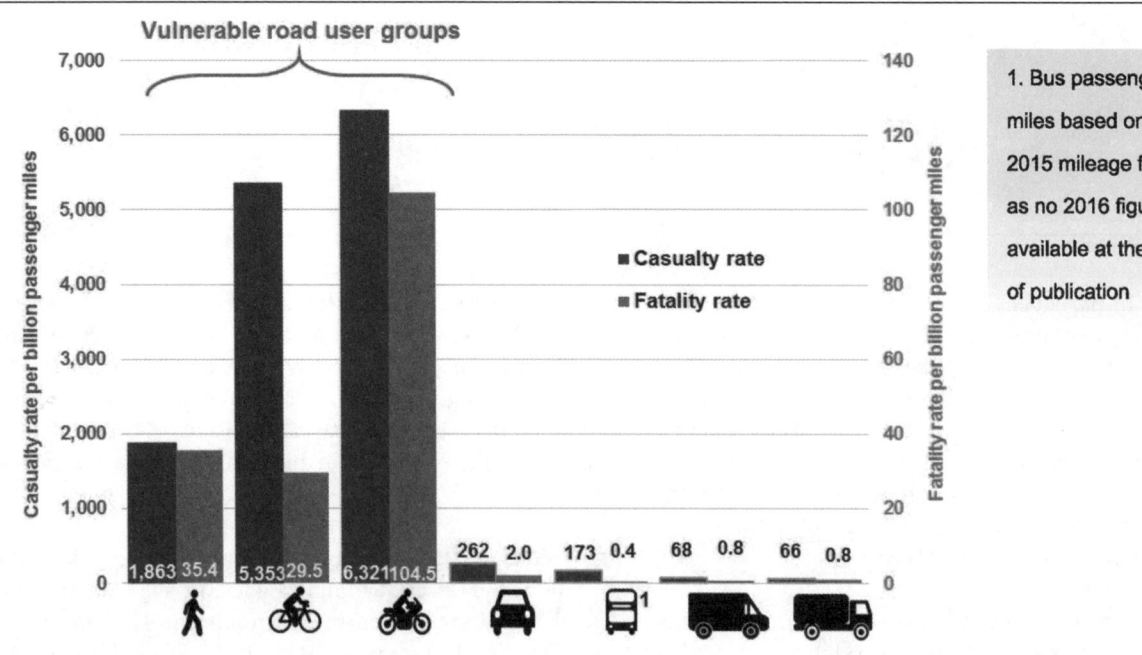

Figure I.4 Pedestrians' share of all road deaths (European Commission, 2015)

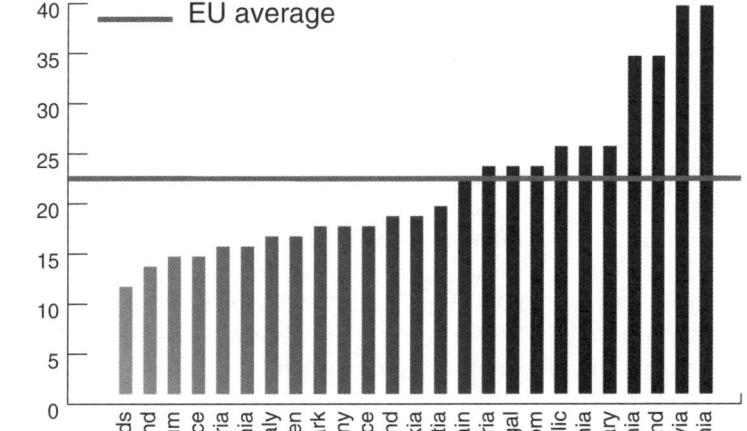

NL	11%	HR	19%
FI	13%	ES	22%
BE	14%	BG	23%
FR	14%	PT	23%
AT	15%	UK	23%
SI	15%	CZ	25%
IT	16%	EE	25%
SE	16%	HU	25%
DK	17%	LT	34%
DE	17%	PL	34%
EL	17%	LV	39%
IE	18%	RO	39%
SK	18%	EU	22%

Table I.2 Criteria for pedestrian facilities design

Item	Importance of item to pedestrian activity	Design characteristics	Example of essential features for pedestrians
Comfort and speed (travel time)	The travel experience, including that in inclement weather, should not cause undue discomfort or damage to clothes, etc.	Design and maintenance features of footpaths to afford protection from inclement weather and safety from falls	Prompt clearance of ice and snow, surface maintenance, adequate drainage at junctions and other locations
	Unless a trip can be made within a reasonable time to allow for other essential activities, it is less likely to be considered	Acceptable overall speed (travel time) for activity contemplated	Direct route to convenience shops with frequent pedestrian-priority crossings
Safety	Physical and operational elements of the trip must be consistent and compatible with the abilities of the range of users to travel with an acceptable risk level	Design features that minimise pedestrians' risk of injury when negotiating the footway and road crossings when walking to a destination	Clear, sufficiently wide footways available, with no parked vehicles, adequately maintained surfaces and sufficient crossing opportunities
Reliability	Route features that cause obstructions or design features that cause intermittent delays or danger can render the route unreliable and reduce its likelihood of use by pedestrians	Provide unhindered access to destination, including lighting and all-weather access	Avoidance of junction crossings, narrow footpaths, any road crossing having no pedestrian priority, temporary parking and inadequate temporary facilities at repair works
Security	Route must be free from real or perceived threat of crime or intimidation from others	Neighbourhood environment should offer good visibility to services, dwellings and destinations	Narrow, long alleyways, obscured entrances and exits should be avoided in the overall design

Additionally, a pleasing environment and a route of interest will greatly add to the likelihood of a route being used. Of the matters mentioned here, appropriate geometric design of the facilities is a key factor in assuring convenience and safety. Clearly, concerns about reliability and security (freedom from crime) are also of importance and matters related to these issues will often overlap those of safety and convenience.

Information sources – geometric design details and guidance

Traditionally, the analysis and geometric design of pedestrian facilities within the boundaries of highways has mainly been addressed from the perspective of drivers and motor vehicles. Although pedestrian facilities and related infrastructure within the highway boundaries are also included, the related characteristics and design procedures are addressed in much less quantitative detail. For example, numerical analysis methods for avoiding inter-vehicle collisions at junctions are provided, based on driver characteristics. Yet no equivalent analysis is detailed for avoiding collisions between vehicles and pedestrians based on relevant pedestrian characteristics.

In addition to publications that describe the philosophy and approaches to design of pedestrian facilities, several publications provide engineering design parameters and their quantitative values. These latter publications are used most extensively here to assist in describing current analysis and design process.

Design Manual for Roads and Bridges (DMRB) (DfT, 2018)

This manual is issued by the Department for Transport (DfT) and is applicable to trunk roads in urban and rural areas, including motorways. The manual is issued in 15 volumes and sections, which are updated when necessary to reflect new guidance. Volume 1, Section 0, Part 1 provides a full definition of its scope and application. An alphanumeric index lists the documents in the *DMRB*, their references and date of issue. A more complete list is presented in the appendices to the *DMRB*. However, the series most relevant to geometric design include

- *HA Advice Notes – Highways*
- *HD Standards – Highways*
- *SH Scottish Technical Memoranda – Highways*
- *TA Advice Notes – Traffic Engineering and Control*
- *TD Standards – Highways.*

Design Bulletin 32 (DB32): Residential Roads and Footpaths: Layout Considerations (DETR, 1992)

This publication, issued by the Department of the Environment and Department of Transport in 1992, provided advice for local planning and highway authorities in preparing local guidance and in specifying their requirements for adoption. *DB32* has been superseded by the *Manual for Streets (MfS)* but is referred to here because many existing streets have been designed in accordance with its guidance, and coordination with future design will usually be necessary. *DB32* was accompanied by a companion guide entitled *Places, Streets and Movement: A Companion Guide to Design Bulletin 32 – Residential Roads and Footpaths* (DETR, 1998).

Manual for Streets (DfT, 2007)

This document, published by the Department for Transport (DfT) supersedes *DB32* (DETR, 1992) and its companion guide *Places, Streets and Movement* (DETR, 1998), which are now withdrawn in England and Wales. The *Manual for Streets (MfS)* comprises technical guidance and does not set out any new policy or legal requirements.

MfS focuses on lightly trafficked streets, but many of its principles may be applicable to other types of street, for example high streets and lightly trafficked lanes in rural areas. *MfS* does not apply to the trunk road network, although many of the principles may be applicable in individual cases. It is the responsibility of users of *MfS* to ensure that its application to the design of streets is appropriate. The policy, legal and technical frameworks are generally the same in England and Wales, but the manual makes clear where differences are made.

Inclusive Mobility (DfT, 2005)

Detailed design configurations and dimensions for a wide range of transport features in order to foster the concept of inclusive mobility for disabled people throughout the transport system are described in *Inclusive Mobility* (DfT, 2005). Part III of the Disability Discrimination Act 1995 gives disabled people a right of access to goods, facilities, services and premises. These rights were phased in between 1996 and 2004. Since 1996, it has been unlawful for service providers to treat disabled people less favourably than other people for a reason related to their disability.

Traffic advisory notes and leaflets and local transport notes

Traffic advisory notes, traffic advisory leaflets (TALs) and local transport notes (LTNs) are numbered and address particular topics. Findings or guidance in such publications may be incorporated into *DMRB* or *MfS* advice and guidance.

Chartered Institution of Highways and Transportation (CIHT)

Formerly the Institution of Highway and Transportation (IHT), the CIHT has produced many publications relating to transport. The book *Transport in the Urban Environment* (IHT, 1997) contains extensive coverage of a wide variety of design issues, including those affecting pedestrians. The publication

Guidelines for Providing for Journeys on Foot (IHT, 2000) addresses detailed aspects of the pedestrian environment, many of which are now further addressed in *Planning for Walking* (CIHT, 2015a) and *Designing for Walking* (CIHT, 2015b).

Other publications

Other publications related to the provision of pedestrian facilities produced in the last twenty years or so, which have guided the design of many existing facilities, include

- *Road Safety Good Practice Guide* (DfT, 2006)
- *Better Streets, Better Places* (ODPM, 2003)
- *Home Zone Guidelines* (IHIE, 2002)
- *Link and Place: A Guide to Street Planning and Design* (Jones *et al.*, 2007)
- *Guidance on the Appraisal of Walking and Cycling Schemes* (DfT, 2017b).

The publication *Link and Place* (Jones *et al.*, 2007) reflects the interaction between street-related activities and through movements. Also, *Street Design for All: An Update of National Advice and Good Practice* (Davis, 2014) provides an overview of many aspects of the street environment and of pedestrian and vehicular traffic interaction, including descriptions of practical applications.

Most aspects of safety sources of information on parameters often overlap and are occasionally at variance with each other. To some extent, the use of a particular design approach and documented guidance depends on the location and jurisdiction of the authority undertaking the design.

Definitions of terms

A number of sources provide definitions of pedestrian-related facilities. Because these can be quite precise in meaning and application they are quoted in the following sections.

Statutory Definitions, England and Wales (CIHT, 2015a)

- *Highway.* A way over which the public have the right to pass and repass.
- *Carriageway.* A highway or part of a highway over which the public have right of way for vehicles. Note that the right of way for vehicles does not detract from the established right of pedestrians to cross the carriageway.
- *Footway.* That part of a highway that also comprises a carriageway over which the public have right of way on foot only.
- *Footpath.* A highway over which the public have right of way on foot only, not being a footway (i.e. not being adjacent to a carriageway). A way that is exclusively for passage on foot is a footpath.
- *Bridleway.* A highway over which the public have right of way on foot and horseback.

- *Cycle track.* A way that is part of a highway over which the public have right of way on pedal cycles, other than pedal cycles that are motor vehicles, with or without the right of way on foot.
- *Road.* In England and Wales, a road is any length of highway or any other road to which the public have access and includes bridges over which a road passes.

DMRB (DfT, 2018)

- *Trunk roads* are roads that connect various points in the United Kingdom. They are divided into A roads, B roads and motorways (designated M roads). Many aspects of design for A and B roads are relevant to local roads, especially at junctions; thus, the *DMRB* standards are of importance to design of pedestrian facilities.

Some further detail on these definitions is as follows (DETR, 1992).

- *Carriageways* are those parts of the access roads that are intended primarily for use by motorised vehicles.
- *Shared surfaces* are those parts of shared-surface roads that are intended for use by pedestrians and vehicles.
- *Footways* are those parts of access roads that are intended for use by pedestrians and that are generally parallel with the carriageways and separated by a kerb or by a verge and a kerb.
- *Footpaths* are those pedestrian routes that are located away from carriageways and not associated with routes for motor vehicles.
- *Cycle tracks* are routes that are intended for use by pedal cyclists with or without rights of way for pedestrians.
- *Segregated cycle tracks* are cycle tracks adjacent to footpaths or footways and separated from them by such features as a kerb, verge or white line.

The various road classifications of trunk, primary, district and local distributor and residential access roads are shown in Figure I.5.

Figure I.5 Urban road network definitions, based on *DB32* (DETR, 1992)

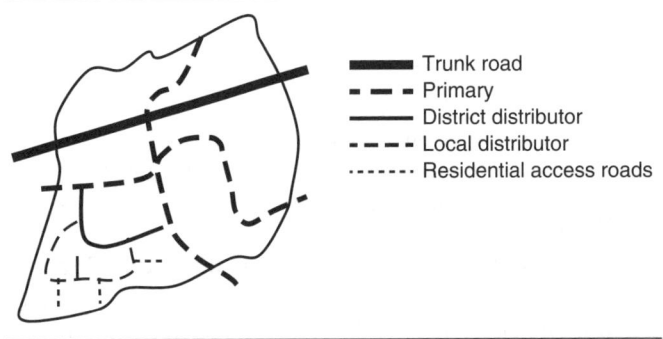

Trunk road
Primary
District distributor
Local distributor
Residential access roads

Definitions of the urban road network, based on definitions in *DMRB* (DfT, 2018) and *DB32* (DETR, 1992)

- *Primary distributors* form the primary network for the town as a whole. Longer-distance traffic are carried through and within the town by such roads.
- *District distributors* distribute traffic between the residential, industrial and principal business districts of the town. They link the primary network with the residential areas.
- *Local distributors* distribute traffic within districts in residential areas and link district distributors with residential roads.
- *Residential access roads* link dwellings and their associated parking areas and open spaces to distributors. These roads are referred to in *DB32* as residential roads.
- Residential roads and driveways take a number of forms.
  - *Access roads* are residential roads with footways that may serve up to around 300 dwellings and provide direct access to dwellings. Where minor or major access roads are referred to, it is assumed that they may serve up to around 100 and 100–300 dwellings, respectively.
  - *Shared-surface roads* are residential roads without footways that may serve up to approximately 50 houses.
  - *Shared driveways* are unadopted paved areas that may serve the driveways of up to five houses.
  - *Driveways* are unadopted paved areas that provide access to garages and other parking spaces within the curtilage of an individual house.

Classification of highways and streets: *MfS* (DfT, 2007)

Consistent with a revised approach to the classification of roads and streets, *MfS* notes that traditional road hierarchies (district distributor, local distributor, access road, etc.) have been based on traffic capacity. It states further that street character types in new residential developments should be determined by the relative importance of both their place and movement functions and gives examples of more descriptive terminology such as

- high street
- main street
- shopping street
- mixed-use street
- avenue
- mews
- lane
- courtyard.

This list is not exhaustive; whatever classification is used, it is important that the street character type is well-defined, whether in a design code or in some other way.

Analysis and design of pedestrian facilities: definitions

The term 'analysis', as used here, means the estimation of physical and operational values used in the geometric design of a location, such that pedestrians may safely interact with the physical and operational infrastructure and vehicular traffic using the roadway. The term 'geometric design' refers essentially to the visible features of the facility, such as the physical configuration and dimensions that affect the way in which users approach and negotiate the roadway (including pedestrian facilities) and its environs.

The geometric design, therefore, affects and is affected by: drivers' characteristics; users' speeds; the abilities of pedestrians moving along the footway and crossing the carriageway; vehicle stopping distances and related safety and capacity aspects; and relative spacing of features affecting visibility of and to the various users. The geometry is an integral feature of the aesthetic aspect of the design and largely affects the safety and attractiveness of a particular part of a pedestrian's journey. However, as presented here, the major emphasis is on the engineering and safety aspects.

For a specified vehicle speed and range of pedestrians' abilities, such as walking speed and reaction times, and the influence area of traffic in the vicinity of a crossing, the geometric characteristics which the analysis should assist in determining include the following.

- *Physical relationship* of elements of the physical design to each other and to the adjacent environment.
- *Width of the crossing*, which is affected by the volume and level of service afforded people wishing to cross, as well as the dimensions of the approaches and exits from the crossing.
- *Dimensions*, such as vertical clearance, gradient and surface configuration of footways and the carriageway; this is largely determined by physical dimensions, movement characteristics, of pedestrians and other non-motorised users (NMUs) and by infrastructure characteristics, such as drainage and emergency vehicle access.
- *Distance and location of sight lines* (may also be referred to as intervisibility lines or forward visibility lines) to and from the crossing from critical pedestrian and vehicle approach locations, to ensure that the pedestrians and drivers negotiating the crossing can see each other in sufficient time for the pedestrian either to cross safely or to decide not to undertake the crossing.
- *Location of the crossing*, with respect to the characteristics of the locality, approach and exit locations and the geometric characteristics (primarily the width, number of lanes, presence of a central reservation, and the horizontal and vertical alignment of the road).
- *Characteristics of design components*, including configuration and dimensions of such elements as pedestrian refuges, build-outs at corners, median islands and raised tables at crossings.

- *Control devices*, characteristics of traffic signals, markings, signs and other operational features.

Regardless of the design components, which may include curve radii, central refuges, build-outs and other features, the analysis provides essential input for determining how and where such features should be placed, as well as their dimensions. Where physical constraints or policy considerations are important, the analysis may be used as a useful tool in objectively establishing appropriate speed limits or other traffic controls.

Pedestrian facilities analysis and design process

Overview

Depending on the location of the particular crossing, the political jurisdictions (such as local authorities, county councils or national governments), the classification of the roads involved, special designations such as 'home zones', and a number of other determinants, the appropriate design guidance documentation varies. However, the principles of dynamic behaviour of pedestrians and drivers will remain the same; these principles govern the geometry of the physical layout and operations.

A flow chart of the major steps in the engineering analysis and design of a specific crossing is shown in Figure I.6. It should be noted that the extensive involvement of community and related issues is not detailed here, but will normally constitute a large part of the total effort. The flow chart shows (1) the location, jurisdiction and highway classification and (2) the characteristics of the related infrastructure as the major considerations in determining the appropriate advice and guidance documents (3) that apply to the design of the element in question. The analysis and geometric design itself (4), which is the emphasis here, comprises an iterative process between analysis and design of the physical and operational elements appropriate to the specific location. Finally, segments of the auditing process (5) provide a review of the completed design, including a Stage 3 audit after the design has been implemented and has been in operation for a representative period, during which its performance can be evaluated and adjustments made, within the context of engineering and geometric design matters. Item 4 is the main focus of the chapters in this book.

Legal framework

An extensive legal framework affecting highways and planning exists. Relevant aspects of legislation affecting geometric design standards are described briefly next (DfT, 2007, Section 3.9). Matters related to new highways, disabled pedestrians and inclusive mobility, and policy, legal and technical frameworks are summarised in the appendix to this chapter.

Assessment and monitoring of the pedestrian environment

Within the framework of providing adequately designed geometric features of pedestrian facilities, the issues of most concern associated with the pedestrian environment are addressed through several approaches. These may be typified by the steps outlined in *Guidelines for Providing for Journeys on Foot* (IHT, 2000) and *Designing for Walking* (CIHT, 2015b), described briefly next.

Background

In addition to monitoring the quality of walking trips, local authorities need to monitor key aspects of the pedestrian environment. Some monitoring functions, such as inspecting footways, are legal requirements or are required for best value. Others will be identified in the objectives and targets of the local transport plan or, if one exists, local pedestrian charter. The costs and practicalities of the monitoring must be considered, yet it should support objectives and not distort priorities.

Quality indicators and walking

The following best value indicators are relevant to walking

- public rights of way that are signposted and easy to use
- pedestrian crossings with facilities for disabled people
- lack of damage to roads and pavements
- street lights working as planned
- street cleaning surveys.

Data requirements

Data are likely to be readily available without additional surveys for the following aspects of the pedestrian environment

- footway conditions
- street lighting defects
- pedestrian casualties
- street crime statistics, available from the local police.

Other indicators

The quality of the pedestrian environment can usefully be monitored at reasonable cost by such indicators as

- length of streets with 20 miles per hour limits
- extent of traffic calming
- number of complaints received about pedestrian conditions
- delays at signal-controlled pedestrian crossings
- response time to repair footway, crossing or lighting effects.

The public should be consulted on the indicators to be monitored in accordance with best practice or guidelines and kept informed of the results, and their views monitored.

Figure I.6 The pedestrian facilities engineering and geometric design process

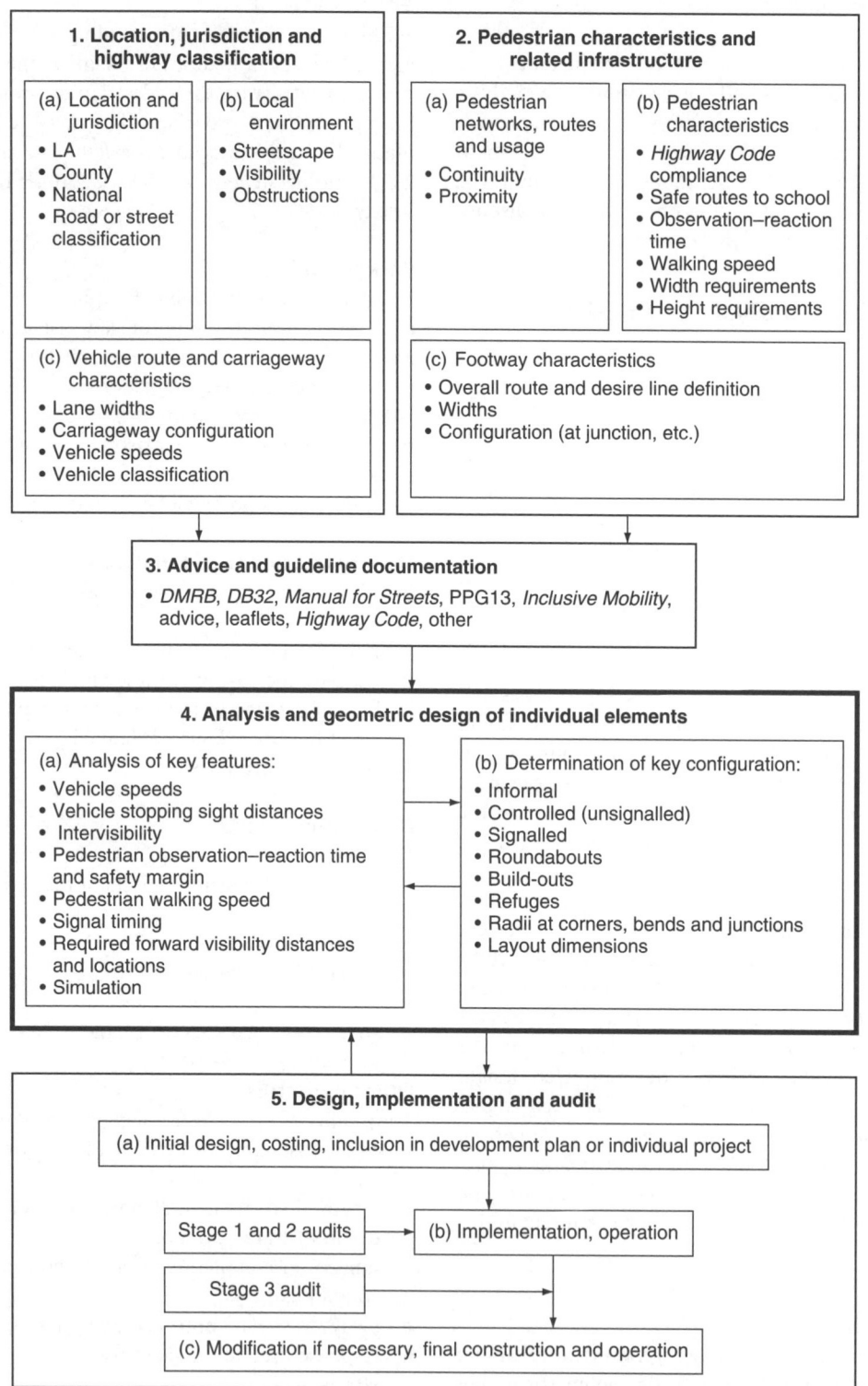

Quantitative methods

In addition to the mostly qualitative methods mentioned, level of service, as detailed in the *Highway Capacity Manual* (TRB, 2016) (see Chapter 13), and other quantitative evaluation methods, including the Pedestrian Environment Review System (PERS) developed as software by the Transport Research Laboratory (TRL, 2009) (see Chapter 14) provide a means of identifying the quality of the pedestrian environment and, hence, act as indicators of how it may be improved.

For illustrative purposes, the criteria used in the PERS (TRL, 2009) modelling system, which ranks many features of a pedestrian's route in terms of attractiveness, provides a useful example. These parameters include surface quality, lighting, conflict with traffic, pedestrian facilities, obstructions, cleaning and drainage, crossing type, deviation from desired route, refuge quality, etc., rest points, public spaces and permeability, as well as other factors such as road safety and public transport waiting areas, bus stops and taxi ranks, public spaces and interchange spaces. Other features can include the aesthetic quality of the walking area, the absence of noise, personal security and the interest value of features along the route. It is up to the designer to provide sound, documented analysis and engineering solutions, while working closely with the many interest groups that combine to encourage walking.

Areas of potential improvement to analysis and design of pedestrian facilities

As will be noted in the various chapters, certain aspects of the analysis and design approach and procedures do not appear to be based on firm evidence and, in some respects, are conflicting. These conflicts are mentioned at the relevant locations in the text. As noted in *MfS*:

> Research carried out in the preparation of *Manual for Streets* indicated that many of the criteria routinely applied in street design are based on questionable or outdated practice.
>
> (DfT, 2007)

Concerns have been expressed about the need for improvement in the resources devoted to the design of pedestrian facilities. Of note are the areas of professional skills, the guidance on analysis and design of the facilities themselves and coordination between designers, formulators of the *Highway Code* and enforcement agencies.

Professional skills

Several reports and studies have addressed the matter of inadequate skills and training of professionals in the design of pedestrian facilities. For example, a report for the UK Parliament (House of Commons Transport Committee, 2009) has addressed many of the perceived problems with the pedestrian environment. This report addresses issues of national targets, planning, the role of local authorities and specific design approaches. A selection of points particularly relevant to the approach and conduct of geometric design of pedestrian facilities is as follows.

- Skills of professionals involved with the design of pedestrian facilities are not consistent nationally and are too often dependent on selected individuals being responsible for the design process.
- Awareness of the knowledge and skills required in providing for pedestrians appears to be lacking.
- There is a chronic shortage of professionals, particularly in senior positions, with the appropriate outlook and technical skills.
- There is an unconscious or conscious bias of senior officers towards car use or, at least, towards concentrating on problems of traffic congestion and vehicle collisions.
- Professionals need to pay much more attention to the needs of pedestrians and to the aesthetics of the street. These matters must be addressed in university and other courses and in continued professional development courses, including the training of officers by local authorities.
- There is a need for consolidated advice, which would be used by the professionals concerned, including highway engineers, planners and designers. A number of professional institutes and bodies have produced guidance on different aspects of access and mobility, but these need drawing together into a coherent *Pedestrians' Digest* that would allow practitioners to understand and develop the technical assessment methods that work well.
- There is a need to address the problems of standards being over-rigorously followed, by challenging and justifying existing practices, and encouraging a body of fully trained professionals, who provide tailored solutions based on their professional judgement.

Apparent deficiencies in design practice

As well as addressing some of the wider issues of designing pedestrian facilities, the report *Paving the Way* (CABE, 2002) lists several areas specifically related to geometric design, ranging from the *Highway Code* to design manuals, such as *DMRB*, local transport notes (LTNs), and traffic advisory leaflets (TALs). Selected areas of concern include the following.

- For priority junctions, there is a lack of emphasis on design for pedestrians and cyclists (*DMRB*, TD 42/95 Part 6; DfT, 2018).
- Pedestrian crossing design provides an underlying assumption that crossings should be made so as not to inconvenience the driver; this should be revised to emphasise pedestrian traffic flow and comfort.

The highway design guidance and specification for national trunk roads given in *DMRB* includes some specific guidance for urban roads. CABE (2002, p. 38) comments on this guidance are that the nature of proposed changes:

… requires greater balance in the geometric and layout presumptions for the design of trunk roads (or other principal routes) in urban areas. Needs to recognise the greater requirements for pedestrian access across and along the street. Express clearly whether and how these guides can be used on local roads and what discretion can be used in their interpretation.

Recommendation 2 of *Paving the Way: How we Achieve Clean, Safe and Attractive Streets* (CABE, 2002, p. 37) states that:

Highway Authorities should, under best value, establish an audit trail for design decisions affecting the streetscape, to show how design guidance, people's needs and vehicle movements have been accommodated. The aim of this audit trail is twofold.

- To enable local authorities to demonstrate that they have acted 'reasonably', if faced with liability claims. Often the main reason cited for the use of a standard (vehicle priority) solution is that such a solution, however inappropriate to the local context, is thought to offer a line of defence in a court of law. […]
- To act as a record of aims and objectives about both how and why decisions were made, and to demonstrate that all users (including the disabled) have been considered in decision-making.

A specific failure in the guidance given in TD 42/95, Paragraph 6, *Geometric Design of major/minor priority junctions* is noted:

[The] assumption is that the movement of traffic is the main concern in junction design: 'advantage is that through traffic on the major road is not delayed.' [There is] no mention of advantages for pedestrians or cyclists.

(CABE, 2002)

A proposed solution is offered:

Revise with a more balanced view highlighting advantages for all road users in urban areas where pedestrians and cyclists needs are higher.

(CABE, 2002)

Regarding visibility splays (clearance of objects to enable a direct view of oncoming traffic), it is stated in *Designing Streets for People* (ICE, 2000) that:

Visibility splays greatly influence the street scene. Guidance on dimensions for visibility splays is given in *Places, Streets and Movement* [(DETR, 1998)] but the rationale behind the figures is not. Therefore, there is no basis for a professional to understand, interpret or question the guidance. However, if the scientific justification behind the figures [were] published, the guidance could be understood and applied to local circumstances.

Attention has also been drawn by Schoon and Hounsell (2006) to a number of apparent deficiencies in documented design practices for pedestrians. These include the lack of coordination with instructions to pedestrians detailed in the *Highway Code*, unrealistic assumptions about the physical movement characteristics of pedestrians, particularly disabled people, and deficiencies in identifying and quantifying intervisibility in the calculation of safe stopping sight distances. These concerns are addressed in greater detail, especially in Chapters 3 to 6.

In terms of university education, although highway engineering is a common topic in most undergraduate and postgraduate courses, the percentage of attention given to pedestrian characteristics and design needs is usually limited. The dynamic characteristics of the pedestrian–vehicle interaction mostly features pedestrians' walking speed and vehicle gap acceptance only. These parameters are usually based on observations of actual behaviour, ignoring the fact that such observations are biased because people who are intimidated and reluctant to walk, or are otherwise affected by the presence of traffic, are not included in the observation samples.

Addressing inconsistencies and shortcomings in the analysis and design of pedestrian facilities – the need for research

Throughout this book, where design guidance appears to be inconsistent or inadequate, the reader's attention is drawn to it. This is intended to alert designers and researchers to the need for further research in particular subject areas – although many additional areas of pedestrian activity also exist. Compared with the quantity of research conducted in relation to vehicular movement and related infrastructure, the amount of research devoted to documenting the characteristics and needs of pedestrians is minimal. The need, in particular, for evidence-based guidance, based on rigorous statistical methods, is paramount, and key issues are addressed in a number of publications, including Hewson (2007).

The context: quality of the pedestrian experience

As mentioned earlier, the engineering, geometric design of individual elements of the pedestrian environment is but one, albeit vitally important, aspect of the pedestrian's transport system –

and indeed of that of transport users in general, for all are pedestrians at some point in their journeys.

It is sometimes easy to forget when conducting a detailed design that many other factors affect a person's decision to walk instead of using a car for a trip. Although not discussed in further detail here, it is essential to be aware of some of the key factors to be considered in the overall effort to encourage walking.

Summary and conclusions

The material in this introduction has attempted to establish a context within which the engineering and geometric design of pedestrian facilities can be viewed. The many references and documents mentioned should indicate the extent of this task as a basis for the more detailed approach adopted in the ensuing chapters. Finally, because of the extremely wide scope of the topic of designing for pedestrians, the material in this book should be taken as an introduction and guide to a highly complex and often not well-defined area of engineering, for which there is much scope for improvement.

REFERENCES
NOTE
Parts of the Design Manual for Roads and Bridges (DMRB) may be referenced here in terms of the DMRB alphanumeric index system (TA, TD, etc.). See the appendices to DMRB for details of this system.

CABE (Commission for Architecture and the Built Environment) (2002) *Paving the Way: How We Achieve Clean, Safe and Attractive Streets.* Thomas Telford, London, UK.

CIHT (Chartered Institute for Highways and Transport) (2015a) *Planning for Walking.* CIHT, London, UK.

CIHT (2015b) *Designing for Walking.* CIHT, London, UK.

Davis CJ (2014) *Street Design for All: An Update of National Advice and Good Practice.* Public Realm Information and Advice Network (PRIAN), London, UK.

DCLG (Department for Communities and Local Government) (2012) *National Planning Policy Framework: An Overview.* DCLG, London, UK.

DETR (Department of the Environment, Transport and the Regions) (1992) *Design Bulletin 32 (DB32) Residential Roads and Footpaths: Layout Considerations*, 2nd edn. DETR, London, UK.

DETR (1998) *Places, Streets and Movement: A Companion Guide to Design Bulletin 32 – Residential Roads and Footpaths.* DETR, London, UK.

DfT (Department for Transport) (2004) LTN 1/04: *Policy, Planning and Design for Walking and Cycling.* DfT, London, UK.

DfT (2005) *Inclusive Mobility: A Guide to Best Practice on Access to Pedestrian and Transport Infrastructure.* DfT, London, UK.

DfT (2006) *Road Safety Good Practice Guide.* DfT, London, UK.

DfT (2007) *Manual for Streets.* Thomas Telford, London, UK.

DfT (2014) *Reported Road Casualties Great Britain: 2013.* DfT, London, UK.

DfT (2015) *Facts on Pedestrian Casualties.* DfT, London, UK.

DfT (2017a) *Reported Road Casualties Great Britain 2016.* DfT, London, UK.

DfT (2017b) *Guidance on the Appraisal of Walking and Cycling Schemes.* DfT, London, UK.

DfT (2018) *Design Manual for Roads and Bridges (DMRB).* DfT, London, UK.

European Commission (2015) *Road Safety in the European Union: Trends, Statistics and Main Challenges.* European Commission, Brussels, Belgium.

Hewson P (2007) Evidence-based practice in road casualty reduction. *Injury Prevention* **13**: 291–292.

House of Commons Transport Committee (2009) *Ending the Scandal of Complacency: Road Safety Beyond 2010. Second Special Report of Session 2008–09.* The Stationery Office, London, UK.

ICE (Institution of Civil Engineers) (2000) *Designing Streets for People.* Thomas Telford, London, UK.

IHIE (Institute of Highway Incorporated Engineers) (2002) *Home Zone Guidelines.* IHIE, London, UK.

IHT (Institution of Highways and Transportation) (1997) *Transport in the Urban Environment.* IHT, London, UK.

IHT (2000) *Guidelines for Providing for Journeys on Foot.* IHT, London, UK.

Jones PM, Boujenko N and Marshall S (2007) *Link and Place: A Guide to Street Planning and Design.* Landor, London, UK.

ODPM (Office of the Deputy Prime Minister) (2003) *Better Streets, Better Places: Delivering Sustainable Residential Environments: PPG 3 and Highway Adoption.* The Stationery Office, London, UK.

Schoon JG and Hounsell N (2006) Access and mobility design policy for disabled pedestrians at road crossings: exploring issues. *Transportation Research Record* **1956(1)**: 76–85.

TRB (Transportation Research Board) (2016) *Highway Capacity Manual.* National Academy of Sciences, Washington, DC, USA.

TRL (Transport Research Laboratory) (2009) *Pedestrian Environment Review System (PERS).* TRL, Crowthorne, UK.

Appendix: A legal framework (based on DfT (2007) and CIHT (2015))

An extensive legal framework affecting highways and planning exists. A selection of relevant aspects of legislation affecting geometric design standards are described briefly in this appendix.

Matters related to new highways, disabled pedestrians and inclusive mobility, and policy, legal and technical frameworks are summarised in this section.

Design standards

The term 'highway', as defined earlier, is a legal term and encompasses roads and streets. The government does not set design standards for new highways – they are set by the highway authority of the roads and streets for which it has responsibility.

Highways England acts as the highway authority for trunk roads within England, and its standard is the *Design Manual for Roads and Bridges* (*DMRB*) (DfT, 2018), the use of which is not mandatory for other highway authorities.

Although some trunk roads could be classed as streets, most are roads, where the movement takes precedent over place-making. *DMRB* does not, therefore, normally provide appropriate design standards for streets, particularly for lower-volume and residential streets. Most local highway authorities have developed their own standards in the past, mainly for use by developers of residential developments. Some of these standards, particularly those that have been published in recent years, have taken on place-making or street issues but many have not.

New highways (DfT, 2007)

Planning legislation encompasses all development of land. The principal act is the Town and Country Planning Act 1990, as amended by the Planning and Compulsory Purchase Act 2004. New highways are included but not maintenance and improvement within existing highway boundaries. Planning policies for new strategic highways and other transport infrastructure of regional significance will be set by regional spatial strategies. Planning policies for most streets will be set by local development plans. Both the regional spatial strategies and the local development plans have been established under the Planning and Compulsory Purchase Act 2004. Supplementary planning documents expand or add details to policies in development plan documents, in the form of design guides, area development briefs, master plans or issue-based documents. Local authority street design guidelines can, therefore, be incorporated into local development plans as supplementary planning documents. Related issues include community strategies, general principles of highways, classification of highways – described earlier – the creation of new highways, and risk and liability.

The duty of the highway authority to maintain the highway is set out in Section 41 of the Highways Act 1980 and case law has clarified the law in several instances, as explained later in this appendix.

Regarding legal aspects of design, few cases are brought for design defects, as it is difficult for claimants to show that a faulty design has caused an accident, rather than another cause – for example, user error, unusual weather, a construction defect or poor maintenance. If faced with such a case, it is a defence for highway authorities to show the following.

- New works were properly designed.
- The authority did not inadvertently trap road users into danger.
- The authority complied with appropriate standards or guidance; however, the emphasis is on the word 'appropriate' – standards and guidance would not be applied without proper consideration of local circumstances.

Disabled pedestrians and inclusive mobility legislative background

The Disability Discrimination Act 1995 was introduced in 1996; Part III gave disabled people a right of access to goods, facilities, services and premises. The original 1995 Act was modified and extended by the introduction of the Disability Discrimination Act 2005 in 2006. This was replaced by the Equality Act 2010, which specifies the public sector's specific duties.

The Acts are also supported by codes of practice issued by the former Disability Rights Commission. The Disability Rights Commission was closed in 2007 and replaced by the Equality and Human Rights Commission. The documents are still available (EHRC, 2016). Further details are contained in Chapter 8.

Policy, legal and technical frameworks

There is a complex set of documentation relating to streets; it is important to distinguish between legal requirements, government policy and technical guidance – whether produced by the government or by various other bodies. Design rules and guidance can be categorised by type, as shown in Table I.3.

Risk and liability (DfT, 2007, Section 2.6)

A major concern expressed by some highway authorities when considering more innovative designs, or designs that are at variance with established practice, is whether they would incur a liability in the event of damage or injury.

This can lead to an over-cautious approach, where designers strictly comply with guidance, regardless of its suitability, and to the detriment of innovation. This is not conducive to creating distinctive places that help to support thriving communities.

In fact, imaginative and context-specific design that does not rely on conventional standards can achieve high levels of safety. This issue was explored in some detail in the publication

Table I.3 Design rules and guidance

Type	Responsible body	Example
Legal framework statutes and regulations	Parliament (interpreted by the courts)	*Statutes* Highways Act 1980 Transport Act 2000 Town and Country Planning Act 1990 Planning and Compulsory Purchase Act 2004 Disability Discrimination Act 2005 *Regulations* Traffic Signs Regulations and General Directions 2002 The Building Regulations 2000
Case law		*Gorringe* v *Calderdale Metropolitan Borough Council* (2004)
Government policy	Department for Transport (DfT)	The Future of Transport White Paper 2004 'Roads' circulars
	Department for Communities and Local Government (DCLG)	Planning Policy Guidance PPG13 Planning Policy Statement PPS3 Planning circulars
Government guidance	Multi-departmental	*Manual for Streets*
	DfT	*Cleaner, Safer, Greener* – 'how to' guides *Guidance on the Preparation of Local Transport Plans* *Traffic Signs Manual* Local transport notes Traffic advisory leaflets Other companion guides to planning policy statements Building regulations approved documents
	Commission for Architecture and the Built Environment (CABE)	*By Design: Urban Design in the Planning System – Towards Better Practice* *Better Places to Live by Design: Companion Guide to PPG3*
	Welsh government	Design Guidance, Active Travel (Wales) Act 2013
Government research reports	DfT	Road safety research reports
	Office of the Deputy Prime Minister (ODPM)	*Better Streets, Better Places: Delivering Sustainable Residential Environments: PPG 3 and Highway Adoption* *Sustainable Communities: Building for the Future*
Local policies	Local authorities	Local offices' local transport plans Public realm strategy Local development framework Local development documents Supplementary planning documents
Design standards and guidance	DfT	Design standards and guidance *Design Manual for Roads and Bridges* (*DMRB*)
	Local authorities	Local design standards and guidance Streetscape manuals (may also have policy function as local development documents)

Table I.3 Continued

Type	Responsible body	Example
Other research and guidance	Various bodies including: UK Roads Board (reports to the UK Roads Liaison Group, and advisory body representing all UK highway authorities)	*Well-maintained Highways: Code of Practice for Highway Maintenance Management*
	The Countryside Agency	*Rural Routes and Networks: Creating and Preserving Routes that are Sustainable, Convenient, Tranquil, Attractive and Safe*
	Institutions (ICE, CIHT, IHIE, IHCSS, etc.)	*Transport in the Urban Environment* *Traffic Calming Techniques* *Home Zone Guidelines*
	Transport Research Laboratory (TRL)	*The Design of Roundabouts* Large library of research reports, e.g. *TRL Report 184 – Accidents at Three-arm Priority Junctions on Urban Single-carriageway Roads*
	British Standards Institution	BS5906: 2005. *Waste Management in Buildings: Code of Practice*
	CABE	*By Design: Urban Design in the Planning System – Towards Better Practice* *Better Places to Live by Design: Companion Guide to PPG3* *Paving the Way: How we Achieve Clean, Safe and Attractive Streets*
	English Partnerships	*Urban Design Compendium*
	Historic England	*Streets for All*

Note: Although relevant to ongoing design matters, the documents and sources of information listed in this table may be updated, and the latest versions should be consulted.

Highway Risk and Liability Claims (UK Roads Board, 2009). Most claims against highway authorities relate to alleged deficiencies in maintenance. The duty of the highway authority to maintain the highway is set out in Section 41 of the Highways Act 1980, and case law has clarified the law in this area.

The most recent judgement of note is *Gorringe* v *Calderdale Metropolitan Borough Council* (2004), where a case was brought against a highway authority for failing to maintain a 'SLOW' marking on the approach to a sharp crest. The judgement confirmed a number of important points, including the following.

■ The authority's duty to 'maintain' covers the fabric of a highway but not signs and markings.
■ There is no requirement for the highway authority to 'give warning of obvious dangers' (Lord Hoffman, para 10).
■ Drivers are 'first and foremost responsible for their own safety' (Lord Scott, para 76).

Some claims for negligence or failure to carry out a statutory duty have been made under Section 39 of the Road Traffic Act

1988, which places a general duty on highway authorities to promote road safety. In connection with new roads, Section 39(3)(c) states that highway authorities,

in constructing new roads, must take such measures as appear to the authority to be appropriate to reduce the possibilities of such accidents when the roads come into use.

The *Gorringe* v *Calderdale* judgment made it clear that Section 39 of the Road Traffic Act 1988 cannot be enforced by an individual, however, and does not form the basis for a liability claim.

Most claims against an authority are for maintenance defects, claims for design faults being relatively rare.

Design of highways

Advice to highway authorities on managing their risks associated with new designs is given in Chapter 5 of *Highway Risk and Liability Claims* (UK Roads Board, 2009). In summary, this advises that authorities should put procedures in place that allow rational decisions to be made with the minimum of bureaucracy, and that create an audit trail that could subsequently be used as evidence in court. Suggested procedures (which accord with those set out in Chapter 3 of *MfS*) include the following key steps.

- Set clear and concise scheme objectives.
- Work up the design against these objectives.
- Review the design against these objectives through a quality audit.

The adequate documentation of design investigations and the matters to be considered and described for possible future review are also of concern during the auditing phase of design and implementation. This is therefore described in greater detail later.

Creating new off-carriageway routes (DfT, 2012)

Local authorities can create new cycle tracks under Section 65(1) of the Highways Act 1980. New footpaths, bridleways or restricted byways can be created under Sections 25 or 26 of the Highways Act 1980, either through agreement or by using compulsory powers. A route might also be dedicated for use as a cycle track if there is a precedent of sustained use by cyclists. Creating a cycle track on a new alignment might require planning approval if it is outside the highway boundary.

Allowing cycling in pedestrianised areas (DfT, 2012)

If cyclists are to be permitted to use a pedestrianised (i.e. vehicle restricted) area, they need to be given legal authority to do so. This can be achieved by amending the order extinguishing the right to use vehicles on a highway under Section 249 of the Town and Country Planning Act 1990 or Section 1 or 6 of the Road Traffic Regulation Act 1984, whichever is appropriate. Further advice is provided in TAL 9/93 *Cycling in Pedestrian Areas* (DfT, 1993).

REFERENCES FOR LEGAL FRAMEWORK

Brown M (1995) *The Design of Roundabouts*. Her Majesty's Stationery Office, London, UK.

BSI (2005) BS5906: 2005. Waste management in buildings: code of practice. BSI, London, UK.

CABE (Commission for Architecture and the Built Environment) (2000) *By Design: Urban Design in the Planning System – Towards Better Practice*. DETR, London, UK.

CABE (2001) *Better Places to Live by Design: A Companion Guide to PPG3*. Department for Transport, Local Government and the Regions, London, UK.

CABE (2002) *Paving the Way: How We Achieve Clean, Safe and Attractive Streets*. Thomas Telford, London, UK.

CIHT (Chartered Institute for Highways and Transport) (2015) *Planning for Walking*. CIHT, London, UK.

DCLG (Department for Communities and Local Government) (2011a) *Planning Policy Guidance 13: Transport*. DCLG, London, UK.

DCLG (Department for Communities and Local Government) (2011b) *Planning Policy Statement 3: Housing*. DCLG, London, UK.

DfT (Department for Transport) (1993) TAL 9/93: *Cycling in Pedestrian Areas*. DfT, London, UK.

DfT (2003) *Traffic Signs Manual*. The Stationery Office, London, UK.

DfT (2007) *Manual for Streets*. Thomas Telford, London, UK.

DfT (2009) *Guidance on Local Transport Plans*. DfT, London, UK.

DfT (2012) LTN 1/12: *Shared Use Routes for Pedestrians and Cyclists*. The Stationery Office, London, UK.

DfT (2018) *Design Manual for Roads and Bridges (DMRB)*. DfT, London, UK.

EHRC (Equality and Human Rights Commission) (2016) https://www.equalityhumanrights.com/en/advice-and-guidance/equality-act-codes-practice (accessed 01/10/2018).

English Partnerships (2007) *Urban Design Compendium*, 2nd edn. Llewelyn-Davies, London, UK.

Gorringe v *Calderdale Metropolitan Borough Council* [2004] UKHL 15.

Historic England (2018) *Streets for All*. Historic England, Swindon, UK.

HMG (Her Majesty's Government) (1980) Highways Act 1980. The Stationery Office, London, UK.

HMG (1984) Road Traffic Regulation Act 1984. The Stationery Office, London, UK.

HMG (1988) Road Traffic Act 1988. The Stationery Office, London, UK.

HMG (1990) Town and Country Planning Act 1990. The Stationery Office, London, UK.

HMG (1995) Disability Discrimination Act 1995. Part III. The Stationery Office, London, UK.

HMG (2000) The Building Regulations 2000. The Stationery Office, London, UK.

HMG (2000) Transport Act 2000. The Stationery Office, London, UK.

HMG (2002) The Traffic Signs Regulations and General Directions 2002. The Stationery Office, London, UK.

HMG (2004) Planning and Compulsory Purchase Act 2004. The Stationery Office, London, UK.

HMG (2005) Disability Discrimination Act 2005. The Stationery Office, London, UK.

HMG (2010) Equality Act 2010. The Stationery Office, London, UK.

IHCSS (Institute of Highways and County Surveyors Society) (2005) *Traffic Calming Techniques*. IHCSS, London, UK.

IHIE (Institute of Highway Incorporated Engineers) (2002) *Home Zone Guidelines*. IHIE, London, UK.

IHT (Institution of Highways and Transportation) (1997) *Transport in the Urban Environment*. IHT, London, UK.

ODPM (Office of the Deputy Prime Minister) (2003a) *Better Streets, Better Places: Delivering Sustainable Residential Environments: PPG 3 and Highway Adoption*. The Stationery Office, London, UK.

ODPM (2003b) *Sustainable Communities Plan*. ODPM, London, UK.

Summersgill I, Kennedy JV and Baynes D (1996) *TRL Report 184 – Accidents at Three-arm Priority Junctions on Urban Single-carriageway Roads*. TRL, Crowthorne, UK.

The Countryside Agency (2015) *Rural Routes and Networks: Creating and Preserving Routes that are Sustainable, Convenient, Tranquil, Attractive and Safe*. Thomas Telford, London, UK.

UK Roads Board (2009) *Highway Risk and Liability Claims: A Practical Guide to Appendix C of The Roads Board Report 'Well-maintained Highways Code of Practice for Highway Maintenance Management'*, 2nd edn. UK Roads Board, London, UK.

UK Roads Liaison Group (2005) *Well-maintained Highways: Code of Practice for Highway Maintenance Management*. The Stationery Office, London, UK.

Welsh Government (2013) Design Guidance Active Travel (Wales) Act 2013. The Welsh Government, Cardiff, UK.

Pedestrian Facilities, Second edition

Schoon, John G
ISBN 978-0-7277-6309-9
https://doi.org/10.1680/pfse.63099.019

Chapter 1
Codes of conduct – the *Highway Code**

Interaction between regulations and the geometric design of highways, including pedestrian facilities, can assist compliance and the general conduct of users. In particular, consideration of rights of way and priorities can emphasise design features that provide adequate warning, sight distances, and related safety features. The *Highway Code* (DfT, 2017) also provides a list of the Acts and Regulations that form the *Code*'s legal basis.

As stated in the introduction to the *Highway Code*:

> This *Highway Code* applies to England, Scotland and Wales. The *Highway Code* is essential reading for everyone.

> The most vulnerable road users are pedestrians, particularly children, older or disabled people, cyclists, motorcyclists and horse riders. It is important that all road users are aware of the *Code* and are considerate towards each other. This applies to pedestrians as much as to drivers and riders.

Despite the existence of a 'code of conduct', road users do not necessarily follow the instructions – either knowingly or through error. Consequently, the design of highway features, including pedestrian facilities, sometimes recognises this fact. Such cases are mentioned throughout the chapters where appropriate.

1.1. Highway user operations: the *Highway Code*

1.1.1 Role of the *Code* in geometric design

The behaviour of drivers and pedestrians in the safe and efficient movement of traffic is beneficially affected by an appropriate code of conduct. In addition, such codes attempt to maintain legal norms of behaviour, such as priority at junctions. The geometric design of the facility, therefore, plays an important role in ensuring that the *Code*'s requirements can be met, although it cannot guarantee compliance. The *Code* applies to all roads and streets; no distinction is made between trunk roads and residential streets or neighbourhood areas, except where specifically noted.

1.1.2 Objectives and practice

A key objective in designing the geometric features of highways is to assist road users in following the rules of the *Highway Code* as follows.

- Assist in safely achieving the mandatory and advisory movement priorities of the various users.
- Indicate or channel the direction of movement of the various users.
- Enable these priorities and directions to take place so that each involved user has adequate time to complete the necessary tasks.

It follows, therefore, that, for the *Code* to be effective, two key factors must obtain.

- The rules of the *Code* must be comprehensive, clear and unambiguous.
- The users must follow the rules laid down in the *Code*.

In practice, neither of these two factors occur; the rules of the *Code* have been shown to have deficiencies and ambiguities, and road users do not always follow the advice and requirements of the *Code*. Consequently, design, as practised by official agencies, has tended to recognise the popularity of motor travel by encouraging its unimpeded flow. The safety of vulnerable users has often been achieved by restricting their movement and encounters with vehicles – often by establishing rules and channelling their movements to encourage their safety and minimise legal liability of an authority in case of accidents.

The deficiencies in the *Code*, the lack of compliance with its provisions by some road users and the response of official agencies and institutions in establishing guidance for geometric design make the task of the responsible engineer and designer challenging in attempting to encourage walking. Regarding implications for the design of pedestrian facilities, the following quotation (CABE, 2002) is relevant to the engineering and design process:

> Recommendation 12: the *Highway Code* should be rewritten to place greater emphasis on the multiple use of streets, rather than mainly vehicle movement.

Also, the *Highway Code* is

… addressed mainly at drivers (or people learning to drive) and says little about what are the rights and duties of pedestrians

… which only increases their sense of vulnerability on the street.

(CABE, 2002)

Furthermore, official coordination of facilities design with the *Code*'s rules is not mentioned in either the *Manual for Streets* (*MfS*) (DfT, 2007) or the *Design Manual for Roads and Bridges* (*DMRB*) (DfT, 2018). Crucially, neither is it mentioned as an aspect of safety auditing for non-motorised users (DfT, 2005), where provision of new or updated facilities should be checked to ensure that provisions of the *Code* for pedestrians can be met. Therefore, the needs and abilities of users related to their actions to conform to the *Code*'s requirements might not have been established, either qualitatively or quantitatively.

In terms of illustrating the advice in the *Code*, several instances exist where graphics are not provided or, if provided, do not clarify and emphasise a specific rule. For example, where turning vehicles are required to give priority to pedestrians crossing the various arms of a junction, only one out of sixteen possible pedestrian movements is illustrated. Also, illustrations of turning movements of vehicles in several instances show no pedestrian facilities or presence, where these could be critical to a driver's decision-making at a junction (Schoon, 2017).

The following sections are intended to highlight a selection of the *Highway Code*'s rules that affect the geometric design of pedestrian facilities (the selection here being subject to additions if necessary by individual designers) and also indicate some of the areas where the *Code* appears to be inconsistent or incomplete, and may warrant clarification.

1.2. *Highway Code* rules

A total of 307 rules are provided in the *Highway Code*, divided into various sections. Those rules that appear to relate directly to the geometric design of pedestrian facilities are listed by section, each rule preceded by its number. For selected rules, a note has been added that relates to the rule itself or to some consideration affecting the understanding of the various road users, primarily pedestrians.

1.2.1   Rules for pedestrians
1.2.1.1   General guidance
1. Pavements (including any path along the side of a road) should be used if provided. Where possible, avoid being next to the kerb with your back to the traffic. If you have to step into the road, look both ways first. Always show due care and consideration for others.

2. If there is no pavement, keep to the right-hand side of the road so that you can see oncoming traffic. You should take extra care and

- be prepared to walk in single file, especially on narrow roads or in poor light
- keep close to the side of the road.

It may be safer to cross the road well before a sharp right-hand bend so that oncoming traffic has a better chance of seeing you. Cross back after the bend.

3. Help other road users to see you. Wear or carry something light-coloured, bright or fluorescent in poor daylight conditions. When it is dark, use reflective materials (e.g. armbands, sashes, waistcoats, jackets, footwear), which can be seen by drivers using headlights up to three times as far away as non-reflective materials.

4. Young children should not be out alone on the pavement or road (see Rule 7). When taking children out, keep between them and the traffic and hold their hands firmly. Strap very young children into pushchairs or use reins. When pushing a young child in a buggy, do not push the buggy into the road when checking to see if it is clear to cross, particularly from between parked vehicles.

NOTE
This reminder, to be careful to avoid pushing a buggy into the path of any oncoming traffic, means that the pedestrian's sight line distance may be restricted – a point raised in later chapters where intervisibility between pedestrians and vehicles is discussed.

5. Organised walks. Large groups of people walking together should use a pavement if available; if one is not, they should keep to the left. Look-outs should be positioned at the front and back of the group, and they should wear fluorescent clothes in daylight and reflective clothes in the dark. At night, the look-out in front should show a white light and the one at the back a red light. People on the outside of large groups should also carry lights and wear reflective clothing.

6. Motorways. Pedestrians **MUST NOT** be on motorways or slip roads except in an emergency (see Rule 271 and Rule 275).

Laws: Road Traffic Regulation Act 1984 Section 17, The Motorways Traffic (England and Wales) Regulations 1982 Regulation 15(1)(b) and The Motorways Traffic (Scotland) Regulations 1995 Regulation 13.

1.2.1.2   Crossing the road
7. The Green Cross Code. The advice given below on crossing the road is for all pedestrians. Children should be taught the Code and should not be allowed out alone until

they can understand and use it properly. The age when they can do this is different for each child. Many children cannot judge how fast vehicles are going or how far away they are. Children learn by example, so parents and carers should always use the Code in full when out with their children. They are responsible for deciding at what age children can use it safely by themselves.

A. First find a safe place to cross and where there is space to reach the pavement on the other side. Where there is a crossing nearby, use it. It is safer to cross using a subway, a footbridge, an island, a zebra, pelican, toucan or puffin crossing, or where there is a crossing point controlled by a police officer, a school crossing patrol or a traffic warden. Otherwise choose a place where you can see clearly in all directions. Try to avoid crossing between parked cars (see Rule 14), on a blind bend, or close to the brow of a hill. Move to a space where drivers and riders can see you clearly. Do not cross the road diagonally.

B. Stop just before you get to the kerb, where you can see if anything is coming. Do not get too close to the traffic. If there's no pavement, keep back from the edge of the road but make sure you can still see approaching traffic.

C. Look all around for traffic and listen. Traffic could come from any direction. Listen as well, because you can sometimes hear traffic before you see it.

D. If traffic is coming, let it pass. Look all around again and listen. Do not cross until there is a safe gap in the traffic and you are certain that there is plenty of time. Remember, even if traffic is a long way off, it may be approaching very quickly.

NOTE
As mentioned in Chapters 5 and 6, the configuration of the road, and the location of dropped kerbs indicating a pedestrian's crossing location, may make it impossible to tell if there is a safe gap in the traffic. This situation must be remedied if a safe crossing place is to be provided.

E. When it is safe, go straight across the road – do not run. Keep looking and listening for traffic while you cross, in case there is any traffic you did not see, or in case other traffic appears suddenly. Look out for cyclists and motorcyclists travelling between lanes of traffic. Do not walk diagonally across the road.

8. At a junction. When crossing the road, look out for traffic turning into the road, especially from behind you. If you have started crossing and traffic wants to turn into the road, you have priority and they should give way (see Rule 170).

NOTE
Although not stated explicitly in Rule 21, pedestrians presumably also have priority over turning vehicles at signalised junctions.

9. Pedestrian safety barriers. Where there are barriers, cross the road only at the gaps provided for pedestrians. Do not climb over the barriers or walk between them and the road.

10. Tactile paving. Raised surfaces that can be felt underfoot provide warning and guidance to blind or partially sighted people. The most common surfaces are a series of raised studs, which are used at crossing points with a dropped kerb, or a series of rounded raised bars, which are used at level crossings, at the top and bottom of steps and at some other hazards.

11. One-way streets. Check which way the traffic is moving. Do not cross until it is safe to do so without stopping. Bus and cycle lanes may operate in the opposite direction to the rest of the traffic.

12. Bus and cycle lanes. Take care when crossing these lanes as traffic may be moving faster than in the other lanes, or against the flow of traffic.

13. Routes shared with cyclists. Some cycle tracks run alongside footpaths or pavements, using a segregating feature to separate cyclists from people on foot. Segregated routes may also incorporate short lengths of tactile paving to help visually impaired people stay on the correct side. On the pedestrian side this will comprise a series of flat-topped bars running across the direction of travel (ladder pattern). On the cyclist side the same bars are orientated in the direction of travel (tramline pattern). Not all routes which are shared with cyclists are segregated. Take extra care where this is so (see Rule 62).

14. Parked vehicles. If you have to cross between parked vehicles, use the outside edges of the vehicles as if they were the kerb. Stop there and make sure you can see all around and that the traffic can see you. Make sure there is a gap between any parked vehicles on the other side, so you can reach the pavement. Never cross the road in front of, or behind, any vehicle with its engine running, especially a large vehicle, as the driver may not be able to see you.

15. Reversing vehicles. Never cross behind a vehicle which is reversing, showing white reversing lights or sounding a warning.

16. Moving vehicles. You **MUST NOT** get onto or hold onto a moving vehicle.

Law: Road Traffic Act 1988 Section 26.

17. At night. Wear something reflective to make it easier for others to see you (see Rule 3). If there is no pedestrian crossing nearby, cross the road near a street light so that traffic can see you more easily.

1.2.1.3 Crossings

18. At all crossings. When using any type of crossing you should

- always check that the traffic has stopped before you start to cross or push a pram onto a crossing
- always cross between the studs or over the zebra markings. Do not cross at the side of the crossing or on the zig-zag lines, as it can be dangerous.

You **MUST NOT** loiter on any type of crossing.

Laws: The Zebra, Pelican and Puffin Pedestrian Crossings Regulations and General Directions 1997 Regulation 19 and Road Traffic Regulation Act 1984 Section 25(5).

19. Zebra crossings. Give traffic plenty of time to see you and to stop before you start to cross. Vehicles will need more time when the road is slippery. Wait until traffic has stopped from both directions or the road is clear before crossing. Remember that traffic does not have to stop until someone has moved onto the crossing. Keep looking both ways, and listening, in case a driver or rider has not seen you and attempts to overtake a vehicle that has stopped.

NOTE
1. *See also Rule 20 regarding central islands on zebra crossings.*
2. *Note that a potential area of confusion and, hence, danger, in this rule is that pedestrians cannot safely start to cross until approaching vehicles have stopped, but vehicles are not required to stop until pedestrians start to cross. Consideration could, therefore, be given in future rule-making for vehicles to stop when pedestrians are waiting to cross.*

20. Where there is an island in the middle of a zebra crossing, wait on the island and follow Rule 19 before you cross the second half of the road – it is a separate crossing.

NOTE
Where there is a central island at signalised crossings, the pedestrian has priority over the entire crossing, as stated in Rule 197 and implied in Rule 199.

21. At traffic lights. There may be special signals for pedestrians. You should only start to cross the road when the green figure shows. If you have started to cross the road and the green figure goes out, you should still have time to reach the other side, but do not delay. If no pedestrian signals have been provided, watch carefully and do not cross until the traffic lights are red and the traffic has stopped. Keep looking and check for traffic that may be turning the corner. Remember that traffic lights may let traffic move in some lanes while traffic in other lanes has stopped.

NOTE
Pedestrians crossing have priority over 'traffic that may be turning the corner' but this is not stated, as in Rule 8, and they do not have priority where specific pedestrian signals indicate otherwise.

22. Pelican crossings. These are signal-controlled crossings operated by pedestrians. Push the control button to activate the traffic signals. When the red figure shows, do not cross. When a steady green figure shows, check the traffic has stopped then cross with care. When the green figure begins to flash you should not start to cross. If you have already started you should have time to finish crossing safely.

NOTE
1. *It is not stated that pedestrians also have priority over the entire crossing (unless it is a staggered crossing) if the crossing is divided by a central island, as stated in Rule 197.*
2. *Pelican crossings are the oldest type of crossing (CIHT, 2015) and are for pedestrians only. However, they are being replaced by 'PedEX' crossings as they exceed their effective economical lives. PedEX crossings are for pedestrians only, and use farside pedestrian signals with red and green figures and standard vehicle signals. They are effectively a junction facility provided as a standard crossing. The steady green 'invitation to cross' figure is followed by a blackout, while drivers see the usual sequence. See also the description of PedEX signals in Chapter 7.*

23. Puffin crossings differ from pelican crossings as the red and green figures are above the control box on your side of the road and there is no flashing green-figure phase. Press the button and wait for the green figure to show.

24. When the road is congested, traffic on your side of the road may be forced to stop even though their lights are green. Traffic may still be moving on the other side of the road, so press the button and wait for the signal to cross.

25. Toucan crossings are light-controlled crossings which allow cyclists and pedestrians to share crossing space and cross at the same time. They are push-button operated. Pedestrians and cyclists will see the green signal together. Cyclists are permitted to ride across.

NOTE
None of Rules 22 to 25 states that these crossings, when divided by an island, give priority to pedestrians for both sides of the crossing, as stated in Rule 197 and implied in Rule 199.

26. At some crossings there is a bleeping sound or voice signal to indicate to blind or partially sighted people when the steady green figure is showing, and there may be a tactile signal to help deafblind people.

27. Equestrian crossings are for horse riders. They have pavement barriers, wider crossing spaces, horse and rider figures in the light panels and either two sets of controls (one higher), or just one higher control panel.

28. 'Staggered' pelican or puffin crossings. When the crossings on each side of the central refuge are not in line they are two separate crossings. On reaching the central island, press the button again and wait for a steady green figure.

29. Crossings controlled by an authorised person. Do not cross the road unless you are signalled to do so by a police officer, traffic warden or school crossing patrol. Always cross in front of them.

30. Where there are no controlled crossing points available it is advisable to cross where there is an island in the middle of the road. Use the Green Cross Code (see Rule 7) to cross to the island and then stop and use it again to cross the second half of the road.

NOTE
This is similar to the case where a central island is present at a zebra crossing but different from signalised crossings where the pedestrian has priority over the entire crossing, as stated in Rule 197 and implied in Rule 199.

1.2.2 Rules for users of powered wheelchairs and mobility scooters (called invalid carriages in law)

1.2.2.1 Powered wheelchairs and mobility scooters
36. There is one class of manual wheelchair (called a Class 1 invalid carriage) and two classes of powered wheelchairs and powered mobility scooters. Manual wheelchairs and Class 2 vehicles are those with an upper speed limit of 4 miles per hour (6 km/h) and are designed to be used on pavements. Class 3 vehicles are those with an upper speed limit of 8 miles per hour (12 km/h) and are equipped to be used on the road as well as the pavement.

37. When you are on the road you should obey the guidance and rules for other vehicles; when on the pavement you should follow the guidance and rules for pedestrians.

1.2.2.2 On pavements
38. Pavements are safer than roads and should be used when available. You should give pedestrians priority and show consideration for other pavement users, particularly those with a hearing or visual impairment, who may not be aware that you are there.

39. Powered wheelchairs and scooters **MUST NOT** travel faster than 4 miles per hour (6 km/h) on pavements or in pedestrian areas. You may need to reduce your speed to adjust to other pavement users who may not be able to move out of your way quickly enough or where the pavement is too narrow.

Law: The Use of Invalid Carriages on Highways Regulations 1988 Regulation 4.

40. When moving off the pavement onto the road, you should take special care. Before moving off, always look round and make sure it's safe to join the traffic. Always try to use dropped kerbs when moving off the pavement, even if this means travelling further to locate one. If you have to climb or descend a kerb, always approach it at right angles and don't try to negotiate a kerb higher than the vehicle manufacturer's recommendations.

1.2.3 Rules for cyclists
1.2.3.1 Overview
62. Cycle Tracks. These are normally located away from the road, but may occasionally be found alongside footpaths or pavements. Cyclists and pedestrians may be segregated or they may share the same space (unsegregated). When using segregated tracks you **MUST** keep to the side intended for cyclists as the pedestrian side remains a pavement or footpath. Take care when passing pedestrians, especially children, older or disabled people, and allow them plenty of room. Always be prepared to slow down and stop if necessary. Take care near road junctions as you may have difficulty seeing other road users, who might not notice you.

Law: Highway Act 1835 Section 72.

1.2.4 General rules, techniques and advice for all drivers and riders
1.2.4.1 Control of the vehicle
1.2.4.1.1 Speed limits
124. You **MUST NOT** exceed the maximum speed limits for the road and for your vehicle [see Table 1.1]. The presence of street lights generally means that there is a 30 miles per hour (48 km/h) speed limit unless otherwise specified.

Law: Road Traffic Regulation Act 1984 Sections 81, 86, 89 and Schedule 6 as amended by The Motor Vehicles (Variation of Speed Limits) (England and Wales) Regulations 2014.

125. The speed limit is the absolute maximum and does not mean it is safe to drive at that speed irrespective of conditions. Driving at speeds too fast for the road and traffic conditions is dangerous. You should always reduce your speed when

- the road layout or condition presents hazards, such as bends
- sharing the road with pedestrians, cyclists and horse riders, particularly children, and motorcyclists

Table 1.1 Speed limits (DfT, 2017)

Type of vehicle	Built-up areas mph (km/h)[a]	Single carriageways mph (km/h)	Dual carriageways mph (km/h)	Motorways mph (km/h)
Cars and motorcycles (including car derived vans up to 2 tonnes maximum laden weight)	30 (48)	60 (96)	70 (112)	70 (112)
Cars towing caravans or trailers (including car derived vans and motorcycles)	30 (48)	50 (80)	60 (96)	60 (96)
Buses, coaches and minibuses (not exceeding 12 m in overall length)	30 (48)	50 (80)	60 (96)	70 (112)
Goods vehicles (not exceeding 7.5 tonnes maximum laden weight)	30 (48)	50 (80)	60 (96)	70[b] (112)
Goods vehicles (exceeding 7.5 tonnes maximum laden weight) in England and Wales	30 (48)	50 (80)	60 (96)	60 (96)
Goods vehicles (exceeding 7.5 tonnes maximum laden weight) in Scotland	30 (48)	40 (64)	50 (80)	60 (96)

[a] The 30 miles per hour limit usually applies to all traffic on all roads with street lighting unless signs show otherwise.
[b] 60 miles per hour (96 km/h) if articulated or towing a trailer.
© Crown Copyright, 2017
This information is licensed under the Open Government Licence v3.0. To view this licence, visit http://www.nationalarchives.gov.uk/doc/open-government-licence/ **OGL**

- weather conditions make it safer to do so
- driving at night as it is more difficult to see other road users.

1.2.4.2   General advice

145. You **MUST NOT** drive on or over a pavement, footpath or bridleway except to gain lawful access to property, or in the case of an emergency.

Laws: Highway Act 1835 Section 72 and Road Traffic Act 1988 Section 34.

1.2.4.2.1 Driving in built-up areas

152. Residential streets. You should drive slowly and carefully on streets where there are likely to be pedestrians, cyclists and parked cars. In some areas a 20 miles per hour (32 km/h) maximum speed limit may be in force. Look out for

- vehicles emerging from junctions or driveways
- vehicles moving off
- car doors opening
- pedestrians
- children running out from between parked cars
- cyclists and motorcyclists.

153. Traffic calming measures. On some roads there are features such as road humps, chicanes and narrowings which are intended to slow you down. When you approach these features reduce your speed. Allow cyclists and motorcyclists room to pass through them. Maintain a reduced speed along the whole of the stretch of road within the calming measures. Give way to oncoming road users if directed to do so by signs. You should not overtake other moving road users while in these areas.

1.2.4.2.2 Vehicles prohibited from using roads and pavements

157. Certain motorised vehicles do not meet the construction and technical requirements for road vehicles and are generally not intended, not suitable and not legal for road, pavement, footpath, cycle path or bridleway use. These include most types of miniature motorcycles, also called mini motos, and motorised scooters, also called go peds, which are powered by electric or internal combustion engines. These types of vehicle **MUST NOT** be used on roads, pavements, footpaths or bridleways.

Laws: Road Traffic Act 1988 Sections 34, 41a, 42, 47, 63 and 66, Highway Act 1835 Section 72, and Roads (Scotland) Act 1984 Section 129.

1.2.5   Using the road
1.2.5.1   Road junctions

170. Take extra care at junctions. You should

- watch out for cyclists, motorcyclists, powered wheelchairs/mobility scooters and pedestrians as they are not always easy to see. Be aware that they may not have seen or heard you if you are approaching from behind

■ watch out for pedestrians crossing a road into which you are turning. If they have started to cross they have priority, so give way.

NOTE

This corresponds to Rule 8 for pedestrians, but is not repeated in Rule 183. As shown in Box 1.1, this rule could be improved by giving more emphasis for drivers on each of the possible turning movements, including diagrams of each combination of vehicle and pedestrian movements.

There are eight different combinations of vehicle and pedestrian movements for a simple T-junction but only one is shown diagrammatically in the Code. *The comments for Rules 171, 172 and 206 should also be noted.*

171. You **MUST** stop behind the line at a junction with a 'Stop' sign and a solid white line across the road. Wait for a safe gap in the traffic before you move off.

Laws: Road Traffic Act 1988 Section 36 and the Traffic Signs Regulations and General Directions 2002 Regulations 10 and 16.

NOTE

This rule does not state specifically that a pedestrian crossing the mouth of the minor junction at any point has priority and that a vehicle approaching the major road (equivalent to a vehicle turning into the minor road from the major road) must give way to that pedestrian. This information is not provided to drivers or to pedestrians.

172. The approach to a junction may have a 'Give Way' sign or a triangle marked on the road. You **MUST** give way to traffic on the main road when emerging from a junction with broken white lines across the road.

Laws: Road Traffic Act 1988 Section 36 and the Traffic Signs Regulations and General Directions 2002 Regulations 10(1), 16(1) and 25.

Box 1.1 Pedestrian priority as mentioned in Rule 170 but not illustrated in the *Highway Code*

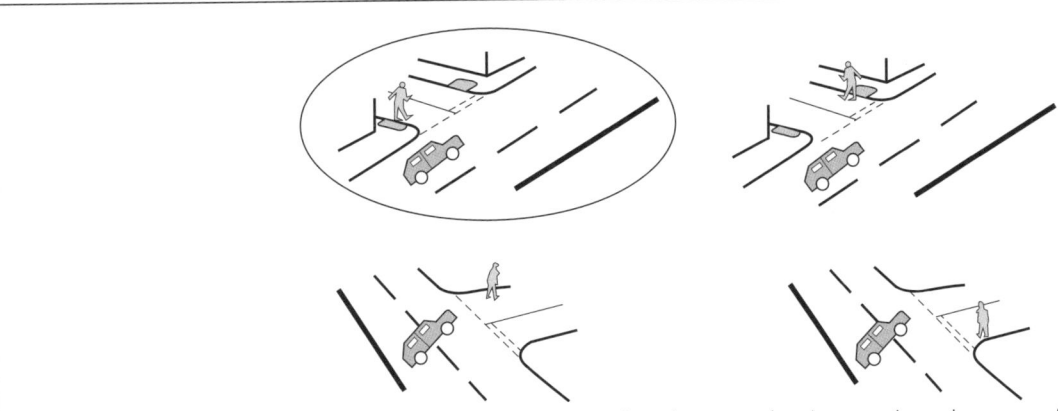

Pedestrian priority over turning vehicles at three-way priority junctions – pedestrian crossing minor arm of junction. Only the situation circled is shown in the *Highway Code*.

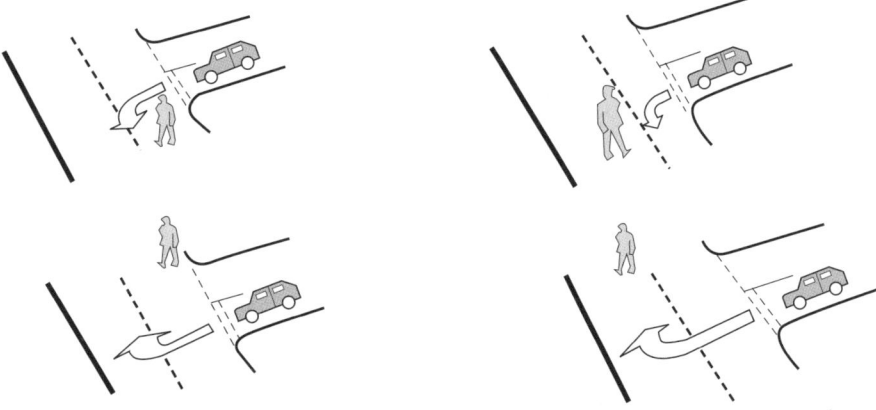

Pedestrian priority over turning vehicles at three-way priority junctions – pedestrian crossing major arms of the junction. None of these situations is illustrated in the *Highway Code*.

Box 1.2 Pedestrian priority movements not addressed in the *Highway Code* in Rules 171 and 172

In Rules 171 and 172, it is not clear that a pedestrian crossing the minor road parallel to the major road must be accorded priority over vehicles approaching the junction along the minor road, as shown in the diagram below. Rule 171 states, 'Wait for a safe gap in the traffic before you move off.' Therefore, because pedestrians are defined as 'traffic' (Traffic Management Act 2004, Section 31) it would seem to be logical and appropriate that priority of pedestrians over vehicles approaching on the minor road should be clearly stated. Also, it is not stated *how* pedestrians are to be given extra care.

Preferably Rules 171 and 172 should state that the vehicles stop before the stop line and also clear of pedestrians crossing because the objective of the *Code* is to assist users in using the roads safely.

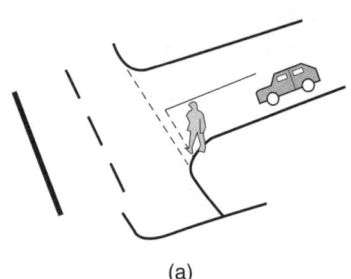

(a)   (b)

Priority of a pedestrian crossing the mouth of a minor road (a) left of vehicle to right, (b) right of vehicle to left. Neither is mentioned or shown in the *Highway Code*.

NOTE
There is no mention in Rules 147 and 148 (addressing road user considerations and concentration, respectively) about the need for drivers to give way also to pedestrians who may be crossing the mouth of the minor road, even though in Rule 206 turning vehicles are required to give way to pedestrians who are crossing.

See Box 1.2, which describes issues related to Rules 171 and 172.

1.2.5.1.1 Turning right
180. Wait until there is a safe gap between you and any oncoming vehicles. Watch out for cyclists, motorcyclists, pedestrians and other road users. Check your mirrors and blind spot again to make sure you are not being overtaken, then make the turn. Do not cut the corner. Take great care when turning into a main road; you will need to watch for traffic in both directions and wait for a safe gap.

NOTE
The meaning of the term 'watch out for' is not defined but would appear to indicate that pedestrians have priority over the entire crossing of the minor road junction.

1.2.5.1.2 Turning left
182. Use your mirrors and give a left-turn signal well before you turn left. Do not overtake just before you turn left and watch out for traffic coming up on your left before you make the turn, especially if driving a large vehicle. Cyclists,

motorcyclists and other road users in particular may be hidden from your view.

NOTE
It would seem that pedestrians should also be mentioned specifically instead of just included in 'other road users'.

183. When turning

- keep as close to the left as is safe and practicable
- give way to any vehicles using a bus lane, cycle lane or tramway from either direction.

NOTE
Pedestrians should be specifically mentioned here, as in other cases of vehicles making left turns, as stated in Rule 170.

1.2.5.2 Roundabouts
185. When reaching the roundabout you should

- give priority to traffic approaching from your right, unless directed otherwise by signs, road markings or traffic lights
- check whether road markings allow you to enter the roundabout without giving way. If so, proceed, but still look to the right before joining
- watch out for all other road users already on the roundabout; be aware they may not be signalling correctly or at all

- look forward before moving off to make sure traffic in front has moved off.

NOTE

It should also be stated that drivers should give priority to pedestrians who may have started to cross at the entrance to the roundabout from the left kerb. Also, the meaning of the term 'watch out for' is unclear and does not convey the requirement that pedestrians who have moved onto the crossing have priority and there is therefore a need for vehicles to stop to afford this priority. Neither is it stated, to pedestrians and drivers, whether such a pedestrian must treat the second half of a crossing as a separate crossing if there is a central island.

187. In all cases watch out for and give plenty of room to

- pedestrians who may be crossing the approach and exit roads
- traffic crossing in front of you on the roundabout, especially vehicles intending to leave by the next exit
- traffic which may be straddling lanes or positioned incorrectly
- motorcyclists.

NOTE

If there is a formal crossing (e.g. zebra), the rules for that crossing should apply. Where there is no formal crossing, it would seem that pedestrians do not have priority.

188. Mini-roundabouts. Approach these in the same way as normal roundabouts. All vehicles **MUST** pass round the central markings except large vehicles which are physically incapable of doing so. Remember, there is less space to manoeuvre and less time to signal. Avoid making U-turns at mini-roundabouts. Beware of others doing this.

Laws: Road Traffic Act 1988 Section 36 and The Traffic Signs Regulations and General Directions 2002 Regulations 10(1) and 16(1).

NOTE

See also the previous notes for roundabouts.

1.2.5.3 Pedestrian crossings

191. You **MUST NOT** park on a crossing or in the area covered by the zig-zag lines. You **MUST NOT** overtake the moving vehicle nearest the crossing or the vehicle nearest the crossing which has stopped to give way to pedestrians.

Laws: The Zebra, Pelican and Puffin Pedestrian Crossings Regulations and General Directions 1997 Regulations 18, 20 and 24, Road Traffic Regulation Act 1984 Section 25(5) and The Traffic Signs Regulations and General Directions 2002 Regulations 10, 27 and 28.

192. In queueing traffic, you should keep the crossing clear.

193. You should take extra care where the view of either side of the crossing is blocked by queueing traffic or incorrectly parked vehicles. Pedestrians may be crossing between stationary vehicles.

194. Allow pedestrians plenty of time to cross and do not harass them by revving your engine or edging forward.

195. Zebra crossings. As you approach a zebra crossing

- look out for pedestrians waiting to cross and be ready to slow down or stop to let them cross
- you **MUST** give way when a pedestrian has moved onto a crossing
- allow more time for stopping on wet or icy roads
- do not wave or use your horn to invite pedestrians across; this could be dangerous if another vehicle is approaching
- be aware of pedestrians approaching from the side of the crossing.

A zebra crossing with a central island is two separate crossings (see Rule 20).

Law: The Zebra, Pelican and Puffin Pedestrian Crossings Regulations and General Directions 1997 Regulation 25.

1.2.5.3.1 Signal-controlled crossings

196. Pelican crossings. These are signal-controlled crossings where flashing amber follows the red 'Stop' light. You **MUST** stop when the red light shows. When the amber light is flashing, you **MUST** give way to any pedestrians on the crossing. If the amber light is flashing and there are no pedestrians on the crossing, you may proceed with caution.

Laws: The Zebra, Pelican and Puffin Pedestrian Crossings Regulations and General Directions 1997 Regulations 23 and 26 and Road Traffic Regulation Act 1984 Section 25(5).

197. Pelican crossings which go straight across the road are one crossing, even when there is a central island. You **MUST** wait for pedestrians who are crossing from the other side of the island.

Laws: The Zebra, Pelican and Puffin Pedestrian Crossings Regulations and General Directions 1997 Regulation 26 and Road Traffic Regulation Act 1984 Section 25(5).

NOTE

In Rule 22, pedestrians are not informed that they have this priority over the entire crossing. New pelican crossings are not permitted and will be replaced by PedEX or puffin crossings as they become inoperable, as discussed in Chapters 3 and 7.

199. Toucan, puffin and equestrian crossings. These are similar to pelican crossings, but there is no flashing amber phase; the light sequence for traffic at these three crossings is the same as at traffic lights. If the signal-controlled crossing is not working, proceed with extreme caution.

NOTE
If it is assumed that toucan, puffin and equestrian crossings afford the same priorities to pedestrians as do pelican crossings, Rules 22 to 25 do not inform them of such priorities.

1.2.6 Road users requiring extra care
1.2.6.1 Pedestrians
206. Drive carefully and slowly when turning at road junctions; give way to pedestrians who are already crossing the road into which you are turning.

NOTE
This would indicate that pedestrians have priority over all turning vehicular traffic when crossing any part of the arm of a junction. But it should perhaps be emphasised that this would also include the crossings of the arms of the major road at a junction as well as the minor arm.

1.3. Status of the *Highway Code* and application to geometric design
1.3.1 Rules for pedestrians
Based on the comments mentioned in the preceding section, it is evident that many areas of ambiguity, apparent inconsistency, partial information and selective provision of information exist in what otherwise is an essential and in many cases informative document. One such area is that although vehicles approaching

a stop line on the minor arm of a junction are required to give way to traffic on the major road, it is not stated that they should also afford priority to pedestrians who may have started to cross or are crossing a part of the junction parallel with the major road.

Instructions to pedestrians related to central islands (refuges), which are often important as design elements in pedestrian crossing locations, are summarised in Table 1.2. From this, several areas where greater clarity could be beneficial include the following.

- It is not apparent why pedestrians who have started to cross a zebra or uncontrolled crossing with a central island must treat the second half of their crossing as a separate crossing. This would seem to be a case where time delays for motorists are considered more valuable than for pedestrians, who often have to wait on the island, owing to vehicles approaching the second half of the crossing.
- It is not stated that if a central refuge is provided at the crossing of the minor arm of a junction whether this means that the pedestrian must treat the crossing as a single movement or as two separate movements – the latter negating a pedestrian's priority (if indeed this is the case) over the entire length of the crossing.

In view of these points, several areas of further investigation appear warranted in order to assist in the design of pedestrian facilities, which can, by their geometric layout and

Table 1.2 *Highway Code* (DfT, 2017) instructions for pedestrians' crossing at selected locations with central islands (refuges)

Location of crossing	Type of crossing	*Highway Code*: pedestrian crossing instructions	Applicable rule(s)
Stand-alone	Uncontrolled	Two separate halves	Rule 30
	Zebra	Two separate halves	Rule 20
	Pelican, puffin and toucan[a]		
	PedEX[b]	One crossing for the entire carriageway	Rules 197 and 199
Priority junction, minor road crossing	Uncontrolled	Not specified	No specific rule
	Zebra	Not specified	No specific rule
	Pelican, puffin and toucan[a]	One crossing	Pedestrian priority over entire crossing is consistent with unsignalised case (Rules 8 and 170)
Roundabouts	Uncontrolled	Not specified	No specific rule

[a] Not including staggered crossings.
[b] PedEXes are replacing pelicans, which will not be replaced when reaching the ends of their working lives.
© Crown Copyright, 2017
This information is licensed under the Open Government Licence v3.0. To view this licence, visit http://www.nationalarchives.gov.uk/doc/open-government-licence/ **OGL**

configuration, assist in improving the safety and convenience of all road users. They include

- the priority of specific road users at each portion of priority junctions, including also the various signalled junctions
- the nature of change in pedestrians' crossing priorities brought about by use of central islands (refuges)
- a clarification of the rules regarding priorities at central islands resulting from their location and function, including free-standing, uncontrolled, zebra and signalled crossings, and at stand-alone, junction and roundabout locations.

1.3.2 Safety, convenience and implications for geometric design

Much of the lack of clarity in the *Code* reportedly encourages road users to be more cautious. This caution is particularly relevant in the case of priority. Here, a vulnerable road user's insistence of right of way or lack of observation combined with a driver's lack of attention or error could result in a collision.

Although this lack of clarity in the *Code* may induce caution, it can also cause confusion. This, in the event of driver–pedestrian interaction, can be equally dangerous. Also, the degree of additional caution that pedestrians must exercise can significantly detract from the utility, convenience and, therefore, attractiveness of walking. As with all codes, individual situations cannot always be predicted, and the ultimate defence against collision is the use of prudence and consideration for others on the part of all road users.

To improve the *Code* and assure that its provisions lead to greater safety and convenience for all road users, it will be essential that concomitant policies and practices in the education, enforcement and conduct of road users will be required. Sensitivity to the needs of road users in the engineering and design of the geometric characteristics of facilities for road users can assist in alerting road users to the need for safety, caution and effectiveness.

1.4. Summary

The code of conduct for road users provides a necessary set of rules by which movements of the various road users are coordinated to assist safety and convenience. The *Highway Code* is based on Acts and Regulations that have evolved over the years, many of which reflect technical characteristics of the various modes of transport, including walking. Also, many of the rules reflect preferences that have developed as a result of political and special interests and, as a result, reflect the undoubted popularity of the car, and motor transport in general. Several areas of the *Code* have been examined and areas of needed

clarification noted, particularly where such clarification would assist in engineering and design of the geometric characteristics of crossings and related pedestrian facilities.

REFERENCES

CABE (Commission for Architecture and the Built Environment) (2002) *Paving the Way: How We Achieve Clean, Safe and Attractive Streets.* Thomas Telford, London, UK.

CIHT (Chartered Institute for Highways and Transport) (2015) *Planning for Walking.* CIHT, London, UK.

DfT (Department for Transport) (2005) HD 42/05: Non-motorised user audits. In *Design Manual for Roads and Bridges (DMRB).* DfT, London, UK.

DfT (2007) *Manual for Streets.* Thomas Telford, London, UK.

DfT (2017) *The Highway Code.* DfT, London, UK.

DfT (2018) *Design Manual for Roads and Bridges (DMRB).* DfT, London, UK.

HMG (His Majesty's Government) (1835) Highway Act 1835. *William IV.* The Stationery Office, London, UK.

HMG (Her Majesty's Government) (1982) The Motorways Traffic (England and Wales) Regulations 1982. The Stationery Office, London, UK.

HMG (1984) Road Traffic Regulation Act 1984. The Stationery Office, London, UK.

HMG (1984) Roads (Scotland) Act 1984. The Stationery Office, London, UK.

HMG (1988) Road Traffic Act 1988. The Stationery Office, London, UK.

HMG (1988) The Use of Invalid Carriages on Highways Regulations 1988. The Stationery Office, London, UK.

HMG (1995) The Motorways Traffic (Scotland) Regulations 1995. The Stationery Office, London, UK.

HMG (1997) The Zebra, Pelican and Puffin Pedestrian Crossings Regulations and General Directions 1997. The Stationery Office, London, UK.

HMG (2002) The Traffic Signs Regulations and General Directions 2002. The Stationery Office, London, UK.

HMG (2014) The Motor Vehicles (Variation of Speed Limits) (England and Wales) Regulations 2014. The Stationery Office, London, UK.

Schoon JG (2017) *Pedestrian Traffic: Walking Safely – Why We Can't and How We May.* Matador, Leicester, UK.

Pedestrian Facilities, Second edition

Schoon, John G
ISBN 978-0-7277-6309-9
https://doi.org/10.1680/pfse.63099.031

Chapter 2
Driver and vehicle characteristics

The interaction between drivers, their vehicles, pedestrians and other non-motorised road users constitutes the essence of geometric design. An extensive body of literature exists on the characteristics of drivers and vehicles; most guidance on the design of highway facilities is based on the needs of motorised traffic. This chapter, therefore, describes the characteristics of essential elements in the design of pedestrian facilities involving interaction with motor vehicles. Characteristics and design features related to other non-motorised users (disabled people, and cyclists and equestrians) are discussed in Chapters 4 and 11, respectively.

2.1.   Parameters describing driver and vehicle characteristics

Safety, in that road users are enabled to avoid collisions and experience convenient travel free of intimidation from traffic, is a universally recognised objective of any functioning transport system – in this case, of establishing a satisfactory geometric design. Essentially, estimation of stopping sight distance, based on driver and vehicle characteristics, is generally the key determinant of the location and much of the individual geometric details of highway and, particularly, pedestrian facilities.

Also of importance in the analysis of proposed facilities are the locations of road users' viewpoints and of objects to be avoided and the dimensions of vehicles when stationary and moving. The matters addressed here, therefore, are

- vehicle speed and its determination
- stopping sight distance
- driver viewpoint location and height of object viewed
- dimensions of vehicles.

For each of these elements, the dimensions and approaches, the major sources of information are the *Design Manual for Roads and Bridges* (*DMRB*) (DfT, 2018), and associated latest design notes, including: TD 9/93 *Highway Link Design, Amendment No 1* (DfT, 2002); TA 12/81 *Traffic Signs and Lighting* (DfT, 1988); TA 22/81 *Vehicle Speed Measurement on All Purpose Roads* (DfT, 1981); TA 12/07 *Traffic Signals on High Speed Roads* (DfT, 2007a). These sources are referred to by their abbreviations in this chapter. Further guidance is contained in: *Manual for Streets* (*MfS*) (DfT, 2007b); *The Manual for Streets: Evidence and Research* (York *et al.*, 2007) and *Manual for Streets 2 – Wider Application of the Principles* (CIHT, 2010). Reference is also made to relevant parts of *Design Bulletin 32: Residential Roads and Footpaths, Layout Considerations* (*DB32*) (DETR, 1992).

2.2.   Vehicle speed

In the UK, approximately 85% of pedestrians died in collisions at impact speeds below 40 miles per hour, 45% at less than 30 miles per hour and 5% at speeds below 20 miles per hour (Richards, 2010). Control of vehicle speeds, therefore, is a crucial factor in pedestrian safety.

Fundamental to any estimate of the ability of a driver to avoid a collision and to conduct appropriate manoeuvres is the concept of stopping sight distance (SSD) – the distance required for a vehicle to stop safely from a given speed. Accordingly, an introduction to how speeds are established, measured and incorporated in the analysis of highway facilities, including pedestrian components, is important in estimating SSDs. Of particular interest in the guidelines are the terms 'speed limits', '85th percentile speeds', 'design speeds', 'journey speeds' and the values established in the various guidelines – namely, *DMRB* and *MfS*.

Apart from consideration of design speeds, in terms of vehicle speed reduction related to collisions, vehicles are increasingly being fitted with autonomous emergency braking. This system can detect other vehicles and pedestrians and warn the driver to take avoiding action. If the driver does not react, the brakes will be applied to reduce the effects of, or eliminate, a collision (Euro NCAP, 2018). Studies (Rosén *et al.*, 2010) indicate that for accidents involving pedestrians, with an activation rate of 65%, the average collision speed decreased from 32 km/h without autonomous emergency braking to 22 km/h with the system. This yielded estimated reductions in the numbers of fatally and severely injured pedestrians by 48% and 42%, respectively. Maximum speeds at which autonomous emergency braking is effective may vary and research is continuing. Along with this system, vehicles are being designed with vulnerable road user protection, which mitigates the effects of a collision (Euro NCAP, 2018).

2.2.1 *Design Manual for Roads and Bridges (DfT, 2018)*

Various definitions of speeds are in use in highway design. Definitions are given in TA 22/81 *Vehicle Speed Measurement on All Purpose Roads* (DfT, 1981); Chapter 3 gives an explanation of the terms and concepts listed here. The various highway carriageways and their designations are as follows.

- S2 designates single carriageway, two lanes.
- WS2 designates wide single carriageway, two lanes.
- D2AP and D3AP designate dual-carriageway all-purpose roads with two and three lanes per carriageway, respectively.
- D2M and D3M designate dual-carriageway motorways with two and three lanes per carriageway, respectively.

2.2.1.1 Spot speed and journey speed

The spot speed is the instantaneous vehicle speed measured at a point, as distinct from the journey speed, which is measured along a length of the road. Spot speeds are measured using such devices as radar speed meters or inductive loops. Journey speeds are measured using moving-observer methods over a given length of highway or by recording and matching registration numbers at times of passing.

2.2.1.2 Speeds – speed limits

To determine speed limits, the 85th percentile dry weather spot speed of cars is used. This is the speed exceeded by only 15% of vehicles. When the 85th percentile's spot speed has been arrived at, it is used to determine the speed limit in the manner described in *Setting Local Speed Limits* (DfT, 2013).

2.2.1.3 Speeds – improvement of alignment and junctions

For improvements of alignments and major or minor junctions or accesses, and for new alignments and junctions or accesses on existing roads, the normal design methods are based on the 85th percentile wet-weather journey speed of vehicles. The precise point and timing of the measurements are important. A point just before the scheme begins and a time of free-flow traffic are suitable. Measurements must be taken at both ends of the scheme so that traffic approaching from each direction is covered. If different values are obtained, the higher speed value should be used in the design process. To convert the dry weather spot speed of vehicles measured to the wet-weather journey speed used in the design, one of the following correction factors should be used.

- For all-purpose dual carriageways, deduct 8 km/h.
- For all-purpose single carriageways, deduct 4 km/h.

2.2.1.4 Speeds – traffic signal design

For areas outside 30 miles per hour speed limits, two types of signal equipment are in use, related to the following conditions.

(a) 85th percentile dry weather spot speed of vehicles approaching between 55 km/h and 72 km/h; double vehicle extension with speed discrimination.
(b) 85th percentile dry weather spot speed of vehicles approaching between 72 km/h and 105 km/h; either triple vehicle extensions with speed discrimination or double vehicle extension with speed assessment.

For condition (a), measurements should be taken at not less than 80 m in advance of the stop sign (as seen by traffic – not more than 100 m). For condition (b), the values should be 150–200 m. To ensure accuracy, certain conditions are necessary; these are listed in TA 22/81, Appendix 1 (DfT, 1981).

For junctions on high-speed roads, typically the strategic road network, the use of microprocessor-optimised vehicle actuation (MOVA) is mandatory (DfT, 2006). MOVA can also be used for junctions on slower speed roads.

2.2.2 Major highways and trunk roads

The design speed of vehicles for major highways and rural trunk roads is determined based on the alignment and layout constraints specified in TD 9/93 (DfT, 2002). This is based on the premise that a road alignment shall be designed to ensure that standards of curvature, visibility, superelevation, etc., are provided for a design speed that shall be consistent with the anticipated vehicle speeds on the road. This approach to determining design speed considers the following variables.

- *Alignment constraint*, which incorporates values of the alignment in terms of bendiness, measured in degrees per kilometre and the harmonic mean visibility (measured in metres).
- *Layout constraint* (measured in kilometres per hour), as presented in Table 2.1, which shows the values of layout constraint relative to cross-section features and density of access, expressed as the total number of junctions, lay-bys and commercial accesses per kilometre, summed for both sides of the road, where
  - L = low access, numbering 2 to 5 per km
  - M = medium access, numbering 6 to 8 per km
  - H = high access, numbering 9 to 12 per km.

2.2.2.1 Mandatory speed limits

On rural derestricted roads, i.e. with national speed limits for

- motorways and dual carriageways: 70 miles per hour (112 km/h)
- single carriageways: 60 miles per hour (96 km/h).

Table 2.1 Layout constraint (DfT, 2002): km/h

Road type	S2				WS2	D2AP	D3AP	D2M	D3M
Carriageway width (excluding metre strips)	6 m	7.3 m			10 m	Dual 7.3 m	Dual 11 m	Dual 7.3 m and hard shoulder	Dual 11 m and hard shoulder
Degree of access and junctions	H	M	M	L	M L	M L	L	L	L
Standard verge width	29	26	23	21	19 17	10 9	6	4	0
1.5 m verge		31	28	25	23 There is no research data available for four-lane single-carriageway roads between 12 m and 14.6 m width (S4). In the limited circumstances of their use described here, design speed should be estimated assuming a normal D2AP with a layout constraint of 15 km/h to 13 km/h				
0.5 m verge		33	30						

Vehicle speeds are constrained only by the physical impression of the road alignment, as described by the alignment and layout constraints. The use of mandatory speed limits (together with more confined urban cross-sections), however, restricts speeds below those freely achievable and will act as a further constraint on speed in addition to that indicated by layout constraint.

2.2.2.2 Selection of design speed

This is described for three categories of road: new rural roads, existing rural road improvements and urban roads.

2.2.2.2.1 New rural roads

The design speed is determined using Figure 2.1, which shows the variation in speeds for a given layout constraint against alignment constraint. The design speeds are arranged in bands, within which suffixes A and B indicate the higher and lower categories of each band, respectively. An initial alignment to a trial design speed should be drawn and alignment constraints measured for each section of the route that demonstrates significant changes over a minimum length of 2 km. The design speed calculated from the ensuing alignment and layout constraints should be checked against the initial choice to identify locations where elements of the initial trial alignment may be relaxed to achieve cost or environmental savings or, conversely, where the design should be upgraded, according to the calculated design speed. If any changes to road geometry are made, the design speed should be recalculated to check that it has not changed.

2.2.2.2.2 Existing rural road improvements

This includes short diversions or bypasses up to about 2 km long. The design speed shall be derived in a similar way to that described for rural roads, with the alignment constraint measured over a minimum length of 2 km, including the improvement, provided there are no discontinuities, such as roundabouts. The strategies for contiguous stretches of roads, however, must be considered when determining the alignment constraint and the cross-section design.

2.2.2.2.3 Urban roads

Low speed limits (30–40 miles per hour) may be required, owing to the amount of frontal activity, but also where physical restrictions on the alignment make it impractical to achieve geometry suitable for a higher design speed. The design speed shall be selected with reference to the speed limits envisaged for the road, so as to permit a small margin for speed in excess of the speed limit, as shown in Table 2.2. The minimum design speed for a primary distributor is 70 km/h.

Table 2.2 Speed limits and design speeds for urban roads (DfT, 2002)

Speed limit		Design speed
miles per hour	km/h	km/h
30	48	60B
40	64	70A
50	80	85A
60	96	100A

Figure 2.1 Selection of design speed for rural roads (DfT, 2002). B, bendiness; VISI, harmonic mean visibility
© Crown Copyright, 2002
This information is licensed under the Open Government Licence v3.0. To view this licence, visit http://www.nationalarchives.gov.uk/doc/open-government-licence/ **OGL**

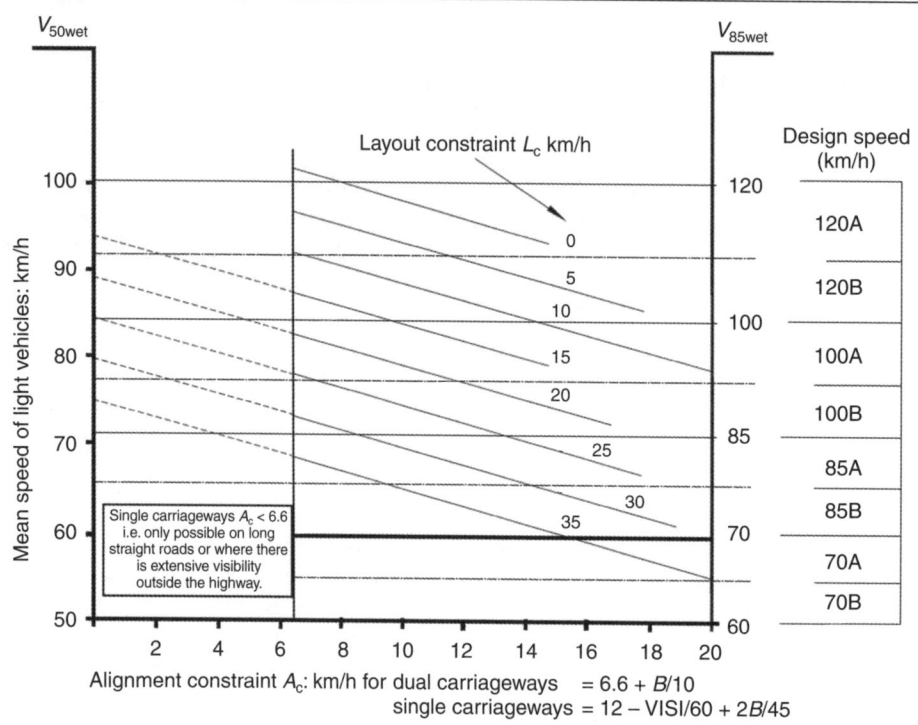

2.2.2.2.4 Design speed-related parameters

The design speed bands – 120, 100, 85, etc. – dictate the minimum geometric parameters for the design, according to Table 2.3, which shows desirable minimum (absolute minimum for sag curves only) values and values for certain design speed steps below the desirable minimum. Desirable minimum values represent the comfortable values dictated by the design speed. In designing pedestrian facilities, the design speeds associated with the horizontal and vertical curvatures of the location concerned would be indicated by the values in this table.

2.2.2.2.5 Changeovers of design speed standards and connections to existing roads

These situations require special consideration. With the former, design speed differences should not present the driver with sudden changes in alignment. In connecting with existing roads, care should be taken to ensure that the curvature and sight distance are adequate for the transition between segments of the road with different design speeds.

The V^2/R values shown in Table 2.3 represent a convenient means of identifying the relative levels of design parameters, irrespective of design speed.

2.2.2.2.6 Minima, relaxations and departures

A three-tier hierarchy of geometric design criteria is related to design speeds. The essential features are as follows.

■ *Desirable minima.* These produce a high standard of road safety and should be the initial objective.
■ *Relaxations.* The type of road (motorway or all-purpose) and the design speed band (A or B) may warrant relaxation of the minimum design standard, at the discretion of the designer and based on local factors. Safety and environmental factors remain important and a cost–benefit analysis with a comparison with minimum standards would help guide decisions. A list of principles to follow when preparing options and considering the merits of a particular design subject to relaxations would consider whether, or to what degree, the site is

– isolated from other relaxations
– isolated from junctions
– one where drivers have minimum stopping sight distance
– subject to momentary visibility impairment only
– one that would affect only a small proportion of traffic
– of a straightforward geometry, readily understood by drivers

Table 2.3 Minimum geometric parameters for design (DfT, 2002)

Design speed: km/h	120	100	85	70	60	50	V^2/R
Stopping sight distance: m							
Desirable minimum	295	215	160	120	90	70	
One step below desirable minimum	215	160	120	90	70	50	
Horizontal curvature: m							
Minimum R without elimination of adverse camber and transitions[a]	2880	2040	1440	1020	720	520	5
Minimum R with superelevation of 2.5%[a]	2040	1440	1020	720	510	360	7.07
Minimum R with superelevation of 3.5%[a]	1440	1020	720	510	360	255	10
Desirable minimum R with superelevation of 5%	1020	720	510	360	255	180	14.14
One step below desirable minimum R with superelevation of 7%	720	510	360	255	180	127	20
Two steps below desirable minimum radius with superelevation of 7%	510	360	255	180	127	90	28.8
Vertical curvature							
Desirable minimum crest K value[a]	182	100	55	30	17	10	
One step below desirable minimum crest K value	100	55	30	17	10	6.5	
Absolute minimum sag K value	37	26	20	20	13	9	
Overtaking sight distances							
Full overtaking sight distance: m	*	580	490	410	345	290	
Full overtaking sight distance, overtaking crest K value	*	400	285	200	142	100	

[a] Not recommended for use in the design of single carriageways.
© Crown Copyright, 2002
This information is licensed under the Open Government Licence v3.0. To view this licence, visit http://www.nationalarchives.gov.uk/doc/open-government-licence/ **OGL**

- on a road with no frontage access
- one where traffic speeds would be reduced locally, owing to adjacent road geometry (e.g. uphill sections, approaching roundabouts and major or minor junctions where traffic has to give way or stop), or where there are speed limits.

Other factors affecting use of relaxations include risk of skidding, signage, sight distance limits, overtaking sight distances and roundabouts. The values associated with these and related features are given in TD 9/93 (DfT, 2002).

- *Departures*. Where exceptionally difficulties cannot be overcome by relaxations, it may be possible to overcome them by means of a departure, the third tier of the hierarchy. Proposals to adopt departures from standards must be submitted to the overseeing organisation for approval before incorporation into a design layout to ensure that safety is not significantly impaired.

2.2.3 *Manual for Streets (MfS)*

Essentially replacing *Design Bulletin 32* (DETR, 1992) regarding residential streets, *MfS* addresses measures that can tend to reduce speeds and contribute to the multiple uses of residential areas. It indicates that conflicts among users can be minimised by reducing the speed and flow of vehicular traffic and that, for residential streets, the following holds.

- A maximum design speed of 20 miles per hour is appropriate; traffic calming schemes suggest that speed control features at intervals of 70 metres or less will help achieve this speed.
- Links can be broken up by means of physical features, changes in priority, street dimensions, reduced visibility and psychological and perceptive measures.

Of particular relevance to road geometry are the effects of visibility (discussed in more detail later) on carriageway width. As shown in Figure 2.2, speeds tended to increase substantially where forward visibility and highway width are increased.

2.3. Stopping sight distance (SSD)

The location and geometric design of any highway element subject to vehicular traffic is affected by the need for drivers to stop within a safe distance when required. A key dimension, therefore, in analysing and establishing a safe design, is the distance that a vehicle will travel after the driver has identified a reason for stopping as quickly as possible. As briefly mentioned earlier, this distance is known as the stopping sight distance (SSD). Examples are the distance to the location of a crossing or other pedestrian facility, such that drivers approaching it can stop if a pedestrian steps onto the crossing or if a signal requires the driver to stop.

Figure 2.2 Correlation between visibility and carriageway width (DETR, 1992)
© Crown Copyright, 1992
This information is licensed under the Open Government Licence v3.0. To view this licence, visit http://www.nationalarchives.gov.uk/doc/
open-government-licence/ **OGL**

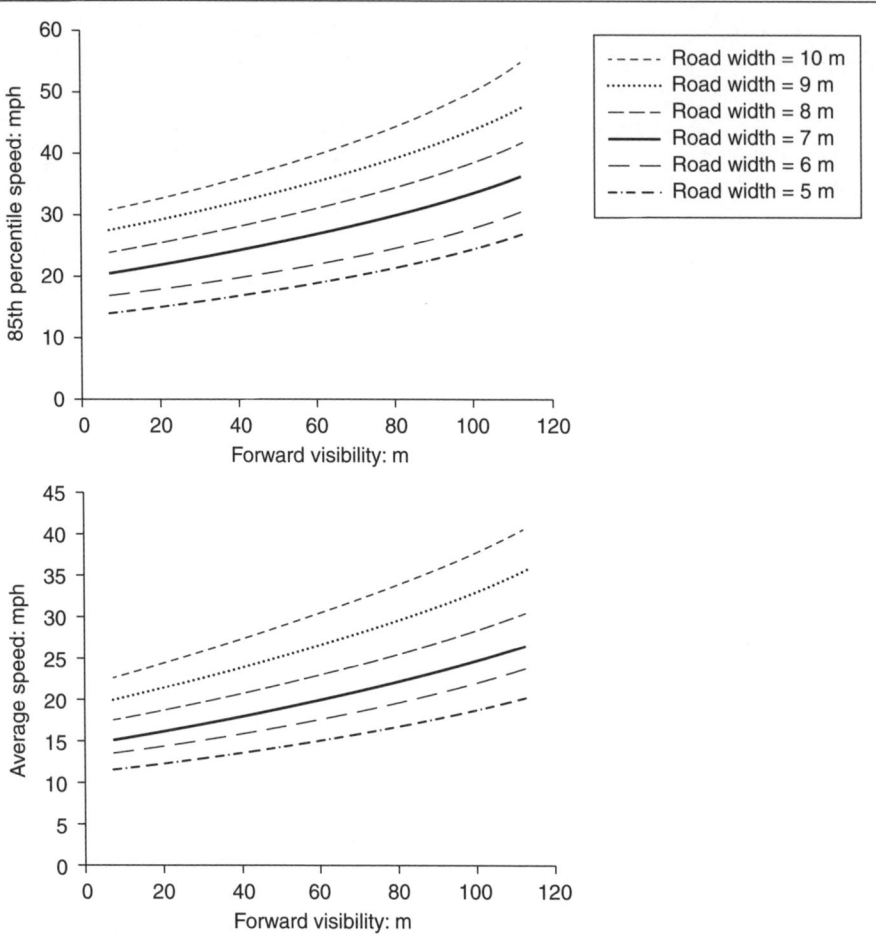

NOTE

Although the SSD may be considered 'safe' in terms of a driver's ability to stop, many collisions are caused by the driver not seeing an object, owing to distraction or as a result of the 'looked but did not see' phenomenon. This matter is addressed with respect to pedestrian sight distance and associated design matters in Chapters 3, 6, 8 and 12.

It should be noted that the term 'stopping sight distance' (SSD) is sometimes used to identify visibility (the distance that can be seen in a desired direction), and is also sometimes referred to as 'forward visibility' or 'stopping distance'. The term 'intervisibility' relates to the ability of drivers to see pedestrians and pedestrians to see approaching vehicles. It is therefore important to identify the context in each of these terms in order to ensure an appropriate interpretation.

2.3.1 Elements of stopping sight distance (SSD)

Several variables contribute to the estimation of SSD: first, the speed of the vehicle affects the distance travelled during the next two elements; next, the driver's reaction time (or 'thinking time') from the time that the object or feature is first seen to the time taken for the driver to apply the brakes or take other action; and, finally, the vehicle's braking distance – the distance required for the vehicle to stop after the brakes have been applied. A driver's perception and reaction time converted to a distance plus the vehicle's braking distance comprises the SSD. Although a driver's reaction time is essentially independent of speed, the braking distance increases with the speed at which the vehicle is travelling when the brakes are applied. Also of importance in practical terms, and described later, are the dimensions related to the position from which most drivers view the road ahead, and the heights of objects that they might be expected to identify.

2.3.1.1 Driver perception–reaction time

A driver's 'perception–reaction' time is a basic parameter, on which most geometric design values depend. A considerable body of literature exists about drivers' perception–reaction times – for example, AASHTO (2011) and Wright (1996).

2.3.1.1.1 Definition

A driver's perception–reaction time is defined by Olson (1989) as the time taken for a driver in a moving vehicle to first see an object or situation and then, through a mental and physical process, which requires a finite amount of time, to react and take physical avoiding action – or not take such action. A summary description of the major sequential features of drivers' perception–reaction times on this basis is as follows.

- *Detection.* The perception–response time begins when some object or condition of concern enters the driver's field of vision. The first step concludes when the driver develops conscious awareness that 'something' is present.
- *Identification.* In this step, sufficient information is acquired about the subject or conditions to be able to reach a decision as to what action, if any, is required. Identification need not be complete in detail.
- *Decision.* At this point the driver must decide what action is appropriate. Assuming that some action is decided on, the choice is whether or not to change speed or direction.
- *Response.* In this step, commands are issued by the motor centre of the brain to the appropriate muscle groups to carry out the required action.

Within these times, it has been found that a driver requires time to shift the direction of vision and also to focus on the particular object or situation. Values for vision shift and focus time in a driving context, estimated by Greenshields (1936) are a 'vision shift' time of 0.15–0.33 s and a 'fixation' (focus) time of 0.1–0.3 s.

2.3.1.1.2 Practice in the UK: *DMRB* (DfT, 2018)

For geometric design in UK practice, based on *DMRB*, drivers' perception–reaction time is taken to be 2.0 s for all classes of road. However, as discussed next, recent guidance in the *MfS* suggests that this value should be reduced in accordance with more recent findings related to residential streets. As a comparison, design policy in the USA is based on a perception–reaction time, for geometric design purposes, of 2.5 s (AASHTO, 2011). This is based on a mean perception–reaction time for an alerted driver of 0.64 s, and a non-alerted driver's mean and maximum values of 1.64 s and $\geqslant 3.5$ s, respectively. A driver's 'start-up' time for crossing a major road is given as 2.0 s, for geometric design purposes.

2.3.1.1.3 *Manual for Streets* (*MfS*)

An investigation (York *et al.*, 2007) made in conjunction with the *Manual for Streets* states that drivers are normally able to stop much more quickly than the 2.0 s mentioned previously in response to an emergency, and comments that values in the *Highway Code* (DfT, 2017) assume a driver reaction time of 0.67 s and a deceleration rate of 0.67*g*. It is further stated that although it would not be appropriate to design junctions for these emergency values, it is considered that some reduction in the key SSD parameters can be accepted for streets with a design speed of 60 km/h or less, without compromising road safety. This advice is based on the following findings.

- A review (Harwood *et al.*, 1995) of practice in other countries has shown that previous UK values are significantly higher than those used elsewhere.
- Olson (1997) found that 85% of drivers will react in less than 1.4 s to a clear and obvious stimulus.
- Research in TRL Report 332 (Maycock *et al.*, 1998) indicates that the 90th percentile reaction time of drivers confronted with a side-road hazard in a driving simulator is 0.9 s.

NOTE
Regarding Olson's findings, many cases where an emergency stop is required do not involve a 'clear and obvious stimulus', especially when visibility is inadequate or if the driver is momentarily distracted when the emergency occurs. Also, regarding the research findings of TRL Report 332, although a simulator may give useful information, the subject is prepared for the need to react and may be expected to do so much quicker than if the event is unexpected, as in an actual road accident.

2.3.1.1.4 The *Highway Code*

As mentioned, *MfS* reports that the *Highway Code* assumes a driver perception–reaction time of 0.67 s.

NOTE
In examining the considerably greater values mentioned above, this value would seem to be the minimum required in an emergency stop. Its use, therefore, in terms of road safety policy, may bear some re-examination if it is considered that drivers should endeavour to stop in a steady and controlled manner.

2.3.1.1.5 Vehicle deceleration and braking distance

Several factors affect the rate of deceleration of vehicles, and therefore the braking distance, starting from activation of a brake pedal. These include the efficiency of the vehicle's brake system, the friction between the tyre and the road surface, the weight of the vehicle and the gradient on which the braking occurs.

Vehicle deceleration may be expressed as the rate of change of speed, for example m/s², or in *g* forces. Some typical values from various sources are 0.25*g* (*MfS*) and 0.67*g* (*Highway*

Code). The basic formula for calculating the braking distance is

$$\text{braking distance} = \frac{v^2}{2d}$$

Where
v = speed (km/h)
d = deceleration rate (m/s^2)

For most design purposes, friction factors are those that would apply for wet roads, thus giving longer braking distances. Deceleration rates stated in guidance documents do not consider ice, snow or other conditions that may similarly affect the surface, and therefore the frictional resistance, between the tyres and the road.

Because a vehicle will decelerate to a stop in a shorter distance if the braked wheels do not skid, most vehicles are fitted with an anti-lock brake system (ABS). This system effectively releases the brake at the point at which the wheel is about to skid. This automatic high frequency (up to 60 cycles per second) release and re-application is achieved by means of electronic sensors. Because the ABS essentially obviates the effects of excessive brake pressure, it may effectively reduce the vehicle's braking distance in an emergency situation when a skid would otherwise have been induced. However, the effects of ABSs are not included in current guidance because many vehicles do not have this system, and there is no assurance that the ABS system on any vehicle so equipped is functioning satisfactorily.

NOTE
It must also be noted that the stated values of the braking distance on wet surfaces provided in guidance documents assume that the vehicle will not skid when stopping. Therefore, it would appear to be open to question whether a vehicle fitted with an ABS system will stop in a lesser distance than the tabulated values; any geometric design should recognise this fact.

2.3.2 Total stopping sight distance (SSD) values – general case

This general case is outlined here to illustrate the values of SSD applicable on straight roads and bends where no other road joins the main one. Although the principles are the same for an SSD as applied to junctions, the geometry of the junction is also important. This is discussed in greater detail in Chapter 6.

As indicated, the SSD is the sum of the distances travelled by the vehicle while the driver reacts to the situation plus the distance covered by the vehicle during braking to a stop. Summing the distance travelled during the driver's perception–reaction time and the vehicle's braking distance under wet road conditions results in a maximum total stopping distance, and, therefore, a distance that a driver needs to see ahead in order to stop at a specific point – the SSD.

2.3.2.1 *DMRB*

Controlled crossings, for which SSDs are specified, may be divided into unsignalled crossings and signalised crossings. Controlled unsignalled crossings primarily include the zebra crossing, which requires a driver to stop whenever a pedestrian steps onto the carriageway. Signalised crossings can include any crossing controlled by a traffic signal, whether actuated by a fixed-time or vehicle- or pedestrian-actuated device. For 85th percentile speeds of up to 50 km/h, visibility requirements for all types of crossing (including zebras) are listed in Table 2.4. In this table, the term 'visibility' is equivalent to the SSD. In this case, both 'desirable' and 'absolute' minimum distances are specified.

NOTE
There are, however, no readily available documented data to support these values.

Signalled crossings may be placed on roads of various categories, including trunk roads that have speed limits of up to 110 km/h (70 miles per hour) (DfT, 2002).

Table 2.4 All types of crossing – visibility requirements, based on TA 12/81 (DfT, 1988)

Speed						
85th percentile approach speed: miles per hour	25	30	35	40	45	50
85th percentile approach speed: km/h	40	48	56	64	70	80
Visibility						
Desirable minimum visibility: metres	50	65	80	100	125	150
Absolute minimum visibility: metres	40	50	65	80	95	115

Table 2.5 Stopping sight distances (DfT, 2002)

Design speed: km/h	120	100	85	70	60	50
Stopping sight distance (SSD): m						
Desirable minimum: m	295	215	160	120	90	70
One step below desirable minimum: m	215	160	120	90	70	50

Stopping sight distances are shown in Table 2.5 for speeds up to 120 km/h. Again, two values are shown for each approach speed: a 'desirable minimum' and a 'one step below desirable minimum' for the various design speeds.

NOTE
Again, no evidential data appear to be documented in the guidelines to support these values.

A further tabulation, based on use of the 85th percentile speeds of drivers, is shown in Table 2.6. The values in this table are equal to or close to those in the preceding table.

2.3.2.2 *DB32* – residential areas
Although essentially superseded by the *MfS*, certain points described in *DB32* are incorporated in many existing streets. These points, therefore, have relevance to street design in residential areas where change may be appropriate. For roads in residential areas with speeds of up to 48 km/h (30 miles per hour), key features of guidelines in *DB32*, include those outlined here.

> The horizontal distance over which unobstructed visibility should be maintained will depend upon the stopping distance of vehicles. This in turn will depend upon vehicle speeds, deceleration rates and drivers' reaction times… The distances are intended to cater for the majority of vehicles and drivers in most weather conditions and may therefore safely be used as general guidance in the design of the residential road network.
>
> (DETR, 1992)

The range of stated stopping distances commensurate with various vehicle speeds is shown in Table 2.7. In this table, 'stopping distance' is equivalent to the SSD.

NOTE
No research is cited in the guidance as a basis for these values, nor are descriptions supplied of how they may vary depending on a driver's ability and environmental conditions, such as lighting, that may be confusing and uncertain to the driver.

Further comments in *DB32*, Section 2.28, relevant to SSD include:

> Restricted visibility in the absence of other precautions cannot be considered a safe method of reducing vehicle speeds. For safety, drivers must be able to see a potential hazard in time to slow down or stop comfortably before reaching it. It is necessary therefore to consider the driver's line of vision, in both the vertical and horizontal planes, and the stopping distance of the vehicle.
>
> Visibility distances must be adequate for the expected speed of vehicles. If measures have been taken to keep the

Table 2.6 Minimum visibility distances to the primary signals required by drivers (DfT, 1988)

85th percentile approach speed (to nearest 10 km/h)	Visibility distance: m
50 km/h (30 miles per hour)	70
60 km/h (37 miles per hour)	95
70 km/h (43 miles per hour)	125
85 km/h (53 miles per hour)	165
100 km/h (62 miles per hour)	225
120 km/h (75 miles per hour)	300

Table 2.7 Visibility distances for selected speed in residential areas, based on *DB32* (DETR, 1992)

Speed: miles per hour	0	5	10	15	20	25	30
Speed: km/h	0	8	16	24	32	40	48
Stopping distance: metres	0	6	14	23	33	45	60

speed of most vehicles below 30 miles per hour, it may be possible to base visibility on these lower speeds. In such cases, likely vehicle speeds along each stretch of carriageway and at each junction and bend should be considered separately, together with the location of any potential obstructions to visibility such as buildings, planting and summits. Section 3 gives detailed guidance on visibility considerations in general and particular requirements for visibility at junctions, on bends and along the edges of carriageways.

(DETR, 1992, Section 2.27)

2.3.2.3 Manual for Streets (MfS) (DfT, 2007b)

New research and approaches to establishing stopping sight distances (York *et al.*, 2007) undertaken during preparation of *MfS* has resulted in considerably smaller SSDs being established for the design of residential streets.

MfS provides guidance on SSDs for streets where the 85th percentile speeds are up to 60 km/h. At speeds above this, the recommended SSDs in *DMRB* may be more appropriate. For new streets, the design speed is set by the designer. For existing streets, the 85th percentile wet-weather speed is used. The basic formula for calculating the SSD is

$$SSD = vt + \frac{v^2}{2d}$$

Where

v = speed
t = driver perception–reaction time
d = deceleration rate (m/s^2)

The relevant sections of *MfS* outlining the basis and recommendations for the SSDs now considered appropriate for residential streets are given in Box 2.1.

2.3.3 *Highway Code SSDs*

The SSDs presented in the *Highway Code* (DfT, 2017) indicate 'typical stopping distances', which include 'thinking distances' and 'braking distances' for vehicle speeds that are all considerably less than the distances given in the design guidance. Table 2.8 shows, for example, that for a speed of 50 km/h, the total stopping distance is 23 m. This is clearly much less than the design total stopping distance given in *DMRB* of 70 m for the 85th percentile of drivers.

Note that many commercial vehicles, including trucks and buses, require a greater stopping distance than do cars. Therefore, if significant numbers of these vehicles are anticipated on a particular segment of highway, appropriate values should be considered. Tests of various trucks braking from 30 miles per hour indicated stopping distances of between approximately 7 m and 21 m and from 45 miles per hour between 18 m and 34 m (Birch, 2001).

2.3.4 Comparison of SSDs

It is important to carefully identify the appropriate road category and use for which a design for a pedestrian crossing is to be applied because the stated SSDs vary depending on the application. The SSDs extracted from the tabulations presented here are shown in Table 2.9 for comparisons only.

NOTE
The values are shown in kilometres per hour; some are approximate because their original values are stated in miles per hour. It could be further noted that, because driver and motor vehicle characteristics affecting SSDs are essentially the same, regardless of the location, some rationalisation between the various sources would appear warranted and could considerably assist in the design process.

A further comparison of drivers' perception times and rates of deceleration is shown in Table 2.10.

Table 2.8 Typical stopping distances (DfT, 2017)

Speed		Distances			
km/h	miles per hour	Thinking distance: m	Braking distance: m	Total stopping distance: m	Total stopping distance: car lengths
32	20	6	6	12	3
48	30	9	14	23	6
64	40	12	24	36	9
80	50	15	38	53	13
96	60	18	55	73	18
112	70	21	75	96	24

Box 2.1 Summary of basis for revised SSDs (*MfS*, Sections 7.5.4 to 7.5.9, DfT, 2007b)

7.5.4 The desirable minimum SSDs used in the *DMRB* are based on a driver perception–reaction time of 2.0 s and a deceleration rate of 2.45 m/s^2 (equivalent to 0.25g where g is the acceleration due to gravity (9.81 m/s^2)). *DB32* adopted these values.

7.5.5 Drivers are normally able to stop much more quickly than this in response to an emergency. The stopping distances given in the *Highway Code* assume a driver reaction time of 0.67 s and a deceleration rate of 6.57 m/s^2.

7.5.6 While it is not appropriate to design street geometry based on braking in an emergency, there is scope for using lower SSDs than those in *DB32*. This is based upon the following

■ a review of practice in other countries has shown that *DB32* values are much more conservative than those used elsewhere
■ research shows that the 90th percentile reaction for drivers confronted with a side-road hazard in a driving simulator is 0.9 s (Maycock *et al.*, 1998)
■ carriageway surfaces are normally able to develop a skidding resistance of at least 0.45g in wet-weather conditions. Deceleration rates of 0.25g (the previously assumed value) are more typically associated with snow-covered roads.

7.5.7 The SSD values used in *MfS* are based on a perception–reaction time of 1.5 s and a deceleration rate of 0.45g (4.41 m/s^2). [The table] uses these values to show the effect of speed on SSD.

7.5.8 Below around 20 m, shorter SSDs themselves will not achieve low vehicle speeds: speed-reducing features will be needed. For higher speed roads, i.e. with an 85th percentile speed over 60 km/h, it may be appropriate to use longer SSDs, as set out in the *DMRB*.

7.5.9 Gradients affect stopping distances. The deceleration rate of 0.45g used to calculate the figures in [the table] is for a level road. A 10% gradient will increase (or decrease) this amount by around 0.1g.

Derived SSDs for streets (figures rounded) (DfT, 2007b)

Speed: km/h	16	20	24	25	30	32	40	45	48	50	60
Speed: miles per hour	10	12	15	16	19	20	25	28	30	31	37
SSD: metres	9	12	15	16	20	22	31	36	40	43	56
SSD adjusted for bonnet length: metres	11	14	17	18	23	25	33	39	43	45	59

Note: For speeds of 16–25 km/h, additional features will be needed to achieve low speeds.
© Crown Copyright, 2007
This information is licensed under the Open Government Licence v3.0. To view this licence, visit http://www.nationalarchives.gov.uk/doc/open-government-licence/ **OGL**

Table 2.9 Summary of SSDs tabulated in *MfS*, *DMRB* and the *Highway Code*, based on foregoing tabulations

Source	Stated speed: km/h							
	30	40	50	60	70	85	100	120
MfS, Table 7.1	20	31	43	56	–	–	–	–
TA 12/81, Section 1 (DfT, 1988)	–	50	65	–	125	–	–	–
Note: Upper value is desirable minimum and lower value is absolute minimum		40	50		95			
TD 9/93, Chapter 1, Section 1.9, Table 3 (DfT, 2002)	–	–	70	90	95	160	215	295
Note: Upper value is minimum, lower value is one step below minimum			50	70		120	160	215
Minimum visibility distances to primary signals required by 85th percentile of drivers, TA 12/81 (DfT, 1988)	–	–	70	95		125	165	225
Highway Code	11	18	24	32	41	60	79	–

Table 2.10 Comparison of drivers' perception–reaction times and vehicle deceleration rates (based on *MfS*) (DfT, 2007b)

Source	Driver's perception–reaction time: s	Deceleration rate: m/s²
DMRB	2.0	2.45
MfS	1.5	4.41
Highway Code	0.67	6.57

2.4. Driver's viewpoint

2.4.1 General principles

Vehicle dimensions and driver location within the vehicles affect a driver's visibility of important objects. For analysis and design, assuming that the driver has the required vision capabilities, estimation of the vertical and horizontal distance along which a driver can see also depends on the height of the object to be seen and the presence of any obstruction of the sight line to the object. Horizontally, it is important that a driver can see ahead and sufficiently to each side to detect any movement that may be relevant to a potential collision.

The vertical dimensions will also affect the visible horizontal distance and so are relevant to the analysis of locations where vertical curves exist in the road or its environs and where possible obstructions of a certain height must be considered. Drivers must, for example, be able to see over, around or through guardrails that might obstruct their vision to places where pedestrians might cross.

Guidance on drivers' sight lines is contained for trunk roads in *DMRB* and for residential areas and streets in *MfS*. A summary of the main features related to driver eye position is provided next.

2.4.2 DMRB

2.4.2.1 Stopping sight distance (SSD)

This is measured assuming the driver's eye height as follows:

Stopping sight distance shall be measured from a maximum driver's eye height of between 1.05 m and 2.0 m, to an object height of between 0.26 m and 2.00 m both above the road surface [as shown in Figure 2.3]. It shall be checked in both the horizontal and vertical plane, between any two points in the centre of the lane on the inside of the curve (for each carriageway in the case of dual carriageways). (DfT, 2002)

2.4.2.2 Full overtaking sight distance

Visibility will also be affected by the sight distance associated with vehicles overtaking, although guidance about pedestrians' crossing facilities along such stretches of road is not provided.

For each design speed:

full overtaking sight distance shall be available between points 1.05 m and 2.00 m above the centre of the carriageway [as shown in Figure 2.4], and shall be checked in both the horizontal and vertical planes.

It should be noted that full overtaking sight distance is considerably greater than SSD. This and related factors associated with its provision are discussed further in TD 9/93 (DfT, 2002).

2.4.3 Manual for Streets (MfS) (DfT, 2007b)

Based on guidance on sight lines associated with residential streets, Figure 2.5 illustrates the key dimensions. It should be noted that the SSD figure relates to the position of the driver. However, the distance between the driver and the front of the vehicle is typically up to 2.4 m, which is a significant portion of the shorter stopping distances. It is therefore recommended that an additional allowance be made by adding 2.4 m to the SSD. Other points of note include the following.

■ Eye levels can vary from 2.00 m, in commercial vehicles, to 0.60 m, for the height of a child.

Figure 2.3 Measurement of stopping sight distance (DfT, 2002)
© Crown Copyright, 2002
This information is licensed under the Open Government Licence v3.0. To view this licence, visit http://www.nationalarchives.gov.uk/doc/open-government-licence/ **OGL**

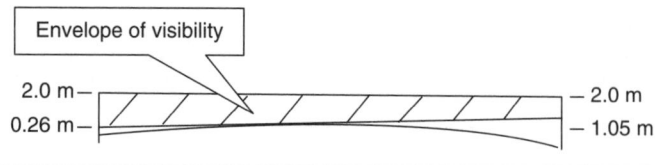

Figure 2.4 Measurement of full overtaking sight distance (DfT, 2002)
© Crown Copyright, 2002
This information is licensed under the Open Government Licence v3.0. To view this licence, visit http://www.nationalarchives.gov.uk/doc/open-government-licence/ **OGL**

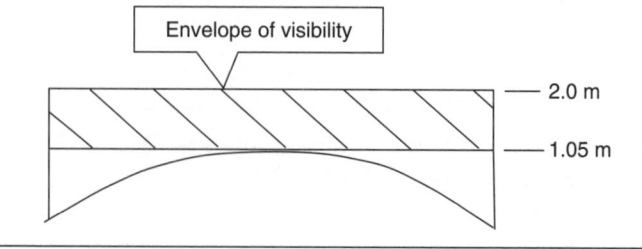

Figure 2.5 Vertical visibility envelope (based on DfT, 2007b)
© Crown Copyright, 2007
This information is licensed under the Open Government Licence v3.0. To view this licence, visit http://www.nationalarchives.gov.uk/doc/open-government-licence/ **OGL**

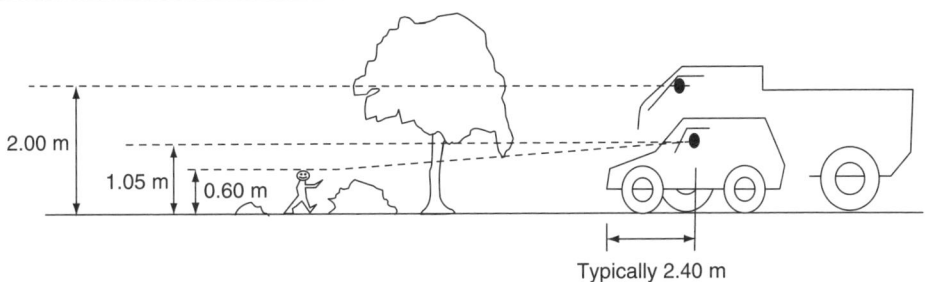

- Obstructions to visibility include summits, adjacent buildings, including bus shelters, screen walls, densely planted trees and parked cars.
- Sight lines should be clear between a height of 600 mm and 2 m above ground level.

2.5.   Vehicle dimensions

From the perspective of geometric design of pedestrian facilities, the principal dimensions of vehicles are those involved when the vehicle is manoeuvring at locations such as junctions and where the carriageway and road width are restricted. In such cases, the design of footways and junction configuration may be affected by the needs of various categories of vehicle. As an indication of the range of dimensions to be expected, Figures 2.6 and 2.7 indicate the major dimensions of typical vehicles. Further details are available in proprietary software,

which provides wheel and overhang pathways for a variety of vehicles when turning and otherwise manoeuvring.

2.6.   Summary

Important aspects of the characteristics of drivers and vehicles, which affect the location, analysis and design of pedestrian facilities, have been outlined in this chapter. Knowledge of the important variables, including vehicle speed and, in particular, stopping sight distance, involved in the siting and configuration of pedestrian facilities is an integral part of the geometric design of pedestrian facilities.

For any specific location, the designer must evaluate the suitability of the values presented and apply them in the context of the trunk road facilities described in the *DMRB* or *MfS*. Moreover, because of the importance of SSD to the safety

Figure 2.6 Vehicle dimensions (DfT, 2007b)
© Crown Copyright, 2007
This information is licensed under the Open Government Licence v3.0. To view this licence, visit http://www.nationalarchives.gov.uk/doc/open-government-licence/ **OGL**

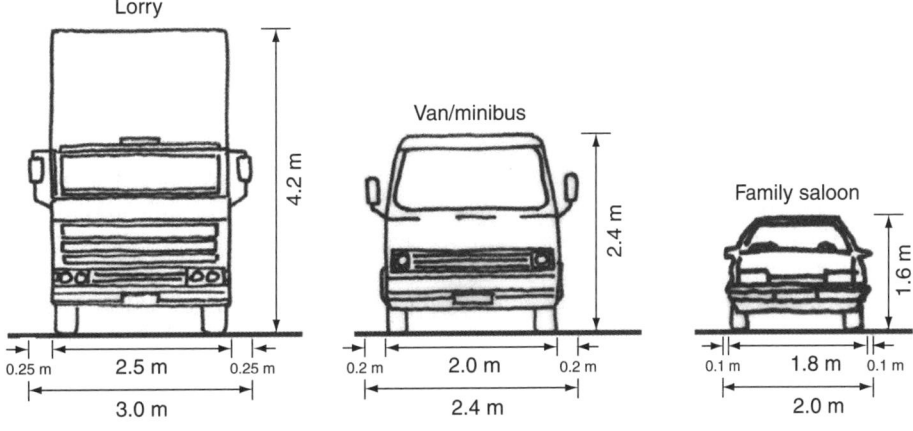

Figure 2.7 Typical bus dimensions (DfT, 2007b)
© Crown Copyright, 2007
This information is licensed under the Open Government Licence v3.0. To view this licence, visit http://www.nationalarchives.gov.uk/doc/open-government-licence/ **OGL**

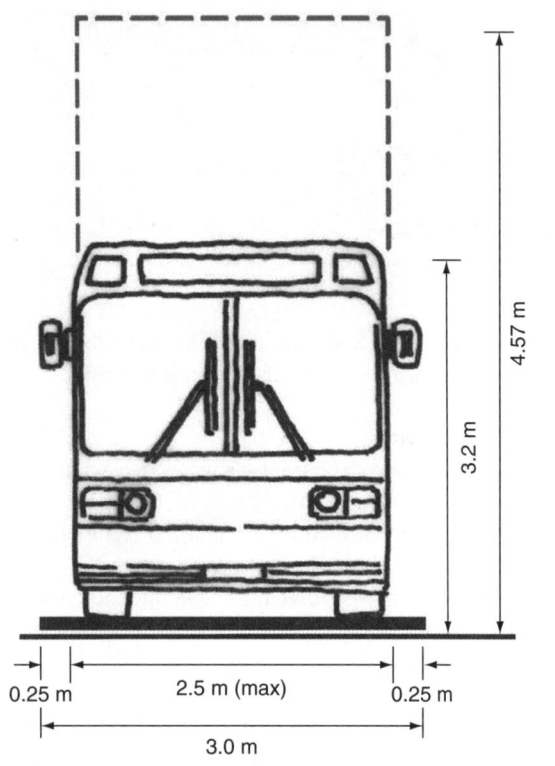

aspects of pedestrian facilities, the theoretical basis for the tabulated values should be an essential consideration in any engineered solution.

REFERENCES

AASHTO (American Association of State Highway and Transportation Officials) (2011) *A Policy on the Geometric Design of Highways and Streets*. AASHTO, Washington, DC, USA.

Birch K (2001) Truck Braking Systems and Stopping Distances, Appendix 1. http://www.ukmotorists.com/hgv%20braking%20distances.asp (accessed 03/10/2018).

CIHT (Chartered Institution for Highways and Transportation) (2010) *Manual for Streets 2 – Wider Application of the Principles*. CIHT, London, UK.

DETR (Department of the Environment, Transport and the Regions) (1992) *Design Bulletin 32 (DB32) Residential Roads and Footpaths: Layout Considerations*, 2nd edn. DETR, London, UK.

DfT (Department for Transport) (1981) TA 22/81: Vehicle speed measurement on all purpose roads. In *Design Manual for Roads and Bridges (DMRB)*. DfT, London, UK.

DfT (1988) TA 12/81: Traffic signs and lighting. In *Design Manual for Roads and Bridges (DMRB)*. DfT, London, UK.

DfT (2002) TD 9/93: Highway link design, Amendment no. 1. In *Design Manual for Roads and Bridges (DMRB)*. DfT, London, UK.

DfT (2006) TAL 1/06, Part 2: *General Principles of Traffic Control by Light Signals*. DfT, London, UK.

DfT (2007a) TA 12/07: Traffic signals on high speed roads. In *Design Manual for Roads and Bridges (DMRB)*. DfT, London, UK.

DfT (2007b) *Manual for Streets*. Thomas Telford, London, UK.

DfT (2013) *Setting Local Speed Limits*. Circular 01/2013. DfT, London, UK.

DfT (2017) *The Highway Code*. DfT, London, UK.

DfT (2018) *Design Manual for Roads and Bridges (DMRB)*. DfT, London, UK.

Euro NCAP (2018) Vulnerable road user (VRU) protection. https://www.euroncap.com/en/vehicle-safety/the-ratings-explained/vulnerable-road-user-vru-protection/ (accessed 01/05/2018).

Greenshields BD (1936) Reaction time in automobile driving. *Journal of Applied Psychology* **20(3)**: 335.

Harwood DW, Fambro DB, Fishburn B *et al.* (1995) International sight distance design practices. *International Symposium on Highway Geometric Design Practices, Boston, MA*. Transportation Research Board, Washington, DC, USA.

Maycock G, Brocklebank PJ and Hall RD (1998) *Road Layout Design Standards and Driver Behaviour*. Report 332. TRL, Crowthorne, UK.

Olson PL (1989) Driver perception response time. In *Motor Vehicle Accident Reconstruction: Review and Update*. Society of Automotive Engineers, Warrendale, PA, USA.

Olson P (1997) Driver perception–response time. *Proceedings of the 3rd National Conference, Telford, 14–16 November 1997*. Institute of Traffic Accident Investigators, Shrewsbury, UK.

Richards DC (2010) Road safety web publication no. 16. Relationship between speed and risk of fatal injury: pedestrians and car occupants. DfT, London, UK. See https://nacto.org/docs/usdg/relationship_between_speed_risk_fatal_injury_pedestrians_and_car_occupants_richards.pdf (accessed 03/10/2018).

Rosén E, Källhammer JE, Eriksson D *et al.* (2010) *Pedestrian Injury Mitigation by Autonomous Braking*. Paper Number 09-0132. Autoliv Research, Stockholm, Sweden.

Wright PH (1996) *Highway Engineering*, 6th edn. John Wiley, New York, NY, USA.

York I, Bradbury A, Reid S, Ewings T and Paradise R (2007) *The Manual for Streets: Evidence and Research*. TRL Report 661. Transport Research Laboratory, Crowthorne, UK.

SELECTED FURTHER READING

CIHT (Chartered Institute of Highways and Transportation) (1997) *Transport in the Urban Environment*. CIHT, London, UK.

Salter RJ and Hounsell NB (1996) *Highway Traffic Analysis and Design*, 3rd edn. McMillan, Basingstoke, UK.

Schoon, John G
ISBN 978-0-7277-6309-9
https://doi.org/10.1680/pfse.63099.047

Chapter 3
Pedestrian characteristics

This chapter describes the physical dimensions required by pedestrians, as well as key characteristics of individual pedestrian movement. Although other pedestrian characteristics, such as characteristics of disabled people, and characteristics pertaining to junctions, signalised locations, roundabouts and locations where pedestrians share facilities with other non-motorised users have similarities to those described here, more specific details of the relevant characteristics are described in the chapters on those topics. Pedestrian movements in groups and the capacities of pedestrian facilities are also of concern; these are addressed in Chapter 13.

The emphasis here is on the characteristics of pedestrians that most affect safety and, by association, the comfort and attractiveness of walking as a major transport mode. It is when they are crossing a carriageway that pedestrians are most at risk of collisions with motor vehicles. Pedestrians' physical and dynamic characteristics reflect a wide range of abilities and are not restricted by legal requirements on access and mobility. Anyone, regardless of physical condition, is entitled to use the public road or highway unless this is prohibited by law; for example, on motorways, access is restricted to motorised traffic.

3.1. Design emphasis
Most current design procedures for pedestrian facilities are based on driver and vehicle characteristics, on the basis that drivers will take the initiative to stop in order to avoid collisions with pedestrians. There is much less emphasis on how designs can provide pedestrians with the means of avoiding collisions with motor vehicles, apart from restricting pedestrian movement, such as with barriers, nor is there research into the behaviour and movement characteristics of pedestrians that would support such design approaches.

An example of research and potential applications addressing pedestrians' behaviour is that of observation–reaction time (similar in concept to a driver's perception–reaction time, described in the preceding chapter), which has only been addressed recently. A brief introduction to the key features of this research and how it may affect approaches to design of pedestrian facilities is provided in this chapter.

3.2. Pedestrian categories
Several methods of categorising pedestrians are in use. For example, one such method is categorisation by trip purpose and abilities (DfT, 2004), in which the categories and comments on their requirements are as follows.

- *Commuter* – prefers a fast direct route between home and work or when accessing public transport, regardless of quality of environment.
- *Shopper/leisure walker* – looks for ease of access, attractive retail environments, and attractive routes.
- *Disabled person* – requires level, clearly defined easy access and careful attention in the design and placement of street furniture, including resting points. Satisfying these requirements will also satisfy the needs of all other users, especially older people, people with heavy shopping/young children, and people with temporary impairments or low levels of fitness.
- *Child* – requires a high level of segregation from motorised traffic and/or other measures to reduce the dominance of motor vehicles, such as speed reduction, together with good passive surveillance from other users. These are important factors where children and young people make independent journeys, especially journeys to school.

(DfT, 2004)

As shown in this list, the categories distinguish between the differing priorities assigned to various aspects of the road – for example, safety versus directness – for users with different requirements due to their journey purpose, level of experience or physical ability. The design of the most appropriate infrastructure must take account of the anticipated type of user.

3.3. Pedestrian dimensions
Most dimensions associated with pedestrians, as published in guidance documents, are those of the physical infrastructure associated with pedestrians rather than the statistical distribution of personal dimensions. There is, therefore, a lack of readily available and documented evidential basis for determining infrastructure dimensions.

Table 3.1 Selected footway widths, applications and information sources

Unobstructed footway width	Application	Source of information
>2 m	Lightly used streets, such as residential areas; allows for wheelchairs to pass	*MfS* (DfT, 2007)
>2 m	Heavily pedestrian use at shops and other locations	*MfS* (DfT, 2007)
>3 m	Clear corridor widths for routes that are designed for occasional use by maintenance vehicles	(DETR, 1992)
>3 m	Outside entrances to schools and similar community buildings	(DETR, 1992)

Guidance on dimensions required by individual pedestrians is published in *Manual for Streets* (*MfS*) (DfT, 2007) and *Inclusive Mobility* (DfT, 2005a). Related dimensions, mostly confined to facilities where pedestrians share space with other non-motorised users are given in TA 90/05 *The Geometric Design of Pedestrian, Cycle and Equestrian Routes* (DfT, 2005b), TD 36/93 *Subways for Pedestrians and Pedal Cyclists Layout and Dimensions* (DfT, 1993) and TA 91/05 *Provision for Non-Motorised Users* (DfT, 2005c), which provides background material on wider issues and applications. Additionally, the *Highway Capacity Manual* (TRB, 2016), discussed in Chapter 13, provides typical dimensions used in the capacity analysis of pedestrian facilities.

In this chapter, the major concern associated with geometric design is with pedestrians' safety when crossing a carriageway. More detailed design issues are addressed in the relevant chapters. These include design for the promotion of inclusive mobility (Chapter 8) and design of facilities for shared use with non-motorised users (Chapter 11).

Because pedestrians need to move along a footway and then position themselves *across* it in order to cross, a selection of minimum footway dimensions is listed in Table 3.1; a portrayal of a more general street location, including frontage furniture and kerb zones, is shown in Figure 3.1.

As will be discussed later in this chapter, the length of the crossing unit when positioned on the footway and waiting to cross is important, owing to the time required to move across the footway after a decision is made to cross. Selected approximate dimensions are shown in Figure 3.2.

3.4. Pedestrians' crossing behaviour under vehicular traffic conditions

The actions of pedestrians, as with all road users, depend primarily on their inherent physical and mental abilities and level of well-being. In addition, their behaviour is affected by conditions from which they experience and learn. With road safety, the presence of physical features and motorised traffic has significant effects. As stated earlier, most pedestrian accidents involving motor vehicles occur when a pedestrian is crossing a carriageway – the space between the safety of the opposing footways.

This section, therefore, presents the results of pedestrians' actions resulting from crossing a carriageway – sometimes resulting in hurried, imprudent decisions, owing to heavy vehicle traffic and the pedestrian's abilities. This is followed by consideration of pedestrians' actions where their abilities reflect a safer sequence and time taken to cross a carriageway in accordance with recognised safety codes and conduct.

NOTE
The description of pedestrians' actions described in this section refer to the case where, essentially, the procedure outlined in the Highway Code *is followed. However, a different approach – the notion of 'eye contact' between drivers and pedestrians, can result in significantly different patterns of pedestrian behaviour when preparing to cross and when crossing. This latter area has not been researched extensively and attempts at defining and quantifying the variables involved appear to consist mainly of anecdotal reports. It remains, therefore, an important area for investigation, and various aspects of it are addressed in the later chapters on crossings.*

3.4.1 Head movement and delay

Observations and analyses of adult and child pedestrians crossing carriageways were made under varying traffic conditions at four uncontrolled crossing sites (Grayson, 1975). The number and percentage of head movements at various stages of the approach and crossing were recorded, together with observations of the related delays, method of crossing, and compliance with the Green Cross Code actions described in Chapter 1. A selection of relevant results is shown in Table 3.2, indicating behaviour in terms of head movements, delay time and compliance with the *Highway Code*. Major conclusions and comments related to these results are given next.

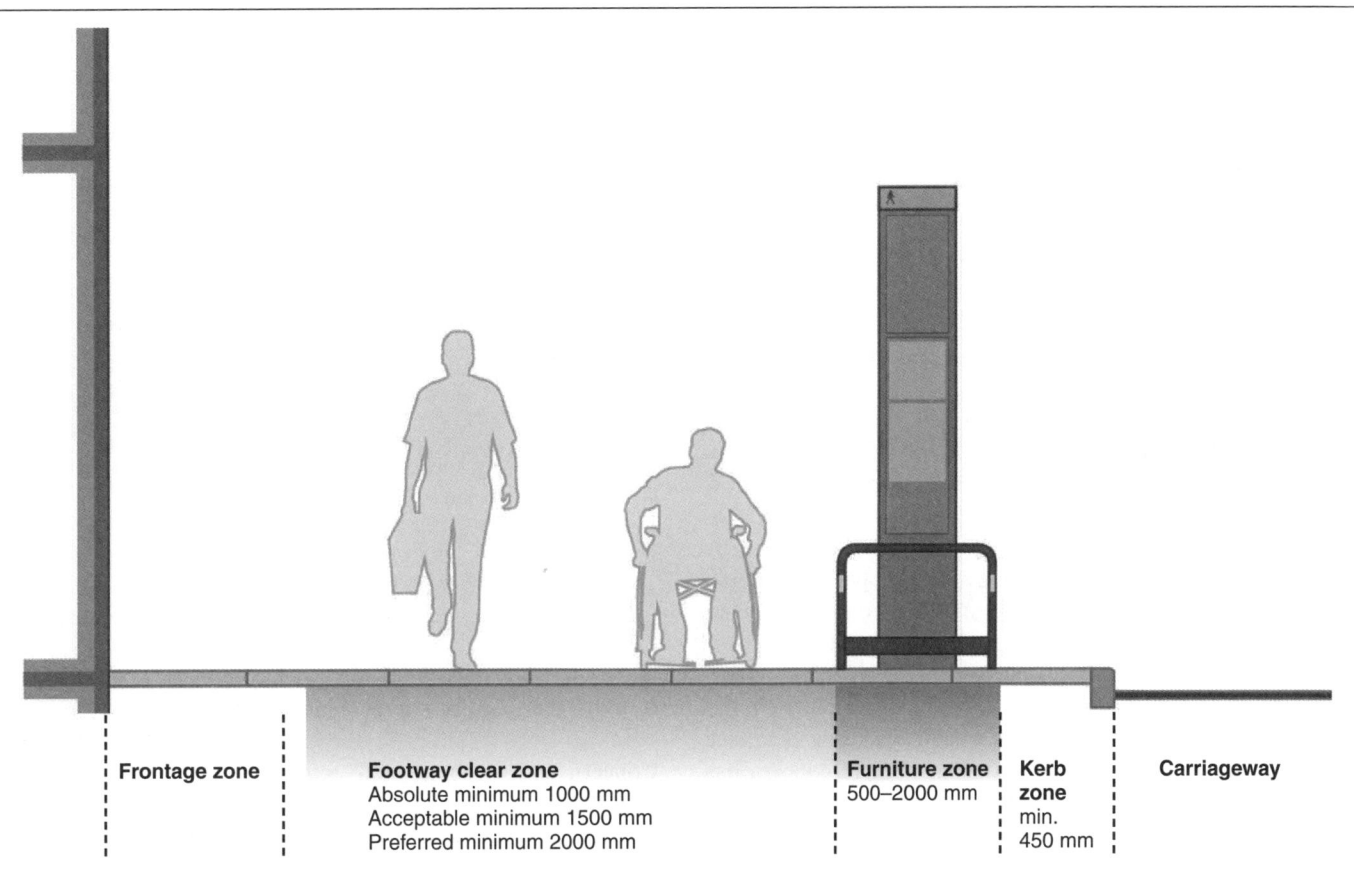

Frontage zone | Footway clear zone
Absolute minimum 1000 mm
Acceptable minimum 1500 mm
Preferred minimum 2000 mm | Furniture zone
500–2000 mm | Kerb zone min. 450 mm | Carriageway

3.4.1.1 Head movements
The average number of head movements, ranging from 1.18 to 2.88, made by the children did not reach the three head movements (look right, look left, look right again) that would be expected from the Green Cross Code. However, many of the pedestrians had made head movements as they approached the location on the kerb from which they crossed. Interestingly, although in most locations the children made more head movements than the adults, for the heavy traffic situation at Camberley, the adults were recorded as making an average of 3.05 head movements, compared with 2.64 for the children.

3.4.1.2 Delay time
Delay times for the children ranging from 3.7 to 7.8 s appear to be considerably greater than those for the adults, again with the exception of the heavy traffic situation at Camberley, where the delay time for adults was 9.5 s compared with 7.8 s for the children.

3.4.1.3 Green Cross Code compliance
Between 19% and 73% of pedestrians stopped at the kerb, and between 5% and 56% looked both ways at the kerb. Yet only between 0% and 5% of pedestrians observed the Code in its entirety.

3.4.2 Pedestrians' delay related to vehicle volumes
In a study of delays to pedestrians at different types of crossing (Goldschmidt, 1977), predictive equations were developed, which linked vehicle volumes to the delays. For example, using some of the relationships developed, for a traffic volume of 500 vehicles per hour at random points on a kerbside, the mean pedestrian delay was 2.4 s. Correspondingly, approximately 40% of pedestrians delayed.

3.4.3 Age-related effects
Age-related differences in crossing behaviour were examined at three roads in busy shopping areas (Wilson and Grayson, 1980).

Figure 3.2 Illustrative approximate unit length requirements for pedestrians waiting to cross

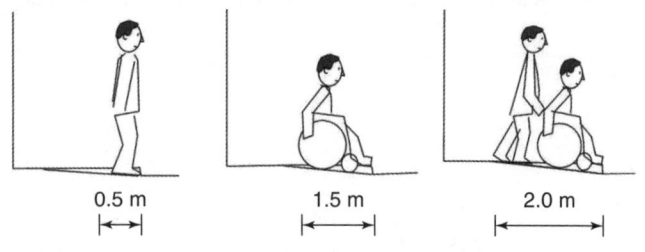

0.5 m	1.5 m	2.0 m

The road segments comprised segments 50–100 m in length, with no pedestrian crossings in the vicinity, shops on both sides of the road and free-flowing traffic in the range of 600–1200 vehicles per hour. The conditions and findings are summarised in Table 3.3.

3.4.4 Walking speeds

Walking speed is generally assumed as a minimum of 1.2 m/s. The origins of this value are unclear. A useful background is provided by Crabtree *et al.* (2014), citing a review of research by LaPlante and Kaeser (2007), which gave a 15th percentile figure of 3.5 ft/s (1.07 m/s) and noted lower speeds for other pedestrian groups, such as older people, with a 15th percentile speed of 3.0 ft/s (0.92 m/s). UK advice specifies that designers of crossings used by a high proportion of slower pedestrians should adopt a slower speed than 1.2 m/s when calculating clearance periods at signalised pedestrian crossings. Recent concern has been expressed about pedestrians crossing while distracted by the use of a mobile phone or other similar device. Reliable data on the extent of this phenomenon have not been published and it is quite possible that records of pedestrian collisions for this

Table 3.3 Summary of pedestrian crossing characteristics, based on Wilson and Grayson (1980)

	Male	Female	All
Mean delay at kerb: s			
Alone	1.6	2.4	2.2
Accompanied	2.3	3.0	2.8
All	1.8	2.7	2.4
Mean number of head movements made before crossing			
Alone	2.4	2.8	2.6
Accompanied	2.8	3.0	2.9
All	2.5	2.9	2.7
Mean number of head movements made during crossing			
Alone	3.2	3.6	3.5
Accompanied	3.5	3.9	3.8
All	3.3	3.7	3.6
Pedestrian walking speed for single pedestrians crossing directly without delay in road: m/s			
Age 15–19	1.37	1.29	1.32
Age 60+	1.10	1.15	1.13
All ages	1.32	1.27	1.28

reason may be categorised under such conditions as failure to see or judge the speed of approaching vehicles.

Concern over mobility and access has been expressed with regard to the increasing numbers of older and disabled pedestrians. Studies show that walking speeds of many of these groups are considerably slower than those of the general population. Walking speeds of older people as reported by

Table 3.2 Summary of pedestrian crossing characteristics, based on Grayson (1975)

	Children, at specific locations				Adults, at specific locations			
	Reading	Windsor	Bracknell	Camberley	Reading	Windsor	Bracknell	Camberley
Traffic/location conditions								
Volume of vehicular traffic	Light	Moderate	Moderate	Heavy	Light	Moderate	Moderate	Heavy
Road characteristics	Straight	T-junction and island	Crossroads	Straight	Straight	T-junction and island	Crossroads	Straight
Number of head movements at kerb								
% with none	55	52	30	32	83	63	48	27
% with	45	48	70	68	17	37	52	73
Average number	1.23	1.18	2.88	2.64	0.24	1.07	1.58	3.05
Delay time at kerb								
% with none	52	47	28	29	81	57	44	27
% with	48	53	72	71	19	43	56	73
Average delay: s	3.7	3.5	7.7	7.8	0.4	3.0	3.6	9.5

Table 3.4 Walking speeds of older people, measured in their homes (Asher *et al.*, 2012)

	65–69	70–74	75–79	80–84	85+	All 65+
Mean walking speed: m/s	1.0	0.9	0.9	0.8	0.7	0.9
25th percentile	0.8	0.8	0.7	0.6	0.5	0.7

Asher *et al.* (2012) are shown in Table 3.4. The results of this study are particularly relevant because the subjects were observed walking where no vehicular traffic was present. Thus the sample was not biased by the presence in the sample of pedestrians who walked hurriedly, owing to intimidation by traffic, or those who did not walk at all because of reluctance to cross, owing to the presence of traffic.

3.4.5 Significance of findings

The values determined in the preceding studies provide a perspective on pedestrians crossing a carriageway under various actual traffic and locational characteristics. From the nature of the studies, the values include those for pedestrians who may adopt unsafe (perhaps owing to feeling pressured in crossing to avoid excessive delay) as well as safe crossing procedures, and exclude potential pedestrians, such as those with a disability or encumbrance, who may be deterred by the traffic and physical conditions at the crossing.

The concept, therefore, of actual crossing time against 'required' time, has relevance to the total crossing process discussed next.

3.5. The pedestrian's crossing process

The process of crossing a carriageway, although typically considered in most official guidance in terms of walking speed only, is considerably more complex. The results of recent studies are presented here in order to illustrate potential considerations in the continuing development of analytical and design techniques.

3.5.1 Elements in crossing a carriageway

The mental and physical task of crossing a road involves a person positioning him- or herself on the footway just behind the kerb before crossing and then sequentially

- observing and reacting to approaching traffic
- crossing the carriageway itself
- gaining the opposite kerb to become fully positioned on the opposite footway.

In addition, a safety margin time before an oncoming vehicle approaches on the side of completing the crossing may also be necessary, especially if a wheelchair or child's buggy has to be tilted to negotiate the kerb. Each of the elements is associated with a time and distance. In turn, these times and distances are key elements in a crossing's analysis and design in terms of determination of visibility distance, prevailing vehicle speeds and, hence, the geometric layout and traffic control methods.

This section addresses the matter of what may be called the notions of 'time required' against 'time accepted' – the latter being the time taken under actual traffic conditions. Key elements of the crossing stages and terminology used are shown in Figure 3.3, and are discussed in greater detail in the ensuing sections.

Figure 3.3 Pedestrian's crossing stages (Schoon and Hounsell, 2006)

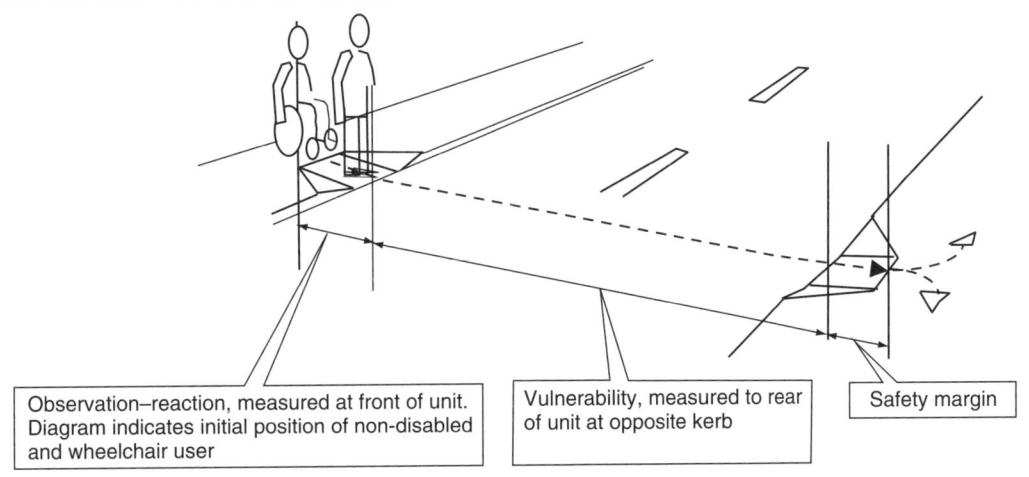

Observation–reaction, measured at front of unit. Diagram indicates initial position of non-disabled and wheelchair user

Vulnerability, measured to rear of unit at opposite kerb

Safety margin

3.6. Pedestrian observation–reaction time

This section presents findings of investigations into the actions of pedestrians in their preparation to cross a road, during their all-important decision-making process, the outcome of which can significantly affect their and other road users' safety.

3.6.1 The concept of observation–reaction

The term 'observation–reaction' for pedestrians was coined by Schoon (2003) for use with pedestrian actions and the total observation–reaction time divided into different components. These components included those times when a pedestrian is 'blind' to vehicles approaching from specific directions when looking each way in turn. The term observation–reaction is used for pedestrians instead of 'perception–reaction' because, whereas a driver in the normal driving task may expect a random appearance (and therefore a need and time to 'perceive' the nature of a reason for stopping), a pedestrian before crossing usually expects to see a vehicle – the approximate form and implications of which he or she is usually aware.

As with drivers who experience perception–reaction times, pedestrians when preparing to cross a carriageway also require time to observe and react to traffic conditions and to take appropriate action – that is, to start to walk across the carriageway or not. This section addresses the matter of what could be called the notions of 'time required' against 'time accepted', the latter illustrated by the results of the studies described in the preceding section.

3.6.2 Examples of pedestrians' actions preparatory to crossing

The exact nature of pedestrians' and drivers' perceptive and cognitive processes is complex. However, it has been implicitly recognised in several cases. For example, a 'comfort time' of 3 s is suggested as a base setting for pedestrians at signal-controlled crossings in the UK (DfT, 2005d). A pedestrian 'start-up' time of 3 s, for design purposes, is documented in the *Highway Capacity Manual* (TRB, 2016) based on extensive studies (Knoblauch *et al.*, 1996) for pedestrians at signalled pedestrian crossings. Other studies (Fugger *et al.*, 2001) found that men at signalled intersections required up to 2.7 s for the time before crossing. However, the conditions for observations at signalled locations are somewhat different from those for uncontrolled crossings. Regarding gap acceptance at unsignalled locations, an assumed time of 2.0 s to allow pedestrians to perceive and react to gaps in vehicle flow before crossing is mentioned (Hunt and Abduljabbar, 1993). In a pilot study, 85th percentile total observation–reaction times of 3.54 s and 5.12 s for a mature male subject and a mature female subject, respectively, were recorded (Schoon, 2003); in further studies of pedestrians in wheelchairs, the corresponding 85th percentile times were 4.2 s and 6 s – the difference appearing to be due to the additional time required for the wheelchair users to cross the footway before entering the carriageway (Schoon and Hounsell, 2005, 2006).

3.6.3 Pedestrians' observational tasks

Some important characteristics of a pedestrian's observations when preparing to cross a carriageway may be summarised as follows.

- The observation angles of up to 180° along a straight road or nearly 270° at a junction that a pedestrian must check require a considerable time to scan. By comparison, the typical driver's scanning angle is up to approximately 48° (Mourant and Rockwell, 1972).
- During the observation stage, the angular distance will require extensive head and some shoulder movement – a movement that some pedestrians, particularly those with physical disabilities, those encumbered and, particularly, those in wheelchairs, may find difficult or, in more extreme cases, impossible.
- In most cases, a pedestrian must be aware of the possibility of vehicles approaching from at least two opposite directions simultaneously. Because a pedestrian can only focus in one direction at a time, he or she has a significant 'blind area' for some time during the crossing process, even if visually scanning to each side during the crossing process.
- Additional time will be required for pedestrians with pushchairs, people in wheelchairs and others who cannot locate themselves, and therefore their viewpoint, just behind the kerb on the footway. In these cases, effective reaction time must include moving their body position some distance across the footway to the kerb, while the pushchair or forward part of the wheelchair precedes the pedestrian into the carriageway.

These points imply that, compared with drivers, the observational tasks of pedestrians have greater complexity of body movement than drivers in turning the head (and usually the torso) and focusing several times during the total observational procedure. In addition, pedestrians with helpers or those accompanied with small children often require considerably more time than others.

The four-stage observation–reaction process for pedestrians (similar to a driver's detection, identification, decision and response) culminates in the pedestrian's start of the first step into the carriageway. If, during the initial observational procedure, a vehicle is approaching that does not permit an acceptable gap to enable the pedestrian to cross, the pedestrian must repeat the entire detection-identification-decision-response process before physically starting to walk.

3.6.4 Analysis of pedestrian crossing actions

As a guide for examining the necessary pedestrian actions when about to cross a road, the Green Cross Code within the *Highway Code* is useful. The instruction

Children learn by example, so parents and carers should always use the Code in full when out with their children
(DfT, 2017)

is also pertinent. Such a procedure also defines a common set of actions for repetition in further investigations. The Code enables the actions to be broken down into finite parts so that the time taken to conduct each one can be identified and analysed.

This approach, using a specified set of actions such as those in the Green Cross Code, contrasts with the values related to pedestrians' gap acceptance – as typified in the studies mentioned in the earlier section. This distinction enables initial attempts at defining a 'minimum required' observation–reaction time to be made for analysis and design purposes. Furthermore, if it is expected that road users abide by the Code, it seems reasonable to design pedestrian (and other) facilities in such a way that the Code can safely be followed. The main headings from the Green Cross Code describing this procedure are listed next, with the original section reference numbers included.

7. The Green Cross Code. The advice given [...] on crossing the road is for all pedestrians.

(*a*) First find a safe place to cross.
(*b*) Stop just before you get to the kerb.
(*c*) Look all around and listen.

NOTE
This action was formerly known as the 'Kerb Drill' and consisted of instructions to 'look right, look left, look right again, and if clear begin to cross' – thereby providing a specific number of actions to be undertaken. These actions are used here in order to provide an identifiable, broadly acceptable, repeatable procedure that may be followed (in possibly modified form) in future investigations for comparison and data-recording purposes.

(*d*) If traffic is coming, let it pass.
(*e*) When it is safe, go straight across the road – do not run.

8. At a junction. When crossing the road look out for traffic turning into the road, especially from behind you.

In this procedure, items (*a*) and (*b*) are considered here to precede the start of the observational process, while item (*c*) incorporates the observation–reaction element during the process of looking (detection, identification and decision)

immediately before the response, or reaction (*d*), i.e. action taken or not taken to start to cross.

A key element is the phrase in subsection (*e*), which states 'When it is safe'. It is assumed here that this means that, in the pedestrian's estimation, a collision with a vehicle will not occur or that a vehicle will not come close enough to cause the pedestrian concern or intimidation.

The Code's actions might be compressed by some pedestrians because they do not stop for each action but conduct them while walking or even while beginning to cross. This, however, is probably difficult or time-consuming for anyone with an encumbrance, a companion or a person with some impediment. Even for a perfectly able pedestrian, this action could lead to an erroneous assessment of approaching vehicles' speeds – especially if lighting conditions are inadequate. However, in the absence of other recognised and generally accepted pedestrian codes, and in order to specify and delineate clearly the steps involved, the Green Cross Code procedure was adopted for use here. This decision was made while also recognising that the Code may have some deficiencies in its content and structure that are not addressed further here.

For two-way roads with refuge islands (which effectively make a two-way road into two one-way roads from a pedestrian's point of view), even though a pedestrian has priority at any part of the crossing, we assume (under 'drive-on-the-left' rules) that the pedestrian looks to the right first on the first part of the crossing and looks to the left first on the second part of the crossing.

It seems reasonable to assume that the combination of head and eye movements in the directions of approaching vehicles amounts to approximately 90° to the left and 90° to the right from the look-ahead position. Each head movement is initiated immediately after the preceding observation direction. Thus, the pedestrian's attention is then aimed at a possible vehicle approaching from the new direction – that is, the pedestrian is no longer focusing on and fully aware of a vehicle approaching from the preceding direction.

3.6.5 Observation–reaction time

Immediately following the end of the 'look all around and listen' procedure (i.e. 'look right, look left, look right again'), the final elements of the pedestrian's observation–reaction process, *and associated time to undertake it*, take place, as follows.

Detection, identification and decision

■ Decide to walk, after estimating that any approaching vehicle can be avoided by a safe margin if it does not slow down.

■ Look back to the ahead position.

■ Focus.

Response

■ React by starting to cross the carriageway by initiating a leaning forward movement of the body in order to take the first step. For someone pushing a pushchair, or in a wheelchair, the first action is to initiate a forward movement to cross part of the footway if the starting point is at the top of a ramp.

3.6.5.1 Observation–reaction times: three concepts

Three forms of observation–reaction time (the directions of looking based on the UK's drive-on-left rule) may be identified as follows.

■ *Total observation–reaction time.* The total time, when standing at the kerb, from beginning to look to the right (right saccade) then looking left and right again, then looking ahead across the carriageway and finally reacting by starting to step from the kerb onto the carriageway.

■ *Penultimate observation–reaction time.* The ending point of the last 'look left' is the latest time when the pedestrian is able to focus on a vehicle approaching from the left. The speed and extent of this loss of focus will depend on the speed of the individual's head-turning movement (left saccade) and on the extent of the pedestrian's peripheral vision and mental processing abilities. From there to the 'start to walk' point, including the last 'look right' (second right saccade), and then across the carriageway, is here called the 'penultimate observation–reaction' time.

■ *Last observation–reaction time.* Starting at the end of the last look to the right (second saccade) and then through the look-ahead across the road in the direction of walking to the 'start to cross' point is here called the last observation–reaction time. The final observation and reaction process should involve observation of the vehicle closest to the pedestrian, which would, therefore, be the first to be involved in a potential collision. Under right-hand drive rules of the road, this would mean vehicles approaching from a pedestrian's right. For left-hand drive rules, the process would be similar but the initial look would be to the pedestrian's left.

The total, penultimate and last observation–reaction times for a pedestrian crossing a straight two-way carriageway and looking first to the right (right saccade) can be shown conceptually in a time–observational angle relationship. Figure 3.4 shows the pedestrian's angle of head (and eye movement) from the 'ahead' position, with saccades to left and back, related to the time taken for the movements during the observations. In this diagram, the focusing time at the end of each head movement is shown as a

horizontal line – that is, there is no head movement during these times. The periods for the penultimate observation–reaction and the last observation–reaction times are also labelled. Relationships between the saccades and fixations are described in the context of drivers' perception of hazards (Velichkovsky *et al.*, 2003). During the observation stage, a pedestrian's angular movement will require extensive head and some shoulder movement and may, for some pedestrians, be difficult or impossible – especially for people in wheelchairs or with encumbrances.

The dashed lines in Figure 3.4 indicate the possible action of most pedestrians in that when turning to look in a certain direction they start to focus on approaching vehicles during the head movement and complete the focusing, observation, etc., while the head is stationary. If no vehicle is approaching from the direction of looking, the head is then immediately turned (in the minimum time case) in the opposite direction.

The exact points at which the focusing begins and ends must be a matter for future investigation, as little information is available on the subject. Therefore, it is assumed that the combination of head and eye movements in the directions of approaching vehicles amounts to a total of approximately 90° from the look-ahead position. It is assumed that the head movement is initiated as a result of the observation–reaction process of the preceding observation direction. This implies that the pedestrian's attention is now aimed at a possible vehicle approaching from the new direction – that is, that the pedestrian no longer is focusing on and fully aware of a vehicle approaching from the preceding direction.

3.6.6 Application of observation–reaction times

In most cases, the pedestrian's observational direction at the beginnings of the penultimate observation–reaction and last observation–reaction times will depend on the initial direction in which the pedestrian looks. This is typically to the right in the UK, because this is the direction of the most immediate threat from approaching vehicles, and to the left in countries with right-hand drive. However, for safety it would be prudent to assume that the penultimate observation–reaction and the last observation–reaction times could apply for movements from looks to the left or the right of a pedestrian's position, although the penultimate observation–reaction time will always be the longer of the two. From the results reported later, a penultimate observation–reaction time of 3 s would not be unusual; in this time, a vehicle travelling at 45 km/h would travel a distance of about 40 m if the driver, for some reason, failed to take avoiding action. From a pedestrian's point of view, then, the following assumptions are appropriate.

■ For a two-way road, the penultimate observation–reaction and last observation–reaction times are applicable in each relevant direction.

Figure 3.4 Pedestrian head movement and observation–reaction time scale: concept diagram for 'drive-on-the-left' rule, based on Schoon (2003)

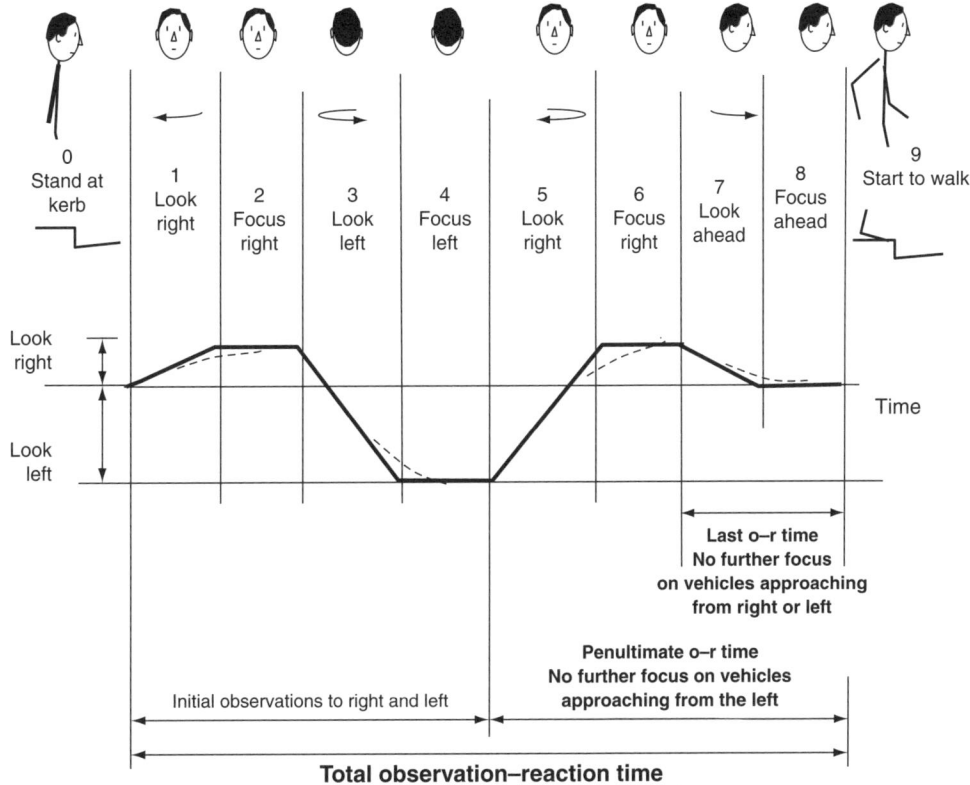

Total observation–reaction time

Summary of pedestrians' actions when following the Green Cross Code in preparing to cross a two-way carriageway

The diagram illustrates the following sequence of actions from the initial standing at the kerb (numbered items correspond with those in the pictograph).

1. First look to the right (right saccade), approximately 90°.
2. Focus to the right (starts at completion of look to the right).
3. First look to the left (left saccade) approximately 180° (starts at completion of focusing right).
4. Focus to the left (starts at completion of look to the left).
5. Second look to the right (right saccade), approximately 180° (starts at completion of focus to the left).
   From the start of this action the pedestrian is 'blind' to vehicles approaching from the left.
6. Focus to the right (starts at completion of second look to the right).
7. Look to the 'ahead' position (left saccade), i.e. looking straight across the carriageway.
   From the start of this action the pedestrian is 'blind' to vehicles approaching from the left and the right.
8. Focus to the 'ahead' position (starts at completion of looking ahead).
9. Start to walk ahead, directly across the carriageway (starts at completion of focusing ahead and represents the reaction stage of the process).

Basis for sequence of actions

The sequence of actions is based on the requirements of the Green Cross Code (formerly the 'kerb drill'): 'Stand at the kerb, look right, look left and look right again. If nothing is coming walk straight across the road.'

■ For a one-way road, a portion of the total observation–reaction time for a two-way road would be omitted. Thus, for vehicles approaching a pedestrian from the right, in Figure 3.4 only Steps 1, 2, 7, 8 and 9 would apply, while for vehicles approaching a pedestrian from the left, after the steps of looking and focusing left, Steps 7, 8 and 9 would apply.

3.6.7 Values of pedestrians' observation–reaction times

In a pilot study featuring the observation–reaction times of mature men and women at a kerb adjacent to a length of straight road, the individual and elapsed times were measured for each of the nine elements of the total observation–reaction time. The results for each subject are summarised in Figure 3.5,

Figure 3.5 Observation–reaction times (s) for men and women, based on Schoon and Hounsell (2005)

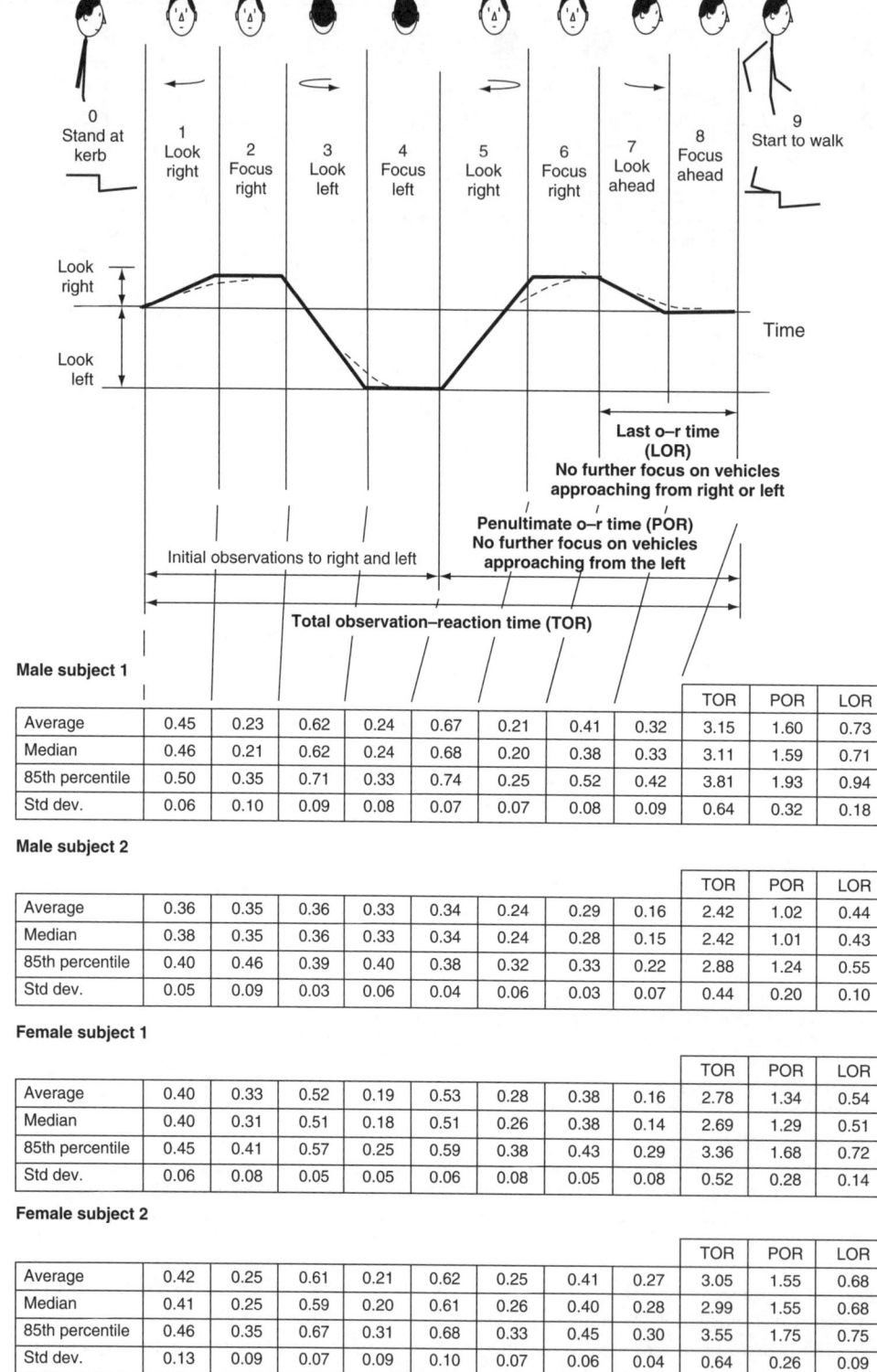

Male subject 1

									TOR	POR	LOR
Average	0.45	0.23	0.62	0.24	0.67	0.21	0.41	0.32	3.15	1.60	0.73
Median	0.46	0.21	0.62	0.24	0.68	0.20	0.38	0.33	3.11	1.59	0.71
85th percentile	0.50	0.35	0.71	0.33	0.74	0.25	0.52	0.42	3.81	1.93	0.94
Std dev.	0.06	0.10	0.09	0.08	0.07	0.07	0.08	0.09	0.64	0.32	0.18

Male subject 2

									TOR	POR	LOR
Average	0.36	0.35	0.36	0.33	0.34	0.24	0.29	0.16	2.42	1.02	0.44
Median	0.38	0.35	0.36	0.33	0.34	0.24	0.28	0.15	2.42	1.01	0.43
85th percentile	0.40	0.46	0.39	0.40	0.38	0.32	0.33	0.22	2.88	1.24	0.55
Std dev.	0.05	0.09	0.03	0.06	0.04	0.06	0.03	0.07	0.44	0.20	0.10

Female subject 1

									TOR	POR	LOR
Average	0.40	0.33	0.52	0.19	0.53	0.28	0.38	0.16	2.78	1.34	0.54
Median	0.40	0.31	0.51	0.18	0.51	0.26	0.38	0.14	2.69	1.29	0.51
85th percentile	0.45	0.41	0.57	0.25	0.59	0.38	0.43	0.29	3.36	1.68	0.72
Std dev.	0.06	0.08	0.05	0.05	0.06	0.08	0.05	0.08	0.52	0.28	0.14

Female subject 2

									TOR	POR	LOR
Average	0.42	0.25	0.61	0.21	0.62	0.25	0.41	0.27	3.05	1.55	0.68
Median	0.41	0.25	0.59	0.20	0.61	0.26	0.40	0.28	2.99	1.55	0.68
85th percentile	0.46	0.35	0.67	0.31	0.68	0.33	0.45	0.30	3.55	1.75	0.75
Std dev.	0.13	0.09	0.07	0.09	0.10	0.07	0.06	0.04	0.64	0.26	0.09

Figure 3.6 85th percentile values for key characteristics: plotted results and diagrammatic summary (Schoon and Hounsell, 2005). LOR, last observation–reaction; POR, penultimate observation–reaction

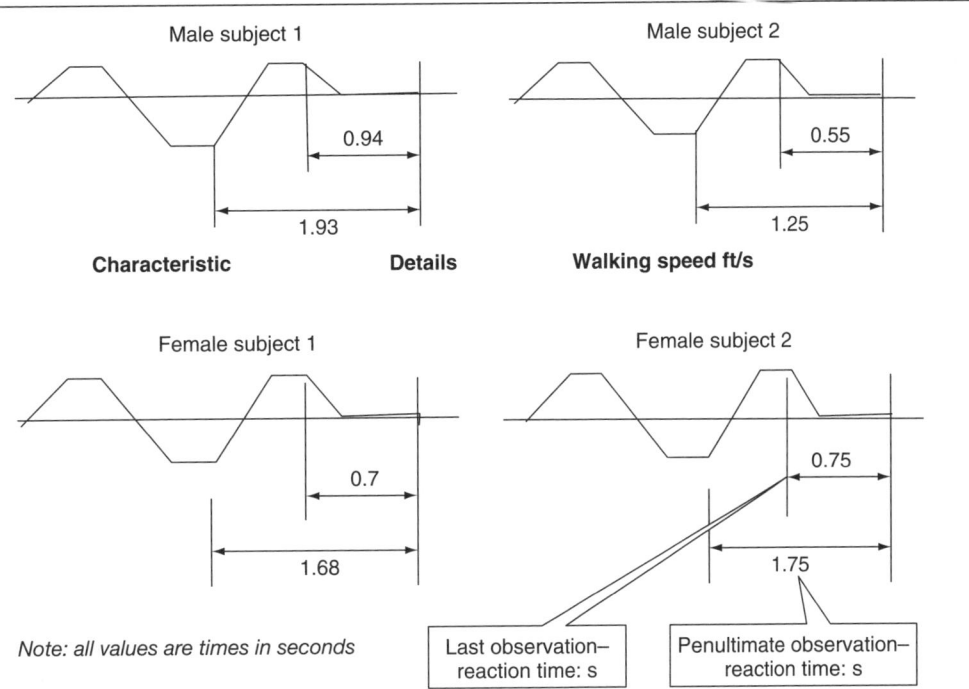

Note: all values are times in seconds

showing the average, median and 85th percentile, as well as the standard deviation and values of the total, penultimate and last observation–reaction times. The 85th percentile times are shown diagrammatically in Figure 3.6. The range of average times for the major head movements and focusing for the four subjects were as follows.

- Range of an average head saccade of 90° was between 0.29 and 0.45 s.
- Range of an average head saccade of 180° was between 0.34 and 0.67 s.
- Range of average focus times to either left or right was between 0.19 and 0.36 s.
- Range of final focus-ahead times before walking was between 0.16 and 0.32 s.
- Range of average total observation–reaction times was between 2.42 and 3.15 s.
- Range of average penultimate observation–reaction times was between 1.02 and 1.60 s.
- Range of average last observation–reaction times was between 0.54 and 0.73 s.

The maximum values, exhibited by male subject 1, showed a total observation–reaction time of nearly 4 s. Within that period, the 85th percentile last observation–reaction time related to vehicles approaching from the pedestrian's right means that this subject would be blind to his right for 1 s plus

the time taken for walking some distance into the carriageway if he simultaneously looked to the left when beginning to cross. If the sum of the last observation–reaction time and the walking time to a point in the carriageway were approximately 3 s, an approaching vehicle moving at 48 km/h (30 miles per hour) would travel approximately 40 m.

Thus if the pedestrian's visibility distance to the vehicle were less than this distance, owing, say, to some obstruction, this could result in an undesirable level of danger of collision if the driver had been inattentive – even if the pedestrian had fully observed the Green Cross Code. Circumstances compounding this situation could include a pedestrian having less than adequate ability to turn his or her head, or a pedestrian crossing with small children. Similarly, the 85th percentile penultimate observation–reaction time of nearly 2 s means that the pedestrians would be essentially unable to see vehicles approaching from their left before walking into the carriageway for this period, and the same principles, but with the relevant values, would apply in terms of pedestrian–vehicle collisions.

Although other factors may come into consideration in this scenario, it points to the fact that the pedestrian's perception-reaction time can be an important element in the calculation of visibility distances. To do this, some minimum time is required to account for the observation–reaction time of the pedestrian in the analysis and design of pedestrian crossings –

whether at controlled or uncontrolled locations. This time is not currently included in official guidance on estimating visibility distances.

3.6.8 Implications of differences between subjects

A comparison between the subjects of this study can be informative for several reasons. For example, it can be seen that all of the observation times differ considerably. This could be expected because each subject represents a small portion of a much wider population of pedestrians, but of essentially the same category (mature adults). As further investigations are made, a more complete picture will emerge about the range of values associated with these and other categories. The investigations would have to be extended to all categories to develop a satisfactory overall spectrum of pedestrians' observation–reaction abilities.

3.7. Pedestrians' crossing speeds and times

Pedestrian walking speeds are the most commonly employed parameters used in current guidance in the analysis of pedestrians crossing a carriageway. The walking speed is particularly important because, together with the width of the carriageway, it determines the amount of time the pedestrian is in the carriageway and thus vulnerable to collision with motor vehicles.

For analysis purposes, current guidelines indicate that a walking speed representing the 85th percentile of pedestrians is 1.2 m/s for use in estimating the timing of signals. Many researchers have investigated pedestrian walking speeds, including walking speeds of people with various disabilities.

Walking speed is an important aspect of crossing analysis and design. Leake *et al.* (1991) adopt an ergonomic approach and Eubanks and Hill (1998) report on a number of sources,

Table 3.5 Walking speeds for selected age groups and disabilities, based on Perry and Burnfield (1992)

Characteristic	Details	Walking speed: ft/s
Average velocity by age group – non-disabled people		
Customary walking speed	Children (6–12)	3.83
	Teenagers (13–19)	3.99
	Adults (20–59)	4.37
	Seniors (60–80)	4.05
Fast walking speed	Children (6–12)	4.81
	Teenagers (13–19)	5.41
	Adults (20–59)	5.80
	Seniors (60–80)	4.92

1 ft/s = 0.303 m/s

including a study addressing matters of gait (Perry and Burnfield, 1992), as shown in Table 3.5, which indicates a range between 3.83 and 5.80 ft/s. These values are compared with speeds of pedestrians with various infirmities in the next chapter.

3.8. Safety margin time

In order that pedestrians should not feel intimidated by having to cross just in time to avoid a vehicle passing the point of crossing, a safety margin of several seconds would seem to be appropriate. However, in current pedestrian facilities guidance analysis, no allowance is mentioned or values indicated. From observations of pedestrians crossing a typical carriageway as vehicles approach, it would seem that most pedestrians prefer at least a 2 s interval between the time when they reach the kerb after crossing to the instant that an approaching vehicle passes that point.

3.8.1 The total crossing process

In estimating the total time taken by a pedestrian in crossing a carriageway, it seems appropriate to include each of the elements experienced by the pedestrian from the point at which he or she ceases looking in the direction of approaching vehicles to the point at which safety is reached at the opposite kerb, i.e. the observation–reaction time, the walking time and the safety margin.

3.8.2 Total crossing time elements

The total crossing time comprises the pedestrian's observation–reaction (including the various head movements during the observation process) and walking times and a safety margin, and is shown conceptually in Figure 3.7. The relationship of the penultimate observation–reaction and last observation–reaction times to the total process is also shown.

3.8.3 Pedestrian's observations during walking

It should be noted that, in Figures 3.4 to 3.7, no indication is shown of the pedestrian's head movement and observations during the walking portion of the total crossing process. Although such movement is considered important from a safety point of view, the focus here is on the crossing design procedure. In practical terms, if a pedestrian sees a previously undetected vehicle approaching, the likelihood of that pedestrian stopping and reversing direction in order to avoid a collision is probably slight at best. It would seem prudent, therefore, to exclude such an action in any analysis or design of pedestrian crossings. Full consideration of the nature, extent and circumstances of the pedestrian's avoidance actions after leaving the kerb is important and can be complex; considerably more research is needed to document it adequately.

3.8.4 Length of crossing unit

Generally, most estimates of crossing time assume a single point movement representing a single person or persons crossing

Figure 3.7 Elements of total pedestrian crossing time

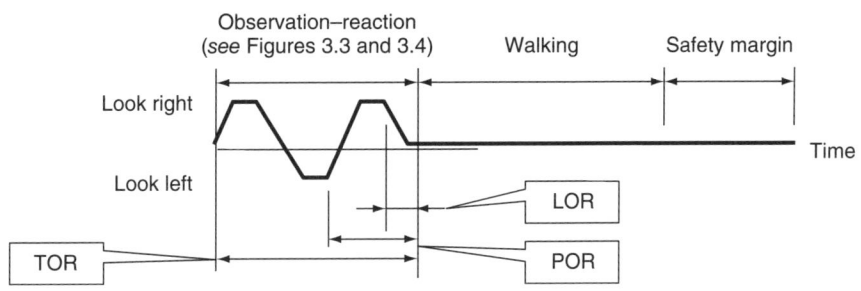

together. However, in cases where a wheelchair is being pushed by a helper or carer or a person is crossing with a buggy and small children, the 'crossing units' have significant length, which may be as much as 2 m. Because of this, the rear of this unit can be several seconds behind the normally estimated point of movement. For example, if the assumed walking speed is 1 m/s and the unit is 2 m long, an additional 2 s should be added to the crossing time in order to adequately provide for the unit's clearance across the carriageway. This matter is illustrated in more detail in Chapter 4.

3.8.5   Example
The inclusion of observation–reaction time, unit length and safety margin in estimating the total time required for a safe and comfortable pedestrian crossing can make a considerable

difference between the current method and the time actually required. In graphic terms, the differences are shown in Figure 3.8.

For a numerical example of the difference in required time, and therefore visibility distance requirements, consider a carriageway of 7 m width, an observation–reaction time of 1.5 s, a safety margin of 2 s and a length of crossing unit of 1.5 m. The difference between the total crossing time calculated using the elements of crossing noted in this chapter and that calculated using current guidance would be as shown in Table 3.6. It can be seen that an additional 5.3 s would be required for pedestrians to cross. The implications for this are that sight distances for drivers should be increased so that pedestrians wishing to cross this carriageway should have a sight distance

Figure 3.8 (a) Current DfT estimation method for pedestrian crossing time at signalled crossings. (b) Method for estimating required pedestrian crossing time at informal crossings

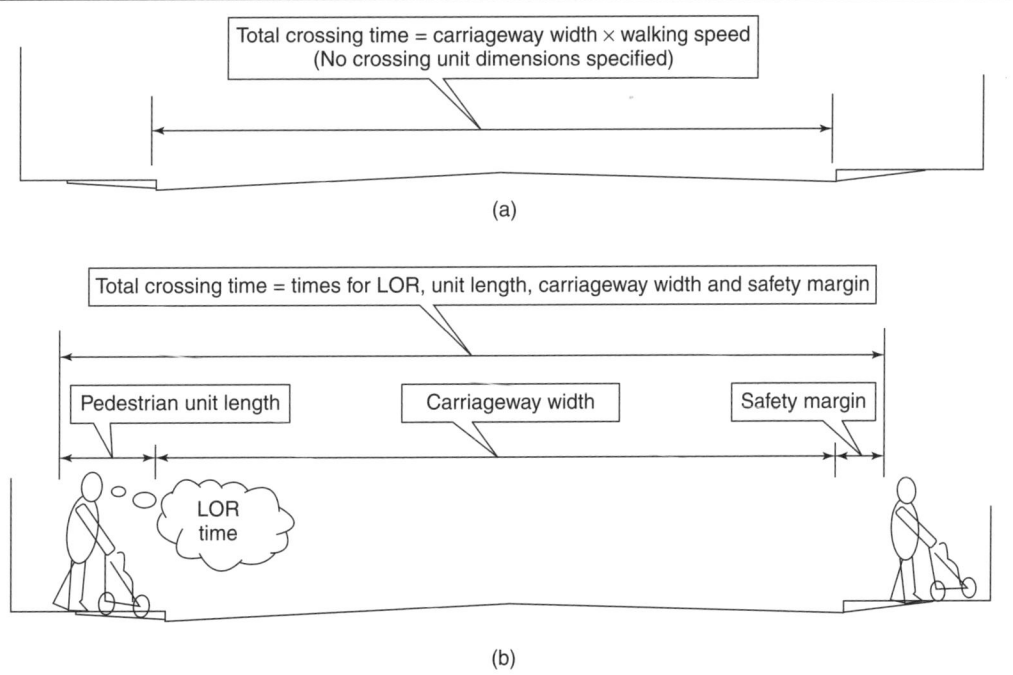

Table 3.6 Comparison of differing calculations between actual conditions and current guidance for pedestrian's total crossing time

Method	Time: s				
	Last observation–reaction time	Crossing the carriageway: 7×1.2 m/s	Length of unit: 1.5 m $\times 1.2$ m/s	Safety margin	Total crossing time
Current guidance	0	8.4	0	0	8.4
Potential total time requirement	1.5	8.4	1.8	2.0	13.7
Difference					5.3

equivalent to the distance associated with the speed of approaching vehicles.

3.8.6 Implications for design of pedestrian facilities

The current method of designing facilities for pedestrian crossings assumes that drivers will take all necessary action to avoid a collision with a pedestrian. The assumption in the analysis process about the pedestrians' actions is that they will cross only when they can see an acceptable gap in traffic to allow them to do so. This, however, ignores the fact that at some locations where pedestrians must cross, such as at some junctions, the visibility distance is insufficient for the pedestrians to cross in the knowledge that, if a driver for some reason does not see them (owing to inattention, distraction or another reason) then the geometric design does not permit the pedestrians to make the crossing without danger of collision. Apart from the obvious danger to pedestrians, this can also have a deterrent effect on the use of the facilities by pedestrians and for walking in general.

Therefore, in addition to the crossing speed, specification of pedestrians' observation–reaction times and a safety margin, as well as consideration of the crossing unit's length, could be beneficial when considering pedestrians' capabilities, especially when calculating intervisibility distances. The concepts of the penultimate and the last observation–reaction times indicated that the observation–reaction times could be a significant portion of the total time required by pedestrians crossing a carriageway. It would appear that inclusion of these times could be an important factor in the analysis and design of uncontrolled (and possibly also controlled) crossing locations and the estimation of appropriate pedestrians' sight line locations and visibility distances. At present, except for the instances mentioned earlier for signalised crossings, no consideration is given to observation–reaction times as an element in a pedestrian's total time required to cross a carriageway, nor for safety margins, nor the crossing unit's length.

3.9. Future research needs

Because of the limited amount of data on which the observation–reaction characteristics are based, further investigations concerning total crossing time would include various other pedestrian categories, types of location, including junctions, and consideration of applications. Considerations in structuring further research would include the following.

- *Pedestrian category.* Before specifying the categories to be included, a policy decision would be required concerning categories of pedestrians to include. Ideally, all pedestrians should be included. However, because little or no data are available on the quantitative observation–reaction times and crossing speeds of people who are disabled or encumbered in some way, the approach to doing this may involve consideration of a possible 'best estimate' based on the location and the abilities of the users.
- *Crossing location.* The nature of crossings can make a considerable difference in complicating a pedestrian's observational tasks and increase the time required to accomplish them. Crossings at or adjacent to a junction or other conditions that require a pedestrian's observation angle of up to 270°, with associated focusing requirements, are of particular concern.
- *Environmental conditions.* These conditions can vary from the ideal to the extremely adverse. Much investigation is needed, particularly about the various levels of lighting and weather conditions, in the context of pedestrians' abilities to observe approaching vehicles.

The concepts and values described in the preceding sections illustrate a part of the total range of values for selected pedestrian categories and locations. Clearly, further, more detailed, investigations would be valuable in the analysis and design of pedestrian facilities.

3.10. Summary

In addition to the physical dimensions, the movement characteristics of pedestrians when exposed to the possibility of collision with motor vehicles are essential inputs to the design process. Current (2018) guidance on estimating crossing times to pedestrians use only the walking speed as the required parameter but, as indicated in this chapter, pedestrian characteristics, such as the observation–reaction time, safety margin and length of crossing unit can be significant inputs to the design of safe, convenient and attractive pedestrian facilities.

REFERENCES

Asher L, Aresu M, Falaschetti E and Mindell J (2012) Most older pedestrians are unable to cross the road in time: a cross sectional study. *Age and Ageing* **41(5)**, 690–694.

Crabtree M, Lodge C and Emerson P (2014) *Review of Pedestrian Walking Speeds and Time Needed to Cross the Road*, TRL published project report PPR700. TRL, Crowthorne, UK.

DETR (Department of the Environment, Transport and the Regions) (1992) *Design Bulletin 32 (DB32) Residential Roads and Footpaths: Layout Considerations*, 2nd edn. DETR, London, UK.

DfT (Department for Transport) (1993) TD 36/93: Subways for pedestrians and pedal cyclists layout and dimensions. In *Design Manual for Roads and Bridges (DMRB)*. DfT, London, UK.

DfT (2004) LTN 1/04: *Policy, Planning and Design for Walking and Cycling*. DfT, London, UK.

DfT (2005a) *Inclusive Mobility: A Guide to Best Practice on Access to Pedestrian and Transport Infrastructure*. DfT, London, UK.

DfT (2005b) TA 90/05: Geometric design of pedestrian, cycle and equestrian routes. In *Design Manual for Roads and Bridges (DMRB)*. DfT, London, UK.

DfT (2005c) TA 91/05: Provision for non-motorised users. In *Design Manual for Roads and Bridges (DMRB)*. DfT, London, UK.

DfT (2005d) TAL 5/05: *Pedestrian Facilities at Signal-controlled Junctions*, Parts 1 to 4. DfT, London, UK.

DfT (2007) *Manual for Streets*. Thomas Telford, London, UK.

DfT (2017) *The Highway Code*. DfT, London, UK.

Eubanks J and Hill P (1998) *Pedestrian Accident Reconstruction*. Lawyers and Judges, Tucson, AZ, USA.

Fugger Jr FT, Randles BC, Wobrock JL *et al.* (2001) Analysis of elderly pedestrian gait and perception/reaction at signal-controlled crosswalk intersections. In *80th Annual Meeting of the Transportation Research Board, Washington, DC.* TRB, Washington, DC, USA.

Goldschmidt J (1977) *Pedestrian Delay and Traffic Management*. Supplementary report 356. TRRL, Crowthorne, UK.

Grayson GB (1975) *Observations of Pedestrians at Four Sites*. Report LR 670, TRRL, Crowthorne, UK.

Hunt J and Abduljabbar J (1993) Crossing the road: a method of assessing pedestrian crossing difficulty. *Traffic Engineering and Control* **34(11)**: 526–532.

Knoblauch RL, Pietrucha MT and Nitzburg M (1996) Field studies of pedestrian walking speed and start-up time. *Transportation Research Record* **1538**: 27–38.

LaPlante J and Kaeser TP (2007) A history of pedestrian signal walking speed assumptions. In *Third Urban Street Symposium, Seattle*. TRB, Washington, DC, USA.

Leake GR, May AD and Parry T (1991) *An Ergonomic Study of Pedestrian Areas for Disabled People*. TRRL contract report CR184. TRRL, Crowthorne, UK.

Mourant RR and Rockwell TH (1972) Strategies of visual search by novice and experienced drivers. *Human Factors* **14(4)**: 325–335.

Perry J and Burnfield JM (1992) Gait analysis: normal and pathological function. SLACK, Thorofare, NJ. New Jersey, USA.

Schoon JG (2003) Pedestrian observation–reaction times: concepts and pilot study. *Proceedings of the Universities' Transport Studies Group, Loughborough University, UK, January 2003*.

Schoon JG and Hounsell N (2005) Observation–reaction times of wheelchair users – a comparison with non-disabled users. In *84th Annual TRB Meeting*. TRB, Washington, DC, USA.

Schoon JG and Hounsell N (2006) Access and mobility design policy for disabled pedestrians at road crossings: exploring issues. *Transportation Research Record* **1956(1)**: 76–85.

TfL (Transport for London) (2017) *Streetscape Guidance*, 3rd edn, revision 1. TfL, London, UK.

TRB (Transportation Research Board) (2016) *Highway Capacity Manual*. National Academy of Sciences, Washington, DC, USA.

Velichkovsky BM, Rothert R, Miniotas D *et al.* (2003) Visual fixation as a rapid indicator of hazard perception. In *Operator Functional State and Impaired Performance in Complex Work Environments* (Hockey GHR, Gaillard AWK and Burov Q (eds)). IOS Press, Amsterdam, Netherlands, pp. 313–321.

Wilson DG and Grayson GB (1980) *Age-Related Differences in the Road Crossing Behaviour of Adult Pedestrians Department of the Environment*, TRRL Report 933. TRRL, Crowthorne, UK.

Schoon, John G
ISBN 978-0-7277-6309-9
https://doi.org/10.1680/pfse.63099.063

Chapter 4
Disabled pedestrians' characteristics

Access and mobility for people with physical and sensory disabilities is often more complex than for the non-disabled. This chapter, therefore, augments the description of pedestrian characteristics given in Chapter 3. It also provides details underlying the guidance on design standards for disabled pedestrians described in Chapter 8.

First, the focus is on the relationship between the *Highway Code* (DfT, 2017) advice on crossing a carriageway and the characteristics of disabled and encumbered people whose physical and movement characteristics differ from non-disabled or non-encumbered pedestrians. This provides a background to key parameters, such as walking speeds and other analysis elements related to vehicle trajectories, which establish a basis for designing safe crossings.

4.1. Disability extent

According to *Disability Facts and Figures* (DWP, 2014), there are over 11 million people with a limiting long-term illness, impairment or disability in the UK. The most commonly reported impairments are those that affect mobility, lifting or carrying. The prevalence of disability rises with age. Around 6% of children are disabled, compared with 16% of working age adults and 45% of adults over the state pension age. Around a fifth of disabled people report having difficulties related to their impairment or disability in accessing transport.

In terms of the road safety of disabled people, including traffic collisions, the problem is summarised by Williams *et al.* (2002), who indicate that between 8 and 12 people in 1000 in the UK are wheelchair users and that they and others may take longer to cross a road. It is also stated that:

> Data on the prevalence of various disabilities among children and adults, their accident involvement and their exposure [to risk] are largely unavailable. It is, therefore, difficult to quantify the extent to which disabled people are at risk of road accident involvement compared with their non-disabled peers [...] In summary, there are insufficient data to identify the policy priorities for disabled children and adults.
>
> Williams *et al.* (2002)

Types of disability may be broadly divided into physical and sensory disabilities. Among the former are limited use of limbs or torso movement; the latter include vision and auditory impairment and cognitive disabilities. Whereas physical disabilities typically require some form of mechanical or human assistance, sensory disabilities can be more difficult to identify and address. Much needs to be done to address such issues adequately, from human and animal assistance to innovative electronic sensor systems.

The design of facilities' layout and associated physical dimensions are addressed in works by Oxley and Alexander (1994), the DfT (2005) and others. The dynamic characteristics of disabled people, i.e. movement characteristics, particularly regarding safety needs when interacting with motor vehicles, have received somewhat less attention. Therefore, the emphasis here is on road crossings, where the danger of collision with motor vehicles is greatest (and, therefore, the possible reluctance to make a specific walking trip at all becomes evident) and where, as shown in Table 4.1, the nature of difficulties in walking for various groups of pedestrians becomes evident. Moreover, the effect of age on distance walked for both sexes is shown in Figure 4.1, indicating a clear decline for both sexes after the age of 65.

Aids to mobility are in general use by disabled people. A selection is shown in Table 4.2; the effects on a person waiting at a kerb to cross are depicted for several cases in Figure 4.2. In particular, this shows some of the more extreme instances of the use of disabled devices in terms of space and positioning – primarily the person's viewpoint, footway space requirements and distance to the kerb's edge. Similar ranges of dimensions and locations on the footway may be expected from people with small children, shopping or other encumbrances, and by the increasing number of motorised 'mobility scooters'.

4.2. Overview of major elements in the crossing process

The mental and physical task of crossing a road, as described in Chapter 3, involves a person positioning him- or herself on the footway just behind the kerb (in accordance with *Highway Code* instructions), before crossing and then sequentially

Table 4.1 Percentage of people reporting difficulties in the pedestrian environment, based on Hitchcock and Mitchell (1984)

Aspect of pedestrian environment	High ← Degree of disability → Low				
	Registered disabled	Elderly, difficulties with walking	Non-elderly, difficulties with walking	Elderly, no difficulty with walking	Non-elderly, no difficulty with walking
Kerbs	12	5	4	4	2
Steps	58	5	4	4	2
Hills or ramps	59	45	30	19	12
Uneven, narrow pavements	21	19	13	14	8
Crowds	50	4	0	5	2
Traffic or crossing roads	35	31	22	16	17
No difficulty	2	23	43	54	67

- observing and reacting to approaching traffic (the observation–reaction time)
- crossing the carriageway itself
- gaining the opposite kerb to become fully positioned on the opposite footway
- allowing a safety margin when reaching the opposite kerb.

Each of these elements is associated with a time and distance. These times and distances are key elements in the crossing's analysis and design, in terms of visibility distances and prevailing vehicle speeds. These, in turn, affect the geometric and control conditions (such as traffic signals and speed limits) at and in the vicinity of the crossing. Key elements of the crossing stages and terminology used are shown in Figure 4.3. It should be noted that blind and visually impaired people, including those accompanied by a helper or guide dog, may take longer

to complete a crossing and are also dependent on clues and other information. Examples of the relevant characteristics, and those associated with hearing impairments and needs, are discussed later in this chapter.

4.2.1 Observation–reaction time
4.2.1.1 Relevance to crossing
The visual and mobility abilities of pedestrians, as discussed earlier, are of major importance in the observation–reaction time of disabled and encumbered people. In particular, during the observation stage a pedestrian's angular movement will require extensive head and some shoulder movement and may for some pedestrians be difficult or impossible – especially for people in wheelchairs or encumbered. For blind and visually impaired people, when not accompanied by a sighted helper, all of the foregoing elements take longer. In such cases, special arrangements, such as audible signals and tactile surfacing, and procedures, such as those described by Bentzen et al. (2000), must then be made for crossings.

Figure 4.1 Number of trips walked by age and sex in 2014 (DfT, 2015)

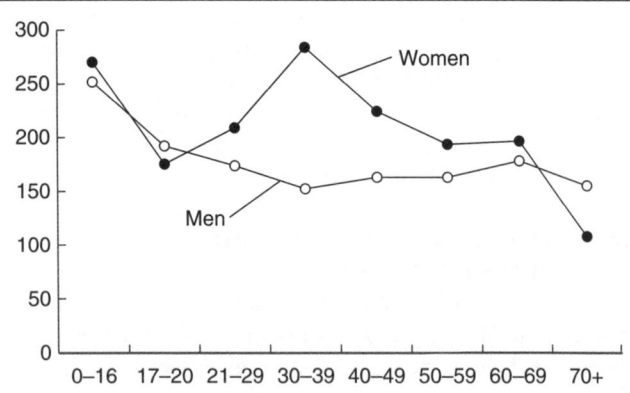

Table 4.2 Type of mobility aid as a percentage of all mobility aids (Office for National Statistics, 2001)

Type of mobility aid	Percentage of all mobility aids
Walking stick	69
Crutches	8
Walking frame, tripod, Zimmer frame	8
Trolley	2
Manual wheelchair	9
Electric wheelchair	2
Mobility scooter	2

Figure 4.2 Examples of people with mobility impairments and encumbered related to position on footway, based on Schoon and Hounsell (2006)

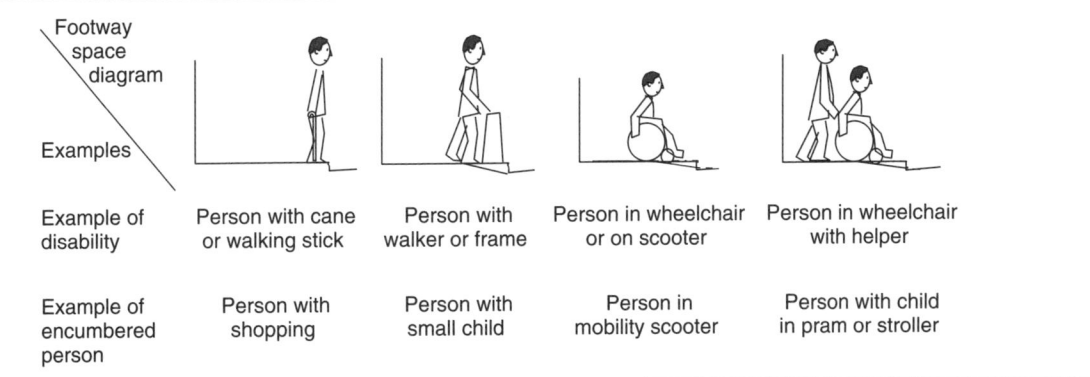

Footway space diagram Examples				
Example of disability	Person with cane or walking stick	Person with walker or frame	Person in wheelchair or on scooter	Person in wheelchair with helper
Example of encumbered person	Person with shopping	Person with small child	Person in mobility scooter	Person with child in pram or stroller

4.2.1.2 Visibility implications

In addition to the horizontal plane, vertical vision shift is important in cases where gradients exist and also in cases where obstructions require the pedestrian to see over or under an obstruction, such as a guardrail, in order to see approaching vehicles. The four-stage observation–reaction process for pedestrians (similar to a driver's detection, identification, decision and response) culminates in the pedestrian's start of the first step into the carriageway or the front of a wheelchair or other device crossing the kerb line. If, during the initial observational procedure, there is not an adequate gap in the vehicle stream, the pedestrian must wait and repeat the entire process. Other, related, characteristics are similar to those described in Chapter 3.

4.2.2 Code requirements related to observation–reaction time

Based on the crossing procedure in accordance with the Green Cross Code described earlier, Figure 4.3(b) summarises the main concepts of the observation–reaction process of a wheelchair user about to cross a two-way carriageway in accordance with the requirements of the Green Cross Code. Note that in the UK, because of the left-hand drive rule, pedestrians are advised to look first to their right.

4.2.3 Observation–reaction time comparisons with non-disabled pedestrians

Measurements of minimum required observation–reaction times for selected wheelchair users, together with those for

Figure 4.3 (a) Example of wheelchair users' crossing stages and associated times, based on Schoon and Hounsell (2006)

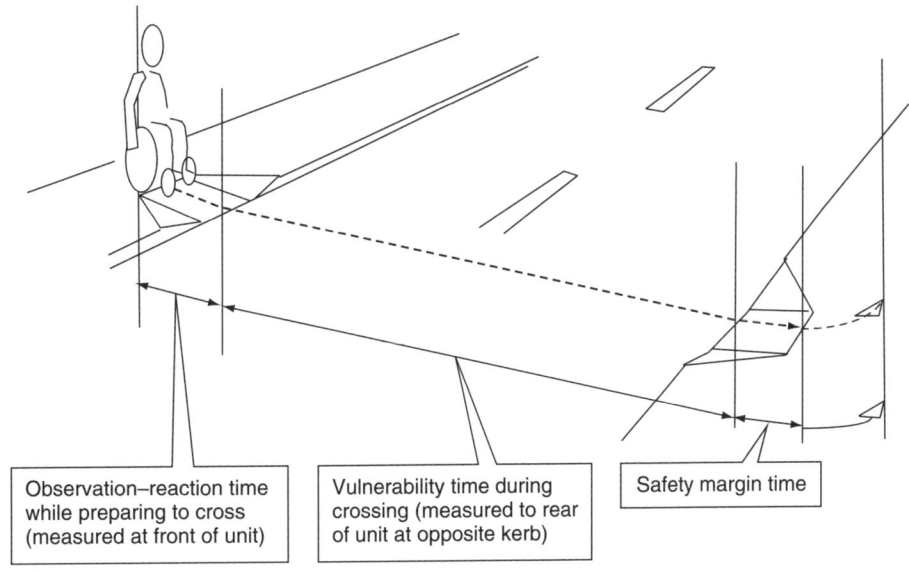

Observation–reaction time while preparing to cross (measured at front of unit)

Vulnerability time during crossing (measured to rear of unit at opposite kerb)

Safety margin time

Figure 4.3 (b) Summary of wheelchair user's actions when following the Green Cross Code for preparing to cross a carriageway, based on Schoon and Hounsell (2006)

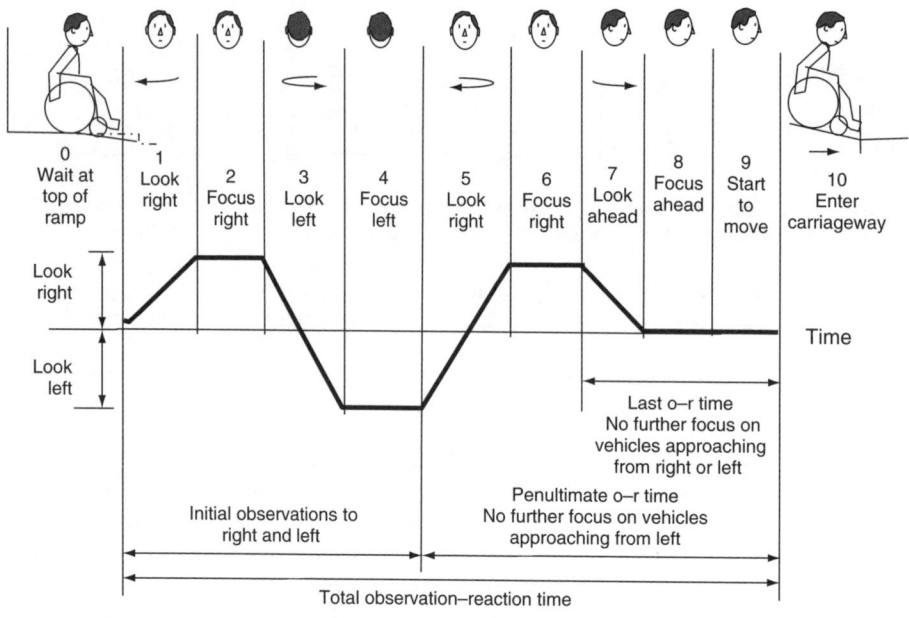

Summary of wheelchair user's actions when following the Green Cross Code in preparing to cross a carriageway

The diagram illustrates the following sequence of actions from the initial waiting with the main wheels at the top of the ramp (numbered items correspond with those in the pictograph).

1. First look to the right, approximately 90°.
2. Focus to the right (starts at completion of look to the right).
3. First look to the left (incl. saccade), approximately 180° (starts at completion of focusing right).
4. Focus to the left (starts at completion of look to the left).
5. Second look to the right (right saccade), approximately 180° (starts at completion of focus to the left).
6. Focus to the right (starts at completion of second look to the right).
7. Look to the 'ahead' position (left saccade), approximately 90°, i.e. looking straight across the carriageway.
8. Focus to the 'ahead' position (starts at completion of looking ahead).
9. Start to move ahead, directly across the carriageway (starts at completion of focusing ahead and represents the start of the reaction stage of the process).
10. Front of wheelchair and occupant just enters carriageway and becomes vulnerable to collision from this point on.

non-disabled users, are summarised in Figure 4.4. These results, shown in the graph of observation angle against elapsed time indicative of the characteristics of a female wheelchair user, show the average, median and 85th percentile times for the observation–reaction process, as well as diagrammatic comparisons of results obtained for male and female pedestrians and wheelchair users.

The significant differences, as might be expected, resulted from the wheelchair users requiring additional time for the wheelchair to move to the kerb line before entering the carriageway. In this respect, it can be seen that the additional times recorded were 1.18 s for the male subject and 1.12 s for the female subject.

The question of whether the difference between observation–reaction times between non-wheelchair and wheelchair users may be due to cognitive or other factors is important, and points to the need for more extensive studies. It should be noted that 1 s of time can amount to a distance of 14 m (44 ft) for vehicles travelling at 48 km/h (30 miles per hour). Because the 85th percentile of a wider range of wheelchair users could be over 2 s, this means that such a wheelchair user would need a sight line to approaching traffic of nearly 28 m (90 ft) greater than that needed by a non-wheelchair user. Moreover, it will be appreciated that the wheelchair user or helper can only see from a point approximately 1 m (3.3 ft) or more *behind* the kerb – thus further restricting the sight line to and from approaching vehicular traffic.

Figure 4.4 Example of experimental results and diagrammatic comparison between wheelchair users and non-disabled users, based on Schoon and Hounsell (2006)

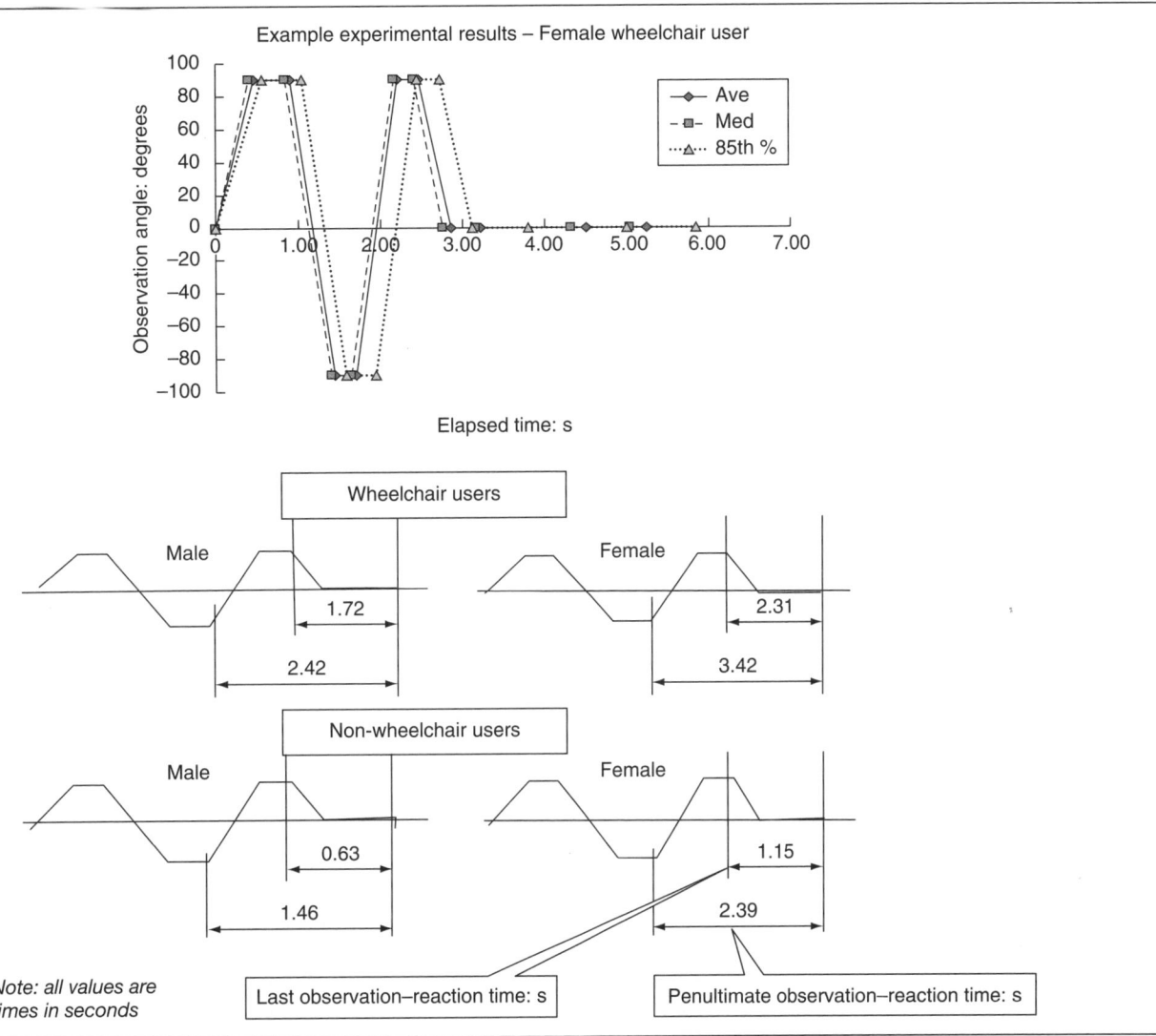

Of particular importance is the penultimate observation–reaction time (the time when pedestrians or wheelchair users are not focusing on vehicles approaching from the left). For the male wheelchair subject, the total mean penultimate observation–reaction time was 2.42 s, compared with 1.46 s for the non-disabled subject. The corresponding times for the female subjects were 3.42 s and 2.39 s, respectively. It should be noted that all observation–reaction times could be considerably longer at locations where the subject is required to look partially backwards, such as at street junctions.

4.2.4   Viewpoint and visibility

Depending on the location of the pedestrian when preparing to start to cross the road, he or she will normally stand just behind the kerb on the footway. This is also in accordance with the Green Cross Code instructions.

However, as mentioned earlier, and as shown in Figure 4.5, it may not be possible for some disabled people to stand immediately behind the kerb, owing to the presence of the assistive devices being used. Their viewpoints will therefore be some distance behind the kerb, thus reducing their visibility distance to less than would be the case for people who are able to stand at the kerb line, as shown conceptually in Figure 4.6.

4.3.   Crossing the carriageway

As soon as the observation–reaction task is completed and the pedestrian (or helper pushing the wheelchair) decides to walk,

Figure 4.5 Examples of disabled person categories and associated mobility devices related to viewpoint and kerb line

Diagram indicating viewpoint
distance behind kerb

Features related to crossing

- Profile of non-disabled or non-assisted pedestrian at kerb
- Viewpoint is approximately just behind kerb as per code instructions

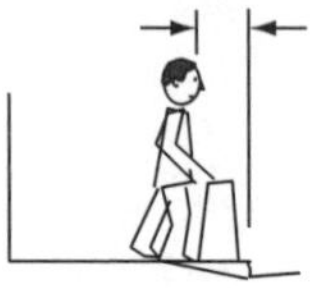

- Profile of disabled pedestrian with walker assistance
- Viewpoint is a significant distance behind kerb

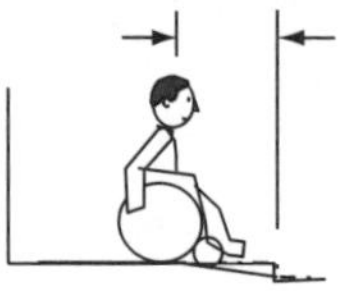

- Profile of manually self-propelled wheelchair
- Similar profile to battery-propelled wheelchair
- Wheels may vary in size but overall front to back dimension is similar
- Safe waiting distance is assumed to be with rear wheels on flat surface

- Profile of helper-assisted wheelchair
- Assumes that helper controls wheelchair and makes movement decisions
- Helper's viewpoint is a considerable distance behind kerb
- Safe waiting distance is assumed to be with rear wheels on flat surface

he or she enters the carriageway and is then vulnerable to a collision with a vehicle. This remains the case until the rear part of the pedestrian, wheelchair or helper crosses the opposite kerb.

Some points particularly relevant to disabled pedestrians include the physical characteristics of the road itself, the slope and width of ramps to and from the footway, the cross-slope of the carriageway, the roughness of the carriageway surface and the height of any 'lip' where the ramp joins the carriageway. For most cases, the access to the ramp and the ramp gradient itself are likely to provide the greatest barriers to a straightforward crossing if they are in any way deficient.

Figure 4.6 Effect of pedestrian's viewpoint on visibility distance

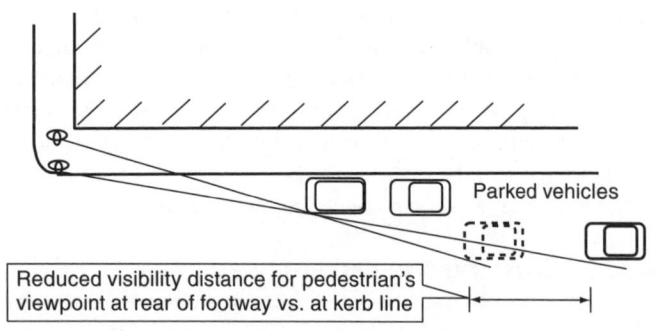

Parked vehicles

Reduced visibility distance for pedestrian's
viewpoint at rear of footway vs. at kerb line

4.3.1 Examples of walking speeds at crossings

Linking of disabilities with corresponding walking speed is an important aspect of crossing analysis and design. Leake *et al.* (1991) adopt an ergonomic approach and Eubanks and Hill (1998) report on a number of sources, including a study addressing matters of gait (Perry and Burnfield, 1992), which reported results of studies on walking and wheelchair speeds of people with various disabilities. Selected excerpts from this study, including normal free gait velocities of non-disabled people for comparison, are shown in Table 4.3. Speeds as low as 1.15 ft/s (approximately 0.36 m/s) were recorded.

Examples of other instances of data on crossing speeds for people with disabilities include the following.

- For puffin (pedestrian user-friendly intelligent) pedestrian-activated crossings in the UK, 'on-crossing' detectors respond to all pedestrians within the crossing area walking at speeds equal to 0.5 m/s (1.6 ft/s) or more (DfT, 2005).
- AASHTO (2011) suggest that pedestrian crossing signal timing be based on a speed of 3 ft/s (0.9 m/s) in areas that are heavily used by older people.
- A study of trends in pedestrians' walking speeds indicates that a walking speed of 0.9 m/s (3 ft/s) would meet the US Access Board recommendation and would accommodate 85% of older or other slower pedestrians.

Table 4.3 Walking speeds for selected age groups and disabilities, based on Perry and Burnfield (1992)

Characteristic	Details	Walking speed: ft/s
Average velocity by age group – non-disabled people		
Customary walking speed	Children (6–12)	3.83
	Teenagers (13–19)	3.99
	Adults (20–59)	4.37
	Seniors (60–80)	4.05
Fast walking speed	Children (6–12)	4.81
	Teenagers (13–19)	5.41
	Adults (20–59)	5.80
	Seniors (60–80)	4.92
Infirmities		
Ambulatory mobility index	<40	1.47
	40–60	1.86
	>60	3.08
Arthritis of the hip		
Arthritis characteristics	Total hip replacement, pre-operation	2.24
	Total hip replacement, post-operation	3.01
	Girdlestone	2.52
	Hip fusion	3.66
Rheumatoid arthritis of the knee		
Assistance devices	Walker	1.15
	Crutches	1.42
	Crutch	1.70
	Cane	1.75
	No assistance device	2.46
Hemiplegia		
	Wheelchair	2.02
	Walking	1.64
Myelodysplasia and swing through vs. wheelchair		
	Reciprocal	1.64
	Swing through	2.30

1 ft/s = 0.33 m/s

- A walking speed of 0.6 m/s (2 ft/s) for disabled people is quoted by Oxley and Alexander (1994), based on studies carried out in Canada.

An important consideration not included in most design guidelines is that a disabled person in a wheelchair may occupy a space of approximately 1.2 m (4 ft) from front to back.

Furthermore, if a helper is pushing the wheelchair, the total length of the wheelchair-and-helper unit is at least 2 m (6.5 ft). Therefore, not only must the speed of the person crossing be taken into account, the fact that a unit moving at, say, 0.6 m/s may itself require up to over 3 s to pass a given point is an important element when estimating the total crossing time.

4.3.2 Gaining the opposite footway

In most cases, pedestrians reaching the opposite kerb, in crossing at a designated crossing location, will progress directly onto the ramp and then turn to whatever direction is required along the footway to continue their journey. Clearly, if no ramp is available, the time to gain the footway will be much greater. No definitive quantitative information appears to be available about this stage of the crossing. However, observations at selected locations indicate that the following may occur.

- There may be a 'lip' at the kerb line of up to 2 cm (1 in) high, sometimes more. This may cause a pause in the progress of a wheelchair to overcome it, especially if the user prefers to avoid jolting, or if it is necessary to tilt the wheelchair at the front and back to overcome this obstruction.
- The effective width of the footway may be inadequate for a wheelchair, or other device, to easily turn into the desired direction along the footway; the wheelchair may therefore have to be manoeuvred until such a direction can be achieved. During this process, the unit may need additional time to fully exit the carriageway and will remain vulnerable during this time.
- The presence of other pedestrians hindering movement may be a significant cause of delay in gaining the opposite footway.

For the time required to complete the process of gaining the opposite footway, it seems reasonable to assume, as a minimum value, the speed at which the main part of the crossing has been undertaken. A brief survey of the time taken for a wheelchair being pushed by an adult to stop briefly at a lip and raise the front of the wheelchair to surmount this obstruction was found to vary between approximately 1 s and 2 s, in addition to the time taken for the front of the wheelchair to reach the opposite kerb line.

4.3.3 Safety margin

Although not addressed in most formal analyses, the inclusion of a safety margin by which the pedestrian clears the carriageway by a certain time period may be appropriate in the overall timing and design of a crossing. The purpose of such a margin would be to provide pedestrians with greater safety and provide a more comfortable, less intimidating experience when crossing, in order to encourage the use of pedestrian facilities (Schoon and Hounsell, 2006).

4.4. Crossing characteristics of people with sensory impairments

So far, in considering people's abilities in crossing the carriageway, the focus has been on people with physical mobility problems, as opposed to mobility problems due to primarily sensory impairments. The latter can include blind or visually impaired people and deaf and partially deaf people. In terms of the geometric characteristics of pedestrian facilities, the lack of visual abilities is perhaps of greatest concern. Key points associated with the needs of blind and visually impaired people are therefore briefly discussed next.

4.4.1 Movement characteristics of blind and visually impaired people

Statistics indicate that there are more than 2 million people in the UK who define themselves as having a sight problem or difficulty in seeing and this number is projected to increase to 4 million by 2050 (RNIB, 2018). This can range from being unable to see a friend across the street or read newsprint, even with the aid of their glasses, to being registered as blind. Many people with sight problems travel independently using either their remaining vision or a mobility aid – a cane or a guide dog.

4.4.2 Canes and mobility

There are different types of white cane: a symbol cane, to indicate to other people that the holder has a sight problem; a guide cane, which is held diagonally across the body and affords some protection from obstacles; and a long cane, which is held out in front and moved from side to side to find and locate objects, such as landmarks and clues, and to identify obstacles in a person's path.

People use a range of landmarks (fixed objects in the environment) and clues, which can help them to orientate themselves in the environment and navigate around – but may not always assist in maintaining a steady speed. These landmarks and clues could be tactile – using the feel of the feet and the cane – auditory, visual or olfactory and may include the following.

- *Tactile*. Kerbs, building lines, slopes or gradients, tactile pavement at crossings and surface textures underfoot.
- *Auditory*. Vehicular traffic, such as buses, goods vehicles and cars; shops and offices, which may have automatic doors, background music or talking sign systems. Note that an emerging problem still to be resolved is that of electrically powered vehicles, which emit almost no sound either when moving or stationary.
- *Visual*. Contrast between the footway and carriageway, yellow and white lines on the carriageway, contrasting coloured tactile pavement, colours of shop fronts and fixation points, such as the skyline.
- *Olfactory*. The smells of the florist, fish and chip shop, etc.

Orientation methods used by a long cane user can include the following.

- Following the inner shore line (shops, garden frontages etc.) using the (cane tactile).
- Following the outer shore line (along the kerb) using the cane (tactile).
- Using the kerb edge to locate carriageway junctions.
- Using the kerb edge to orientate/maintain the straight line.
- Using tactile pavement to orientate safe/safer carriageway crossing points.
- Using building lines to provide clues to the next carriageway.
- Using traffic flow to indicate the straight line of travel; and.
- Using other sound clues as well as smell and auditory information.

4.4.3 Role of the guide dog

The Guide Dogs for the Blind Association currently supports about 5000 guide dog partnerships in the UK. The role of the guide dog is to guide its visually impaired owner in a straight line, unless directed otherwise, avoiding any obstacles. The owner gives encouragement, and commands and informs the dog of the direction to go. The guide dog is trained to stop at steps and kerbs, find doors, carriageway crossings and places that are visited frequently. The guide dog will lead its owner across the carriageway but the owner must determine where and when to cross safely.

4.4.4 Orientation methods used by a guide dog owner

Using the approach and location of the kerb edge to plan routes (a route would involve planning from kerb to kerb until required to turn) – for example, 'first down the kerb, turn right; second up the kerb turn left; 25 paces on the right is the supermarket'.

- Using the location of the kerb edge to orientate or check for a straight line of travel, using sound clues – sound shadows – to aid in locating the objective.
- Using other sound clues to aid orientation – for example, the travel of traffic.
- Use of tactile paving to indicate safe or safer crossing points.
- Following the behaviour of the dog to recognise certain locations in the environment.

All mobility aids require the blind or partially sighted traveller to stay orientated within the immediate environment in order to remain safe and to achieve desired objectives. If the foundations of orientation techniques are removed, safety is compromised.

4.5.   Analysis and design of the total crossing

Incorporating the points presented previously permits estimation of total crossing time, and therefore the required distances and dimensions of the crossing and its environs, including sight lines related to motor vehicle speeds. A useful graphical portrayal of a pedestrian unit crossing a carriageway is provided by the time–space diagram shown in Figure 4.7. This is similar in concept to the time–space diagrams used to portray movement of vehicular traffic streams, where vehicle speed is shown by the slope of the line (distance divided by time). The double diagonal lines show the front and the back of the crossing unit on the distance axis and therefore indicate the space taken up as the unit crosses the carriageway. At any point in time during the crossing, from the stationary position on the footway, the position of the crossing unit is defined by reading up from the time axis to the diagonal lines and then horizontally to the distance axis. The total time for the crossing can then be related to approaching vehicle speeds to determine the necessary sight line distance. The diagram shows, in an analytical sense, the features shown in Figure 4.3.

Figure 4.7 shows the relationship of the elapsed time for each stage of the unit's crossing, beginning with the observation–reaction time, immediately before crossing the kerb. From the point where the front of the unit crosses the kerb into the carriageway, the unit becomes vulnerable to collision with an approaching vehicle and remains so until the rear part of the unit crosses the opposite kerb onto the footway. When this has occurred, it is assumed that the unit then continues, with a safety margin. In the diagram, the unit's progress across the carriageway is shown to occur at a constant velocity – a possible approximation of the actual speed, which may vary in practice with, for example, the cross fall of the carriageway and some acceleration at the start of the crossing and deceleration as the opposite kerb is approached.

An example of the times and speed characteristics of a wheelchair crossing based on investigations by Schoon (2011) using the Global Positioning System (GPS) is shown in Figure 4.8: panel (a) shows the general crossing layout, indicating the actual distance crossed; panel (b) shows the main crossing elements; and panel (c) shows the actual speed of the wheelchair at 0.1 s intervals, the acceleration and deceleration rates, and the delay incurred when the wheelchair slows towards the 'lip' of the far kerb before accessing the opposite footway.

Note from Figure 4.8(c) that the average speed in the carriageway (typically used for signal timing and intervisibility estimates) is greater than the actual average speed needed to cross the entire distance, from starting to finishing positions of the wheelchair. This is primarily due to the inclusion of the wheelchair length and the acceleration and deceleration elements of the crossing process.

Figure 4.7 Conceptual time–space diagram for disabled unit crossing from footway to opposite footway, based on Schoon and Hounsell (2006)

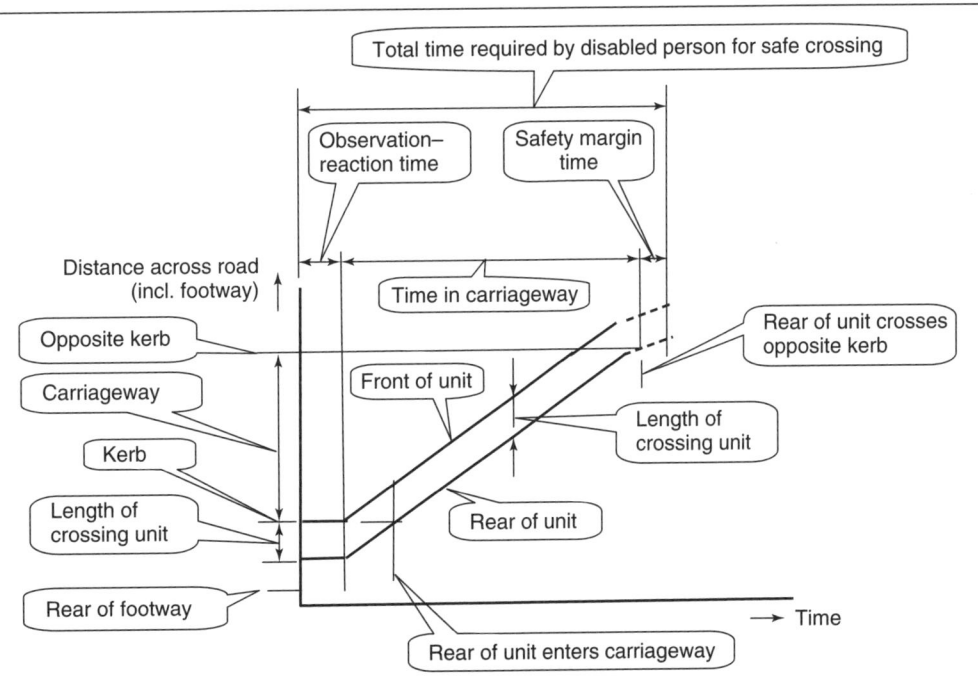

Figure 4.8 Wheelchair crossing speed and distance characteristics, based on Schoon (2011)

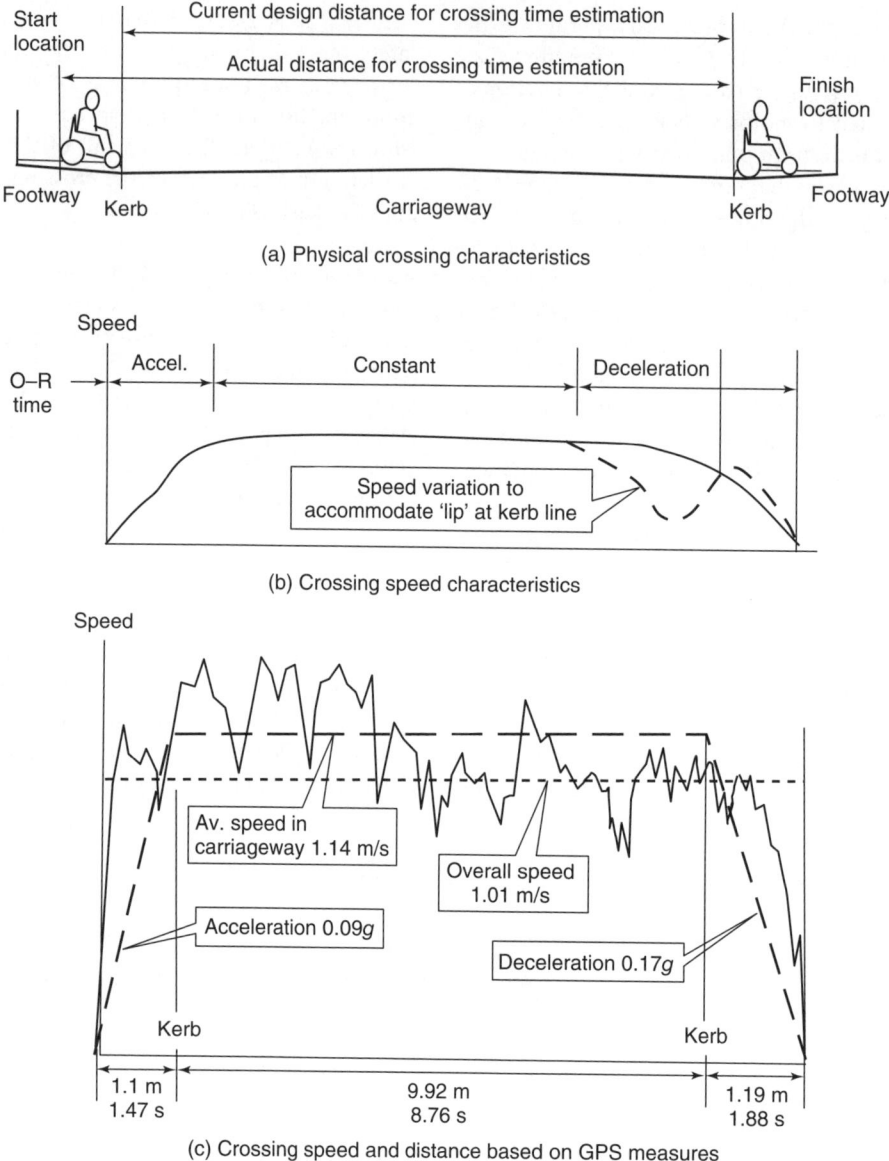

(a) Physical crossing characteristics

(b) Crossing speed characteristics

(c) Crossing speed and distance based on GPS measures

Once the times for the various elements of the crossing have been determined, i.e. the appropriate observation–reaction time, the time to cross the carriageway, including the clearance time until the rear of the unit crosses the opposite kerb, and the safety margin time, the sum of these times can be used as a basis for estimating the length of sight lines to approaching traffic. The required lengths and locations of these lines form the basis for ensuring that the pedestrian can see approaching vehicles and that drivers can see the pedestrian in good time for either or both to take avoiding action or proceed, in the knowledge that the pedestrian is visible.

It is of interest to calculate briefly the difference in total crossing times for a disabled person while addressing the issues with a more usual approach explored in this chapter. Here, we assume an unsignalled intersection of two-way roads, a carriageway width of 7 m, a wheelchair measuring 1.3 m front to back, a penultimate observation–reaction time of 3 s, a crossing speed of 0.7 m/s and a safety margin of 2 s. Thus, the crossing time, including safety margin would be

$$\left(\text{observation–reaction time} + \frac{\text{carriageway width}}{\text{walking speed}}\right.$$

$$\left. + \frac{\text{length of crossing unit}}{\text{walking speed}} + \text{safety margin time}\right)$$

Inserting the appropriate values gives

$$\left(3 + \frac{7}{0.7} + \frac{1.3}{0.7} + 2\right) \text{s} = 17 \text{ s (rounded)}$$

With current analysis methods, it would not be usual to include the penultimate observation–reaction time, the length of the unit or the safety margin time. The total crossing time would therefore be

$$\left(\frac{7}{0.7}\right) \text{s} = 10 \text{ s}$$

Table 4.4 Summary of issues and consideration related to design policy for disabled pedestrians, based on Schoon and Hounsell (2006)

Analysis and design feature	Current approach	Potential additional considerations	Comments
Green Cross Code or similar instructions for users to encourage a safe crossing	Reference to the Code and related functional design requirements are not explicitly described	Integrate the Code with the observation–reaction time before crossing. Reconsider implications for design of the practicality of the Code's 'look all around' instruction	Consistency between instructions to pedestrians and design functionality should improve responsiveness of design to pedestrians' actions
Location of pedestrian's viewpoint relative to kerb	Assumed to be immediately behind kerb, according to Green Cross Code	For disabled people in wheelchairs, helpers, or people pushing strollers, the viewpoint will be some distance behind the kerb – possibly up to 2 m (6.6 ft)	Sight lines to approaching traffic are likely to be inadequate if viewpoint is not considered
Observation–reaction time	May be included as 'start-up time' but often not included	May often be considerable, especially for people with torso, neck and head movement disabilities	Time to observe and react to traffic may result in increased pedestrian crossing time
Travel time across carriageway	Normally included	Particular attention to movement speeds of disabled	Lack of adequate time to cross will increase danger and inhibit use of facilities
Inclusion of time due to longitudinal dimension of a unit comprising people and equipment crossing	Not included	Adding the time taken for the length of the crossing unit to pass a given point recognises the greater time taken to cross	May be a considerable portion of additional time, owing to the front-to-back dimension of the crossing unit
Code requirement to 'look all around' while crossing	Not included in analysis and design, but implies that pedestrian can effectively stop and reverse direction to safety	Analysis and design of crossing should preferably assume that people do not look around after leaving the kerb, even though they may do so if able	For most disabled people, looking around can be difficult or impossible – as with potentially reversing direction
Time to mount opposite footway	Not included	Often may be required and may require up to 2 s or more	If time is required, it should be added to the total crossing time
Safety margin time	Not included	A desirable feature of crossings	Pedestrians can be inhibited by 'near misses'
Length of crossing unit	Not included	May add significantly to the total time for the unit to clear the crossing	Consideration should be given to including the necessary time in the total crossing time

The difference in this case is 7 s. The difference would presumably be somewhat less in the calculations for a signalised intersection, where a start-up or 'comfort' time may be included but nevertheless would still be considerable.

4.6. Summary and conclusions

The key points explored in this chapter are listed in Table 4.4, which indicates, for each analysis feature, the current approaches described in design manuals and guidelines, potential additional considerations in designing for disabled people and comments related to future design policy. The comments recognise that pedestrians' behaviour, particularly when disabled or encumbered, warrants a detailed approach in analysing and designing the geometric and operational characteristics of highways and streets – a level of detail to match at least that given to driver and vehicular movement. In this regard, the issues discussed may assist in complementing existing policies on geometric design of highways and streets.

Many important features have not been detailed, as the intent has been to identify and provide examples of analysis and design inputs in preference to emphasising databases. Regarding the latter, much work needs to be done to establish a base of behavioural characteristics of non-disabled, disabled and encumbered people, specifically aimed at its use in street and highway design.

Supporting evidence related to the response of disabled people to improvements, in this case related to the safety and ease of crossing a carriageway, will warrant further investigations for establishing and adopting preferred analysis and design methods.

Much work needs to be done to expand on the matters discussed in this chapter. In terms of inputs to establishing a 'design pedestrian', much thought needs to be devoted to concepts already developed in this area. Consideration of a more uniform yet convenient image lacking some detail, as opposed to a more time- and effort-consuming approach that accommodates the essential variables associated with a variety of pedestrians' needs will be an essential topic for debate.

As with any feature of infrastructure design, the adoption of appropriate parameters, values and procedures for inclusion in the analysis must remain the responsibility of the designer, including the conduct of possible individual experimentation or observations for specific locations.

REFERENCES

AASHTO (American Association of State Highway and Transportation Officials) (2011) *A Policy on the Geometric Design of Highways and Streets*. AASHTO, Washington, DC, USA.

Bentzen BL, Basleus JM and Frank C (2000) Addressing barriers to blind pedestrians at pedestrian crossings. *ITE Journal* **70(9)**: 32–35.

DfT (2005) *Inclusive Mobility: A Guide to Best Practice on Access to Pedestrian and Transport Infrastructure*. DfT, London, UK.

DfT (2015) *National Travel Survey: 2014*. DfT, London, UK.

DfT (2017) *The Highway Code*. DfT, London, UK.

DWP (Department for Work and Pensions) (2014) *Official Statistics: Disability Facts and Figures*. DWP, London, UK.

Eubanks J and Hill P (1998) *Pedestrian Accident Reconstruction*. Lawyers and Judges, Tucson, AZ, USA.

Hitchcock A and Mitchell CGB (1984) Man and his transport behaviour, Part 2a: walking as a means of transport. *Transport Reviews* **4(2)**: 177–187.

Leake GR, May AD and Parry T (1991) *An Ergonomic Study of Pedestrian Areas for Disabled People*. TRRL contract report CR184. TRRL, Crowthorne, UK.

Office for National Statistics (2001) *General Household Survey 2001: Table 10.9. Type of Mobility Aid as a Percentage of all Mobility Aids*. Office for National Statistics, London, UK.

Oxley PR and Alexander J (1994) *Mobility in London – A Follow-up to the London Area Travel Survey*. TRL Project Report PR 34. Transport Research Laboratory, Crowthorne, UK.

Perry J and Burnfield JM (1992) Gait analysis: normal and pathological function. SLACK, Thorofare, NJ. New Jersey, USA.

RNIB (Royal National Institute for the Blind) (2018) Key Information and Statistics on Sight Loss in the UK. https://www.rnib.org.uk/professionals/knowledge-and-research-hub/key-information-and-statistics (accessed 05/10/2018).

Schoon JG (2011) Crossing time characteristics for wheelchairs and mobility scooters based on Global Positioning System (GPS): a pilot study. In *Transportation Research Board 90th Annual Meeting*, poster P11-1193. TRB, Washington, DC, USA.

Schoon JG and Hounsell N (2006) Access and mobility design policy for disabled pedestrians at road crossings: exploring issues. *Transportation Research Record* **1956(1)**: 76–85.

Williams K, Savill T and Wheeler A (2002) *Review of Road Safety of Disabled Children and Adults*, TRL Report 559. TRL, Crowthorne, UK.

SELECTED FURTHER READING

Carr M, Lund T, Oxley P and Alexander J (1994) *Cross-sector Benefits of Accessible Public Transport*, Project Report 39. TRL, Crowthorne, UK.

Dawson D (2004) Designing accessible facilities in the public right of way. *ITE Journal* **74(9)**: 46–48.

DFID (Department for International Development) (2004) *Enhancing the Mobility of Disabled People: Guidelines for Practitioners*. TRL, Crowthorne, UK.

Schoon JG and Hounsell N (2005) Observation–reaction times of wheelchair users – a comparison with non-disabled users. In *84th Annual TRB Meeting*. TRB, Washington, DC, USA.

Transport Canada (1998) *Canadian Accessible Transportation Guide*. Canadian Department of Transportation. Ottawa, Canada.

Windley S (2004) Toward accessible public rights-of-way. *ITE Journal* **74(9)**: 42–44.

Pedestrian Facilities, Second edition

Schoon, John G
ISBN 978-0-7277-6309-9
https://doi.org/10.1680/pfse.63099.077

Chapter 5
General road layout practice

Although design details of individual elements of vehicle and pedestrian ways are described later, this chapter provides an overview of the purpose, dimensions and configuration of infrastructure in general road layouts. Many of these characteristics have evolved over time and in response to perceived needs. In many cases, documented support for the configurations and dimensions is not available, or may vary between jurisdictions.

Essentially, the geometric layout of roads is based on guidance described in the *Design Manual for Roads and Bridges* (*DMRB*) (DfT, 2018) and the *Manual for Streets* (*MfS*) (DfT, 2007). These two publications predominantly address trunk and major roads and residential streets, respectively, although, as with most cases, some overlap in practice may occur.

As well as practice in the UK, geometric design of pedestrian and cyclists' facilities in other countries has evolved in response to continuing concerns over personal safety and mobility and for urban design in general. In addition to design standards, differences may include: attitudes to road safety; risk-taking; speeding; traffic volumes and mix; characteristics of non-motorised traffic; and traffic law and enforcement (Belcher *et al.*, 2015). These factors must be considered when presenting on or reviewing practices in other countries. This chapter contains analysis specifically related to UK norms of conduct. However, several instance of European practice are included, particularly regarding shared space, to provide a wider perspective.

5.1.   Design principles and approaches

The layouts and characters of many streets have developed over decades or, often, centuries. In many cases, unplanned development and growth in traffic have dictated the predominant street features and dimensions of the different elements. Some basic design principles are described in this section.

5.1.1   Objectives and approaches to designing pedestrian facilities in streets

Road layouts have a wide range of requirements to fulfil: they must cater for the needs of various categories of pedestrian; through and local motor traffic; goods movement and public transport, as well as providing access to adjacent land uses, including shops and transport terminals, and for emergency services. Added to the functional needs are often historical and geographical determinants, which may limit choice of location and form.

In terms of geometric design involving the layout and dimensions of pedestrian facilities, the interaction of pedestrians with motor vehicles is the key concern of this chapter. Official guidance provided by the Department for Transport (DfT) in the *DMRB* (DfT, 2018) and its associated advice and guidance notes and the *MfS* (DfT, 2007) are the key sources of information.

In addition, the importance of the street to both movement and local activity is recognised. Numerous publications, including advice on the planning and design of pedestrian facilities published by the Chartered Institute of Transport (CIHT, 2015a, 2015b), the Public Realm Information and Advice Network (Davis, 2014) and the Commission for Architecture and the Built Environment (CABE, 2000), among others, offer information on trends for potential inclusion into official guidance.

5.1.1.1   Street-space dimensions

Establishing street-space dimensions (a fundamental aspect of geometric design) is embodied in the concept of 'link and place' (Jones *et al.*, 2007). Here, the permutations of 'link' (enabling the users to pass through the street as quickly and conveniently as possible) and 'place' (where users are encouraged to stay as long as desirable and enjoy the street's surroundings) are the two primary functions, competing for space. Additional aspects include active frontages and meeting places (Davis, 2014).

In recognising these two primary street functions, Figure 5.1 summarises some of the main concepts in the design relationships between link and place. This shows how the total street width may be portrayed as a combination of 'link spaces' and 'place spaces'. The diagram may then be used to explore different layouts, including the case where the total width is inadequate and where other choices, possibly network-wide, must be made.

Figure 5.1 Minimum (min) and desirable (des) provision for combined link (L) and place (P) needs, based on Jones *et al.*, 2007. Here, the total available street-space capacity lies between the 'minimum' and 'desirable' levels

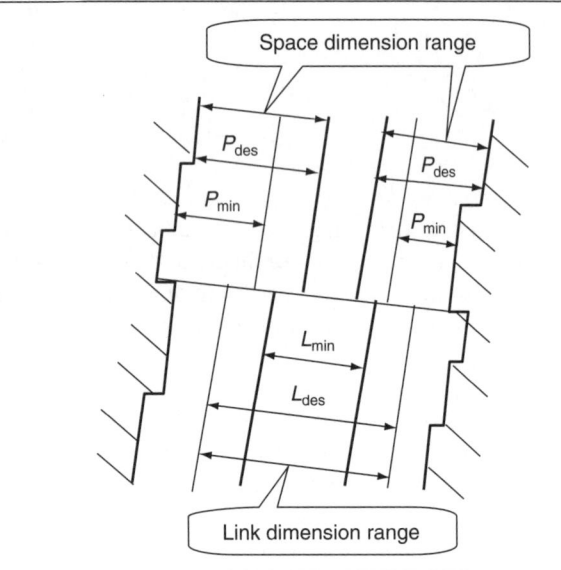

Figure 5.2 A walking and cycling route between St Pancras station and Trafalgar Square (Davis, 2014)
© Crown Copyright, 2014
This information is licensed under the Open Government Licence v3.0. To view this licence, visit http://www.nationalarchives.gov.uk/doc/open-government-licence/ **OGL**

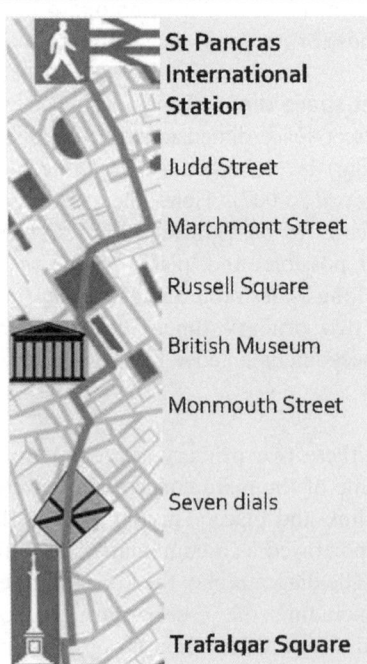

Along with establishing the basic dimensions, interested stakeholders may be involved in exploring alternatives. Plan view and cross-section overlays, such as those developed in the ARTISTS project and reported in Jones *et al.* (2007) enable participants to see whether a street design element can be fitted within the space available along a section of street. An example of a possible route could be as illustrated in Figure 5.2 and put into specific street dimensional terms, as illustrated in Figure 5.3.

5.2. Trunk and other high-vehicular-traffic roads

The geometric components of a particular design are often defined by the required cross-section of the road; whether it is a main trunk road or a residential street, and whether it is in an urban or rural area. This section outlines selected examples of the guidance presented in TD 27/05 (DfT, 2005a), predominantly focusing on trunk roads in urban areas.

5.2.1 Features of cross-sections

Considerations include the neighbourhood location and features, traffic volumes and characteristics, and the configuration and dimensions of contiguous roads. The design organisation's role then focuses on deciding which of the components to include and selection of the appropriate dimensions.

Features included in the cross-section can affect the overall road width. Consequently, their interaction is often complex. The preferred locations of features in verges and the central reserve may often coincide or overlap and the design organisation should be aware of the potential for such conflicts. Generally, there is far more equipment below the surface of verges and central reserves than is apparent above ground level; some underground features must be readily accessible for routine maintenance.

5.2.2 Road classification

Two broad divisions are used to define roads by association with the environment through which they pass; namely rural roads and urban roads. Each of these is further subdivided into motorways and all-purpose roads, with a classification to distinguish between mainline and connector roads.

5.2.3 Variations between rural and urban roads

In urban areas, there will usually be less scope for coordinating features than in rural areas, although every effort should be made to do so wherever economically and environmentally practicable. The design organisation will need to ensure a careful balance between the many competing demands. In urban areas, there are likely to be numerous items of street furniture and underground equipment within the highway cross-section. Further advice is given in TD 9/12 (DfT, 2012) and *Transport in the Urban Environment* (IHT, 1997) on designing urban single- and dual-carriageway roads.

Figure 5.3 Example of street design cross-section (Jones *et al.*, 2007); the suggestion is to convert the central portion of the street (currently shared by trams and general traffic) to a dedicated tramway, and to add cycle lanes. The segregated tram and cycle tracks are to be accommodated in this street design section by reducing the number and width of lanes for general traffic. In this design, the 'link' status is higher than the 'place' status

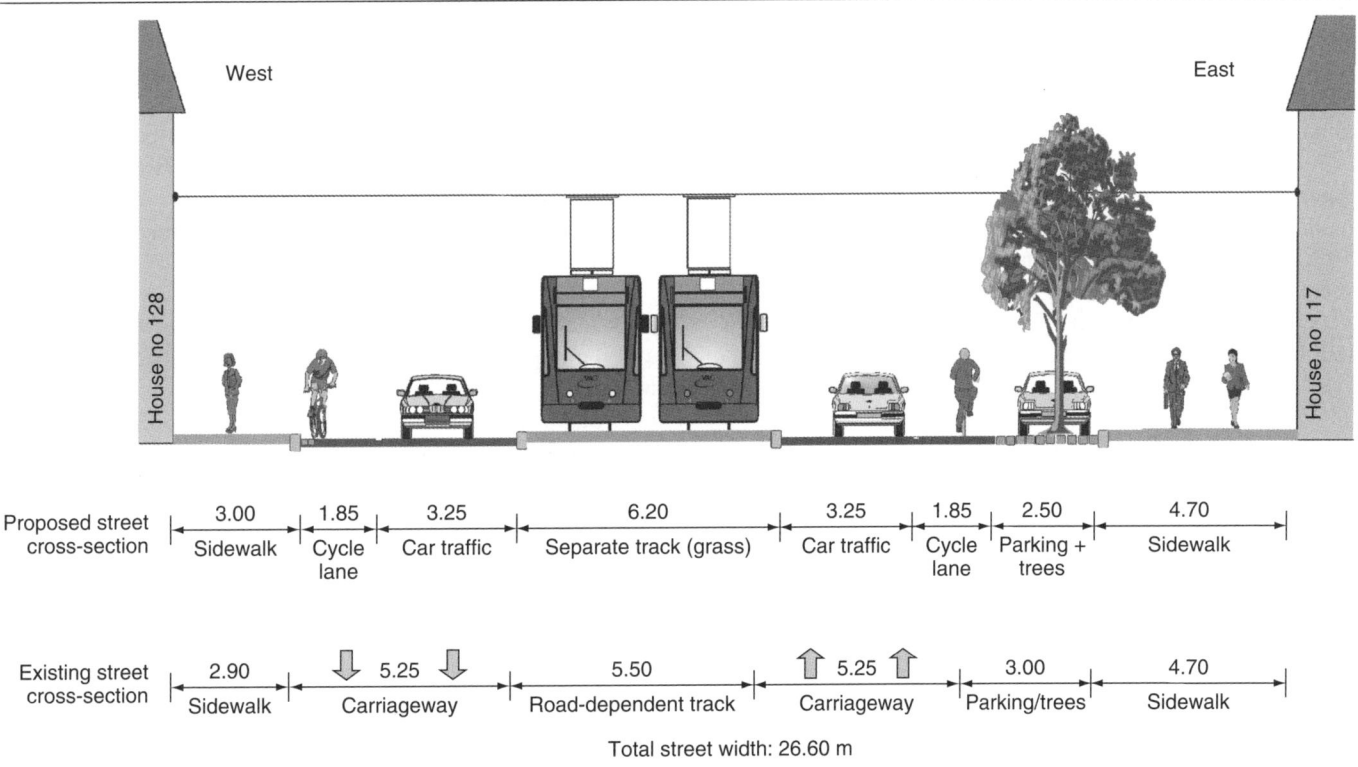

	3.00	1.85	3.25	6.20	3.25	1.85	2.50	4.70
Proposed street cross-section	Sidewalk	Cycle lane	Car traffic	Separate track (grass)	Car traffic	Cycle lane	Parking + trees	Sidewalk

	2.90	⬇ 5.25 ⬇	5.50	⬆ 5.25 ⬆	3.00	4.70
Existing street cross-section	Sidewalk	Carriageway	Road-dependent track	Carriageway	Parking/trees	Sidewalk

Total street width: 26.60 m

Urban roads have lower design speeds and are often more congested than roads in rural areas. Generally, drivers do not expect rural standards in urban areas and the restriction of width can assist with the encouragement of low speeds, which is of safety benefit, owing to the large number of accesses and non-motorised users (NMUs), particularly those crossing the road. On urban roads, the carriageway edge treatment will generally include positive drainage and kerbs, which provide additional edge restraint and support for raised footways and verges.

The size and extent of typical features above ground that may need to be accommodated in the verge and central reserve of an urban road are illustrated in Figure 5.4. The possible need for future features above and below ground should be considered and the design made accordingly. A balance should be struck between safety, environmental impact, cost, construction feasibility, operation and maintenance.

5.2.4 Non-motorised users (NMUs)

It is essential that design organisations integrate facilities for NMUs in the design at an early stage so design organisations must understand the highway environment, in relation to the

Figure 5.4 Typical features to be accommodated in the cross-section (DfT, 2005a)
© Crown Copyright, 2005
This information is licensed under the Open Government Licence v3.0. To view this licence, visit http://www.nationalarchives.gov.uk/doc/open-government-licence/ **OGL**

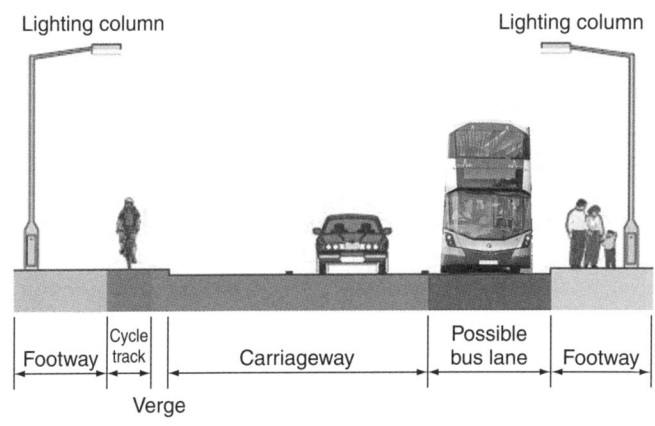

Urban cross-section
Where appropriate, e.g. on lower flow roads, cyclists can often be safely accommodated on the carriageway.

various road design components, from an NMU's perspective. Design organisations must determine and make adequate provision for any NMU requirements; see TA 90/05 (DfT, 2005b), TA 91/05 (DfT, 2005c), HD 42/05 (DfT, 2005d) and *Guidelines for Providing for Journeys on Foot* (IHT, 2000). There is a statutory duty to provide proper and sufficient footways for pedestrians and adequate margins for ridden horses and driven livestock where necessary, or desirable, for their safe accommodation.

Other considerations include health and safety regulations, maintenance and the choice of cross-section features. These latter include paved widths, hard shoulders, slopes, central reserves and the positions of features related to the presence of NMUs. Commonly occurring features within the highway, with corresponding references, are detailed in Appendix A of TD 27/05 (DfT, 2005a).

Reference should be made to HD 42/05 (DfT, 2005d) for guidance on widening the central reserve at priority junctions on dual-carriageway all-purpose roads. Away from junctions, crossing places for NMUs, both controlled and uncontrolled, may require significant space, particularly where equestrians and cyclists are expected.

5.2.5 People with disabilities
The required standard of provision for persons with disabilities must be considered at the early stages of scheme preparation and the appropriate level of facilities must be agreed with the overseeing organisation as part of the reporting procedure. For further advice on designing for the needs of persons with disabilities, see DfT guidelines in *Inclusive Mobility* (DfT, 2005e), many features of which are described in Chapter 8.

5.2.6 Design process
For the purposes of developing initial layouts, the design organisation should determine the appropriate typical width for the highway cross-section and any variation in width required. The type of highway and number of lanes needed for a facility are usually determined during the concept stage of project development. Figure 5.5 provides a flow chart as simplified guidance on the design process.

5.3. Residential streets
As indicated earlier, most of the design practice involving residential streets is contained in *MfS* (DfT, 2007); key features are outlined in this section.

5.3.1 Principal functions
From a residential point of view, streets can be considered to have five principal functions, as described in *Paving the Way* (CABE, 2002).

- *Place.* An important consideration in distinguishing a street from a road, as described in *Places, Streets and Movements* (DETR, 1998).
- *Movement.* Providing for all modes of transport, including pedestrians and cyclists.
- *Access.* To buildings and public spaces; including frontage roads and access to areas not directly facing the street.
- *Parking.* A key function of many streets, it can provide convenient access to frontages and add to the vitality of a street but can also create safety problems and impair visual quality.
- *Drainage, utilities and street lighting.* Essential space for these vital public services.

Movement status can be expressed in terms of traffic volume and the importance of the street, or section of street, within a network, either for general traffic or within a mode-specific (e.g. bus or cycle) network. It can vary along the length of a route, such as where a street passes through a town centre.

5.3.2 Relative importance of routes
Highway authorities assess the relative importance of particular routes within an urban area as part of their normal responsibilities, such as those under the New Roads and Street Works Act 1991. One of the network management duties under the Traffic Management Act 2004 is that all local traffic authorities should determine specific policies or objectives for different roads or classes of road in their road networks. Other guidance, including *Traffic Management Act 2004: Network Management Duty Guidance* (DfT, 2004), also applies. Typically, it is for the authority to decide the levels of priority given to different road users on each road. For example, particular routes may be defined as being important with respect to the response times of the emergency services.

5.3.3 Place and movement matrix
Defining the relative importance of particular streets or roads in terms of place and movement functions should inform subsequent design choices. For example

- motorways – high movement function, low place function
- high streets – medium movement function, medium-to-high place function
- residential streets – low-to-medium movement function, low-to-medium place function.

Therefore, streets can be looked at as a two-dimensional hierarchy. The two-dimensional hierarchy as a way of informing street design recognises the relative importance of traffic flow and place function (DfT, 2007, Figure 2.5) giving a wider range of design options. It is recommended that the design of a scheme should follow a user hierarchy (DfT, 2007, Table 3.2) of

Figure 5.5 Cross-section design flow chart (DfT, 2005a). NMU, non-motorised user
© Crown Copyright, 2005
This information is licensed under the Open Government Licence v3.0. To view this licence, visit http://www.nationalarchives.gov.uk/doc/open-government-licence/ **OGL**

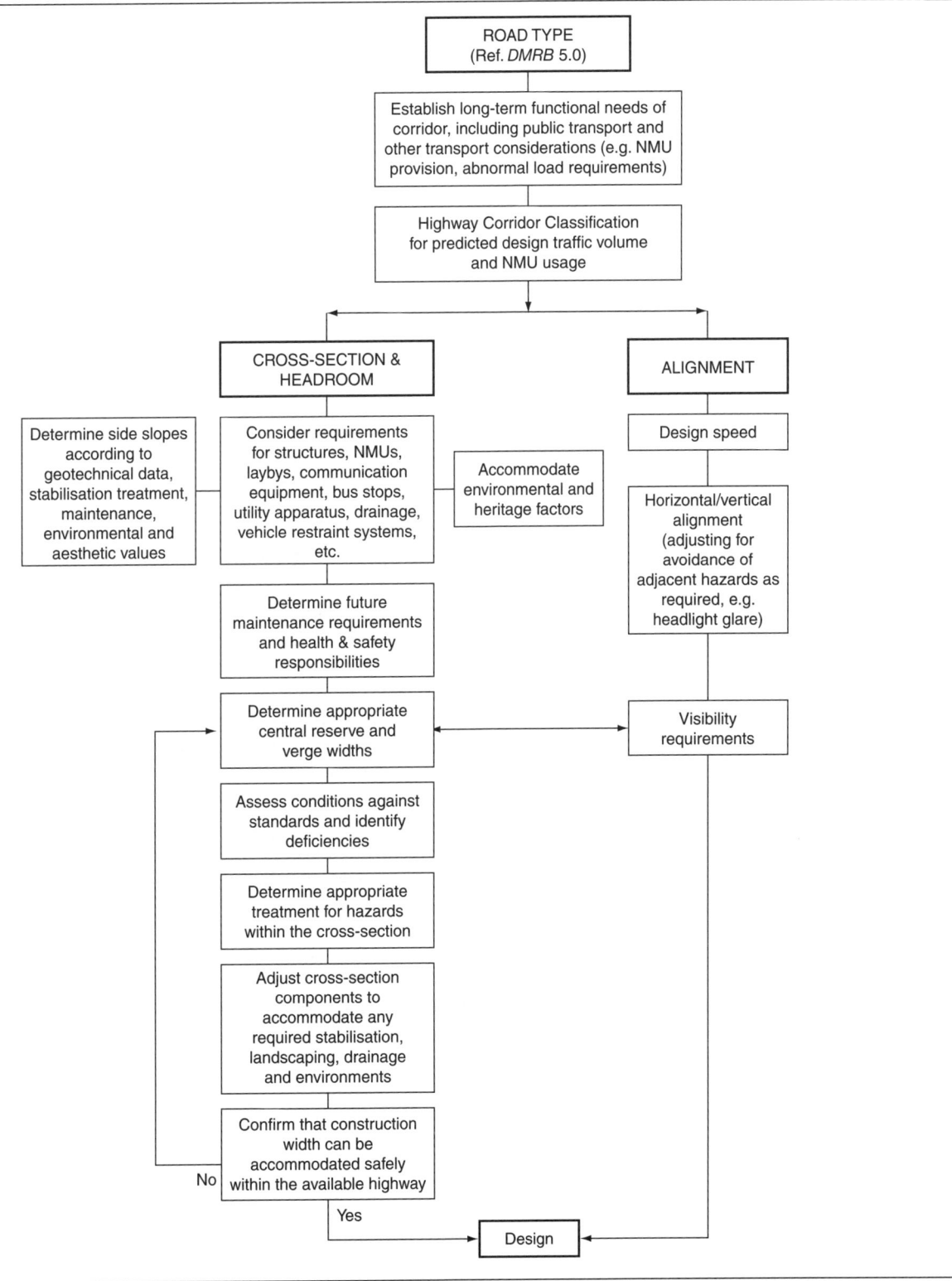

Table 5.1 Hierarchy of provision for pedestrians and cyclists (DfT, 2007)

	Pedestrians	Cyclists
Consider first	Traffic volume reduction	Traffic volume reduction
	Traffic speed reduction	Traffic speed reduction
	Reallocation of road space to pedestrians	Junction treatment, hazard site treatment, traffic management
	Provision of direct at-grade crossings, improved pedestrian routes on existing desire lines	Cycle tracks away from roads
Consider last	New pedestrian alignment or grade separation	Conversion of footways or footpaths to adjacent-* or shared-use routes for pedestrians and cyclists

* Adjacent-use routes are those where the cyclists are segregated from pedestrians.
© Crown Copyright, 2007
This information is licensed under the Open Government Licence v3.0. To view this licence, visit http://www.nationalarchives.gov.uk/doc/open-government-licence/ **OGL**

- pedestrians (which are considered first)
- cyclists
- public transport users
- specialist service vehicles (e.g. emergency services, waste, etc.)
- other motor traffic (which are considered last).

With regard to provision for pedestrians and cyclists, Table 5.1 indicates a useful hierarchy.

The hierarchy is not meant to be rigid. However, this list will ensure that all of the users will be served in a balanced way. Often, the position of a street and the demands on it will inform decisions on its capacity, cross-section and connectivity. Moreover, regarding provision of cycling facilities, people may still want to cycle on a main road where volume reduction is unlikely, so a cycle track may offer assistance.

5.3.4   Design codes
Design codes comprise detailed written and graphically presented rules for building out a site or an area. They are often promoted by local authorities but may be put forward by the private sector. Example dimensions for the design code for a riverside development in Rotherham (DfT, 2007, Figure 3.11) are

- carriageway width, 6.0 m
- footpath width, minimum 2.0 m; minimum 3.0 m along riverfront
- design speed, 20 miles per hour
- traffic calming, carriageway narrowing
- junction radii, minimum 4.0 m
- vehicle type to be accommodated: cars, small service vehicles, fire appliances, cycles

- on-street parking: perpendicular, 5.0 m × 2.5 m
- direct access to plots: no
- street trees, 8.0–12.0 m spacing between trees (adjust to accommodate parking areas).

5.3.5   Geometric choices and street patterns
Straight streets are efficient in the use of land, facilitating direct connections between places for pedestrians who prefer direct routes, but, if too long, may encourage higher vehicle speeds. Shorter and curved or irregular streets contribute to variety and a sense of place, and may be suitable for topographic and activity-related reasons. However, layouts that use excessive or gratuitous curves can be less efficient and may hinder access for non-motorised users. Geometric choices and street patterns should be based on a thorough understanding of context. This can include considerations of crime prevention; for example, dwellings and activity centres should preferably overlook footpaths and other pedestrian areas.

Culs-de-sac may be required because of topographical, boundary or other constraints. They can result in less motor traffic but through connections for pedestrians and cyclists should be visible from active frontages. Culs-de-sac can also assist where through routes are not practical, but they may concentrate traffic impact on a few dwellings and require turning heads that are wasteful of land and lead to additional vehicle travel and emissions, particularly by service vehicles.

Most neighbourhoods include a range of street character types, each with differing characteristics, including type of use, width and building heights. These characteristics dictate how pedestrians and traffic use the street.

Figure 5.6 Typical street widths for different types of street (DfT, 2007)
© Crown Copyright, 2007
This information is licensed under the Open Government Licence v3.0. To view this licence, visit http://www.nationalarchives.gov.uk/doc/open-government-licence/ OGL

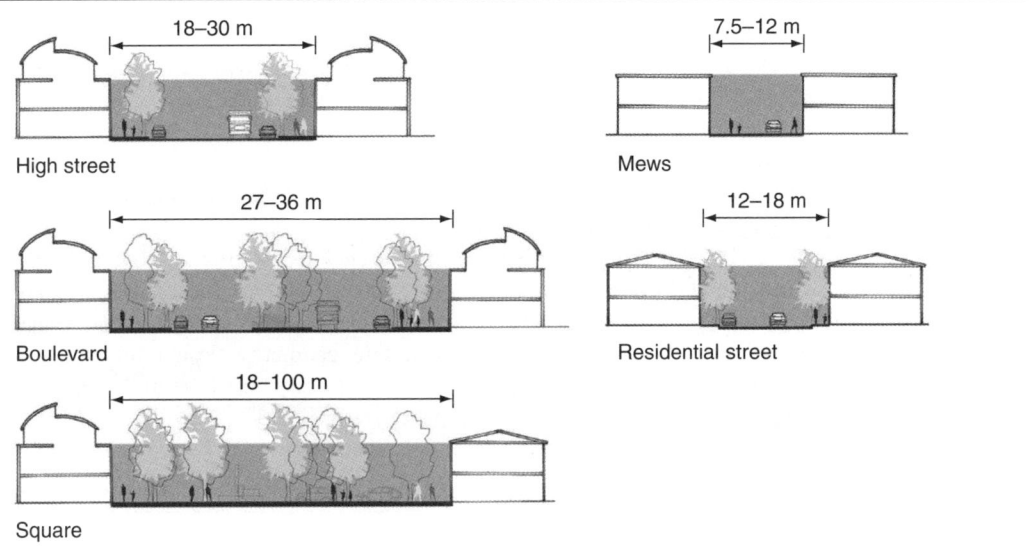

5.3.6 Width

Width between buildings is a key dimension and needs to be considered in relation to function and aesthetics. Figure 5.6 shows typical widths for different types of street. The distance between frontages in residential streets typically ranges from 12 to 18 m, although there are examples of widths less than this that work well. There are no fixed rules but account should be taken of the variety of activities taking place in the street and of the scale of the buildings on either side.

Typical widths of carriageways for various combinations of motorised and non-motorised vehicles are shown in Figure 5.7.

Figure 5.7 Typical dimensions of carriageways to accommodate various combinations of motorised and non-motorised vehicles (DfT, 2007)
© Crown Copyright, 2007
This information is licensed under the Open Government Licence v3.0. To view this licence, visit http://www.nationalarchives.gov.uk/doc/open-government-licence/ OGL

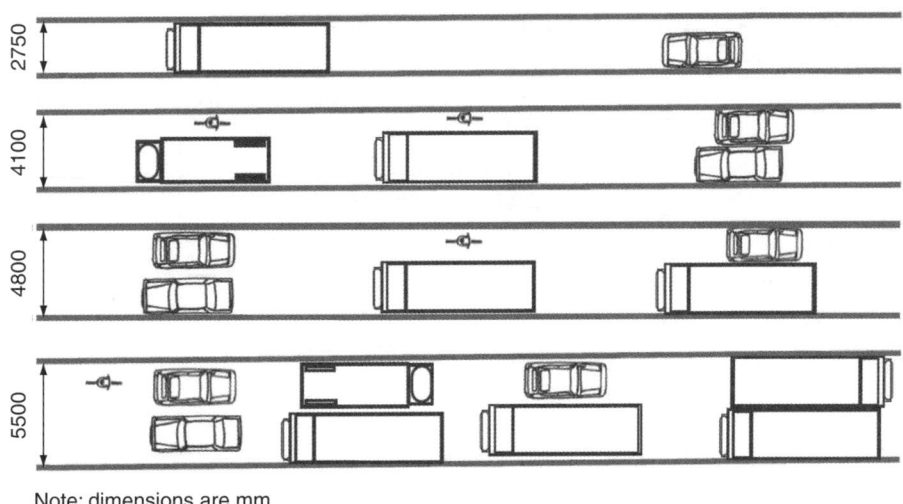

Note: dimensions are mm.

5.3.7 Swept path analysis

Swept path analysis, or tracking, is used to determine the space required for various vehicles and is a key tool for designing carriageways for vehicular movement within the overall layout of the street.

The use of computer-aided design (CAD) tracking models and similar techniques often proves beneficial in determining how the street will operate and how vehicles will move within it. Layouts designed using this approach enable buildings to be laid out to suit the character of the street, with footways and kerbs helping to define and emphasise spaces. Designers have the freedom to vary the spaces between kerbs or buildings. The kerb line does not need to follow the line of vehicle tracking if careful attention is given to the combination of sight lines, parking and pedestrian movements. Figure 5.8 shows a typical representation of a street character type and two street arrangements to accommodate parking and buses. Shallow gradient junction tables should be used for buses. Both examples also show applicable dimensions.

5.3.8 Principles of urban design

The fundamental principles of urban design are described in a wide range of publications, including *By Design: Urban Design in the Planning System – Towards Better Practice* (CABE, 2000). There is a section relating specifically to pedestrians in *Planning for Walking* (CIHT, 2015b).

Based on principles of urban design, roads should be complemented by networks of pedestrian routes; the use of sustainable modes, such as walking, cycling and public transport should be encouraged. Walking destinations up to approximately 800 m are preferred; infrastructure to encourage walking must include adequate geometric design, including footway widths, good access to public transport and crossings that provide effective pedestrian priority.

Figure 5.8 Representation of a street character type showing detail for a minor street junction and examples of on-street parking arrangements (DfT, 2007)
© Crown Copyright, 2007
This information is licensed under the Open Government Licence v3.0. To view this licence, visit http://www.nationalarchives.gov.uk/doc/open-government-licence/ **OGL**

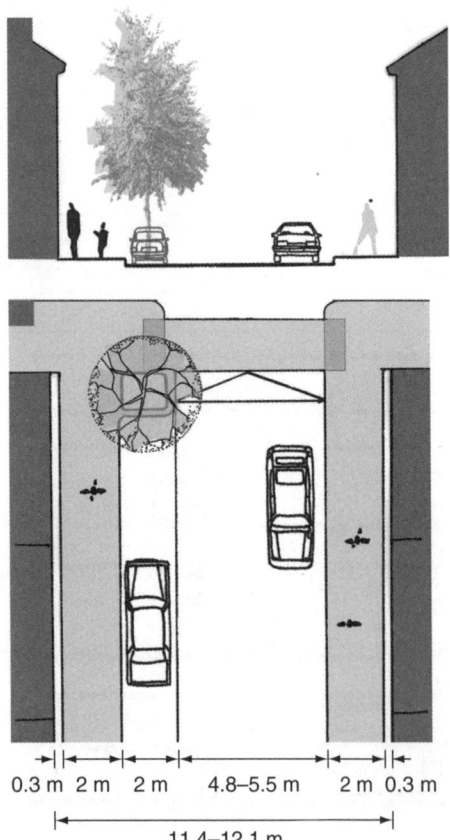

0.3 m 2 m   2 m   4.8–5.5 m   2 m 0.3 m

11.4–12.1 m

Newhall case study demonstrates that adherence to masterplan principles can be achieved through the use of design codes that are attached to land sales and achieved by covenants.

A list of key dimensions was applied:

- Frontage to frontage – min 10.5 m;
- Carriageway width – min 4.8 m, max 8.8 m;
- Footway width – min 1.5 m;
- Front gardens – min 1.5 m, max 3 m;
- Reservation for services – 1 m; and
- Design speed – 20 mph.

The design is based on pedestrian priority and vehicle speeds of less than 20 mph controlled through the street design.

The specific conditions in a street will determine what form of crossing is most relevant. See also other discussions related to unsignalled and signalled crossings in Chapters 6 and 7.

Surface level crossings can be of a number of types.

- *Uncontrolled crossings.* These can be created by dropping kerbs at intervals along a link. As with other types of crossing, these should be matched to pedestrian desire lines. If the crossing pattern is fairly random and there is an appreciable amount of pedestrian activity, a minimum spacing of 100 m is recommended. Dropped kerbs should be marked with appropriate tactile paving and aligned on each side of the carriageway.
- *Informal crossings.* These use paving materials and street furniture to indicate a crossing place, encouraging slow-moving traffic to give way to pedestrians.
- *Pedestrian refuges and kerb build-outs.* These may be used separately or in combination. They effectively narrow the carriageway, reducing the crossing distance. However, they can create pinch-points for cyclists if the remaining gap is still wide enough for motor vehicles to squeeze past them.
- *Zebra crossing.* Of the formal crossing types, these involve the minimum delay for pedestrians when used in the right situations.
- *Signalised crossings.* There are four types: pelican, puffin, toucan (including equestrian) crossings and PedEX crossings. The pelican crossing was the first to be introduced. Puffin crossings, which have nearside pedestrian signals and a variable crossing time, are replacing pelican crossings. They use pedestrian detectors to match the length of the crossing period to the time pedestrians take to cross. Toucan and equestrian crossings operate in a similar manner to puffin crossings, except that cyclists can also use toucan crossings, while equestrian crossings include a separate crossing for horse riders. Signalled crossings are preferred by blind or partially sighted people. PedEX crossings (IHE, 2016) feature farside pedestrian signals with red and green figures, a junction push-button and a blackout period. They have the same markings as a pelican crossing and feature countdown or detection (but not both). As individual pelican crossings reach the end of their useful life they will be replaced by PedEX crossings.

5.3.9 Bus stops
It is essential to consider the siting of public transport stops and related pedestrian routes at an early stage of design to ensure their easy access on foot and, importantly, by disabled people. The width of the footway should be adjusted to ensure that people passing the bus stop have sufficient space to comfortably pass those waiting, boarding and alighting at the stop. See Chapter 8 for details of bus stop layouts and dimensions.

Bus stops should be easily accessed on foot. Their precise location will depend on other issues, such as the need to avoid noise nuisance, visibility requirements and the convenience of pedestrians and cyclists. Routes to bus stops must be accessible by disabled people.

5.3.10 Emergency vehicles
The requirements for emergency vehicles are generally dictated by the fire service requirements; such requirements will also accommodate police vehicles and ambulances. Regulation B5 of The Building Regulations 2000 and the Fire and Rescue Services Act 2004 address these matters. The latter specifies

- a minimum carriageway width of 3.7 m
- vehicle access for a pump appliance within 45 m of single-family houses
- vehicle access for a pump appliance within 45 m of every dwelling entrance for flats or maisonettes
- that a vehicle access route may be a road or other route
- that fire service vehicles should not have to reverse further than 20 m.

The Association of Chief Fire Officers has expanded on and clarified these requirements in terms of distances to dwellings, situations requiring integrated management plans, parking and the use of sprinkler systems. Other relevant publications include the Fire and Rescue Services Act 2004, risk reduction plans required by the Welsh Assembly (National Assembly for Wales (2005, 2007)) The Fire and Rescue National Framework (Wales) 2005 (Revisions) Order 2007.

5.3.11 Service vehicles
Service vehicles must be designed for, although larger vehicles that are only expected to use a street infrequently, such as pantechnicons, need not be fully accommodated; they will have to reverse or undertake multipoint turns to turn around on the rare occasions they will require access. For culs-de-sac longer than 20 m, a turning area should be provided to cater for vehicles that will regularly need to enter the street.

Carriageway widths should be appropriate for the particular context and uses of the street. Key factors to take into account include vehicular and pedestrian traffic volumes, composition of vehicles, demarcation between carriageway and footway, parking arrangements and design speed (recommended to be less than 20 miles per hour in residential areas). Street curvature and one-way systems must also be considered.

In streets with light traffic, carriageways may be narrowed over short lengths to single lanes as a traffic calming feature. In such single-lane working sections of street, to prevent parking, the width between constraining vertical features, such as bollards, should be no more than 3.5 m (DfT, 2007). In particular circumstances, this may be reduced to a minimum value of 2.75 m, which will still allow for occasional large vehicles. However, widths between 2.75 and 3.25 m should be avoided in most cases, since they could result in drivers trying to squeeze past cyclists. The local fire safety officer should be consulted where a carriageway width of less than 3.7 m is proposed.

5.4. Shared space

5.4.1 Shared-space streets

As well as the *MfS*, the concept of making vehicle and pedestrian traffic more compatible has received much attention. The notion of 'sojourn' space, as well as space for movement for pedestrians, and the creation of environmentally attractive features have resulted in related requirements for geometric design sensitive to these needs, although, in most urban areas, traditional street and traffic patterns, and associated road safety concerns, are likely to dominate for the foreseeable future. An extensive literature on shared space has become available from government, non-government and academic sources.

Within the context of overall planning and design for pedestrian facilities in the public realm, a graphical portrayal of the sequence from research on users' needs and desires through the various involvements and responsibilities to conform to the Equality Act 2010 is shown in Figure 5.9 (CIHT, 2018).

A useful description of the key issues is provided in LTN 1/11 *Shared Space* (DfT, 2011), which focuses on shared space in high-street environments, although many of its principles will apply in other settings. Emphasis is placed on stakeholder engagement, inclusive design, sustainability and maintenance. It explains how the scheme development process introduced in LTN 1/08 *Traffic Management and Streetscape* (DfT, 2008) can be applied to shared-space projects. Key parts of LTN 1/11 are summarised next.

5.4.1.1 Shared space – characteristics

Shared space may be defined as:

> A street or place designed to improve pedestrian movement and comfort by reducing the dominance of motor vehicles and enabling all users to share the space rather than follow the clearly defined rules implied by more conventional designs.
>
> (DfT, 2011)

In continental Europe, shared space is often used to smooth traffic flow and reduce delays at major junctions. In the UK, it is usually applied to links and minor junctions with the aim of allowing pedestrians to move more freely within the space.

> In a conventional street, [...] although pedestrians and motorists are equally entitled to occupy the carriageway, pedestrians generally exercise little control over vehicular traffic, except at controlled crossings.
>
> (DfT, 2011)

Figure 5.9 Policy and guidance characteristics from user needs to Equality Act 2010 (CIHT, 2018)

In a shared space, drivers' progress is more dependent on interpreting the behaviour of other users than on facilitating vehicular movement. Key characteristics of shared space are

- pedestrians occupying the carriageway
- increased levels of social interaction and leisure activity
- people spending longer in the street (evidence of an enhanced sense of place)
- drivers and cyclists giving way to pedestrians
- pedestrians crossing the street at locations, angles and times of their choosing
- drivers and cyclists giving way to one another.

(DfT, 2011)

5.4.1.2 Examples

Many historic streets operate as shared spaces, particularly narrow streets in historic core zones and residential mews. [Examples include] Chertsey Road in Woking, Surrey, [and] Seven Dials in Covent Garden, London. Shared space has also been applied to some arterial routes, restoring their traditional place functions. Home zones and some country lanes, particularly those with a quiet lanes designation, tend to operate as share spaces.

(DfT, 2011)

Shared-space schemes have been implemented internationally. Figures 5.10 to 5.12 show instances of shared space in the UK and continental Europe.

5.4.1.3 Physical characteristics

Some shared-space streets omit conventional kerbs – these are often called shared-surface streets. However, the term is

Figure 5.10 Chertsey Road in Woking (DfT, 2011)
© Crown Copyright, 2011
This information is licensed under the Open Government Licence v3.0. To view this licence, visit http://www.nationalarchives.gov.uk/doc/open-government-licence/ **OGL**

not necessarily an accurate description of the way the space operates – not all such surfaces will be truly shared. [...] The term 'level surface' is used to describe this feature. A level surface is defined thus:

'A street surface with no level difference to segregate pedestrians from vehicular traffic.'

A level surface is often intended to remove a physical and psychological barrier to pedestrian movement. It can also indicate to drivers that pedestrians are not confined to the footway and may be encountered in the whole street.

(DfT, 2011)

The characteristic of 'no level difference' has, however, caused much concern, particularly for people with visual impairments; calls for a moratorium on shared space have been made (Holmes, 2015). Pertinent comments relating to LTN 1/11 made by Disability Wales (2014) address kerb locations and heights and are addressed in Chapter 8.

In general, sharing between vehicle users and pedestrians should take place in the street's carriageway area, not the sides of the street, which should mainly be the preserve of pedestrians.

[...] Sharing may be facilitated by [...]

- removing any implied priority of vehicles over pedestrians in the carriageway
- reducing demarcation between pedestrians and vehicular traffic
- introducing features not necessarily limited to the sides of the street, such as seating, public art and cafes, which encourage pedestrians to use the space

The design speed is a target speed that designers intend most vehicles not to exceed and is dictated primarily by the geometry of tracked vehicle paths within the street. For shared space, a design speed of no more than 20 miles per hour is desirable, and preferably less than 15 miles per hour.

(DfT, 2011)

Speeds may be reduced by:

further restricting width and increasing horizontal deflection at key locations such as crossing points, even though this could require large vehicles to negotiate them at speeds well below the design speed.

(DfT, 2011)

The concept of shared space is under continuing review; one way of defining shared-space characteristics (CIHT, 2018) is as follows.

Figure 5.11 Queen's Square, Wolverhampton, before and after installation of shared-space scheme, also showing reduced traffic flows (European Commission, 2004)

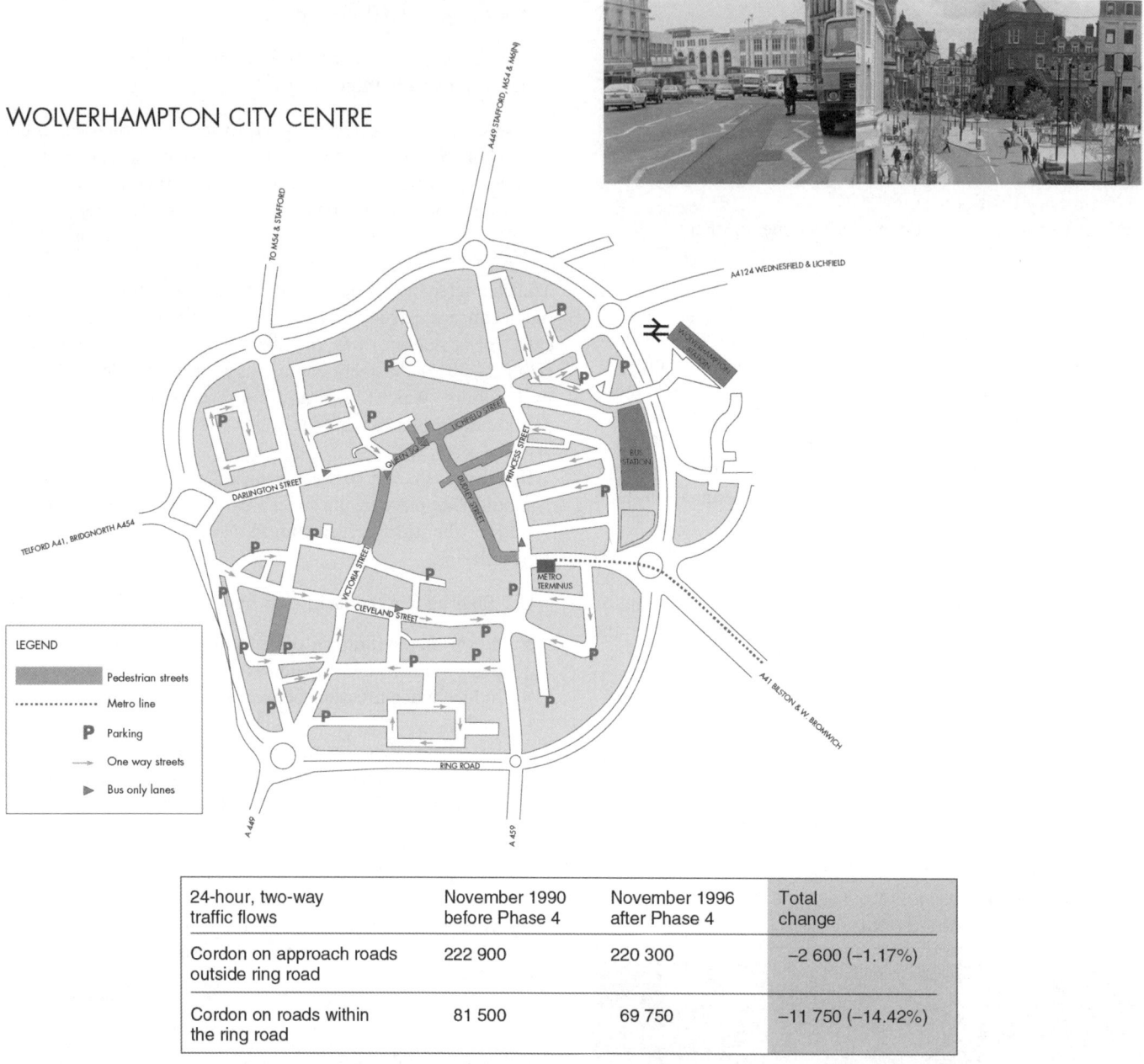

WOLVERHAMPTON CITY CENTRE

24-hour, two-way traffic flows	November 1990 before Phase 4	November 1996 after Phase 4	Total change
Cordon on approach roads outside ring road	222 900	220 300	−2 600 (−1.17%)
Cordon on roads within the ring road	81 500	69 750	−11 750 (−14.42%)

- *Pedestrian prioritised streets.* Streets where pedestrians feel that they can move freely anywhere and where drivers should feel they are a guest.
- *Informal streets.* Streets where formal traffic controls (signs, markings and signals) are absent or reduced but there remains a footway and carriageway, even if with less differentiation.
- *Enhanced streets.* Streets where the public realm has been improved and restrictions on pedestrian movement (e.g.

guardrails) have been removed but conventional traffic controls largely remain.

Regarding other European countries, it is reported (Parkin, 2018) that Switzerland, Belgium and France have meeting zones (or encounter zones) assigned to all users, in which pedestrians have legal priority over vehicles. Design of space is, therefore, important; a key issue in the absence of a clearly differentiated carriageway. It has been commented that such equitable design,

Figure 5.12 Cycle route at the junction of Quai du Général Koenig and Rue de la Brigade d'Alsace Lorraine (left) and Place Kléber (right), with accompanying indicators of lessons learnt (European Commission, 2004)

KEY SUCCESS FACTORS/LESSONS LEARNT(*)
- It is necessary to have strong political vision and commitment to finding more sustainable solutions even in the face of opposition.
- Carry out a comprehensive consultation exercise.
- Provide clear and regular information about the progress of the project.
- Provide tangible 'benefits' when taking away road space from car drivers.

(*) *Source:* Communauté urbaine de Strasbourg.

coupled with the knowledge that the fault would lie with the driver in the event of a collision, would assist and accelerate the introduction of such schemes more widely.

5.4.1.4 Comfort space

While shared space appears to work well for most people, some disabled and older people can feel apprehensive about using the space, particularly where a level surface is used. Comfort space is defined as:

> An area of the street predominantly for pedestrian use where motor vehicles are unlikely to be present.
>
> (DfT, 2011)

In general, comfort space only needs to be considered when designing streets with a level surface.

5.4.2 Home zones

Home zones are residential areas designed with streets to be places for people, instead of just for motor traffic. By creating a high-quality street environment, home zones strike a better balance between the needs of the local community and drivers. Involving the local community is the key to a successful scheme. Effective consultation with all sectors of the community, including young people, can help ensure that the design of individual home zones meets the needs of local residents.

Home zones often include shared surfaces as part of the scheme design; these can create difficulties for disabled people. Research commissioned by the Disabled Persons Transport Advisory Committee (DPTAC) on the implications of home zones for disabled people will demonstrate those concerns. Chapter 8 discusses key elements of these issues in more detail.

Home zones are encouraged in planning and transport policies for new developments and existing streets. They are distinguished from other streets by having signed entry and exit points, which indicate the special nature of the street. Local traffic authorities in England and Wales were given the powers to designate roads as home zones in Section 268 of the Transport Act 2000.

Developers sometimes implement 'home zone style' schemes without formal designation. However, it is preferable for the proper steps to be followed to involve the community in deciding how the street will be used. In existing streets, it is essential that the design of the home zone involves significant participation by local residents and local access groups (DfT, 2007). Further guidance on the design of home zones is given in *Home Zones: Challenging the Future of Our Streets* (DfT, 2005f), *Home Zone Design Guidelines* (IHIE, 2002) and on the website *Home Zone News* (2013).

5.5. Summary

This chapter has briefly addressed a method of understanding the principles of street design and outlined key concerns in the actual design of streets described in current guidance. The topic of street design continues to be one of considerable debate and the matter of street design in rural areas and villages, for example, is only lightly addressed in much of the guidance. Considerably more research and innovative approaches to this important aspect of pedestrian facilities and movement is warranted.

REFERENCES

Belcher M, Proctor S and Cook C (2015) *Practical Road Safety Auditing*, 3rd edn. Thomas Telford, London, UK.

CABE (Commission for Architecture and the Built Environment) (2000) *By Design: Urban Design in the Planning System – Towards Better Practice*. DETR, London, UK.

CABE (2002) *Paving the Way: How We Achieve Clean, Safe and Attractive Streets*. Thomas Telford, London, UK.

CIHT (Chartered Institute of Highways and Transportation) (2015a) *Designing for Walking*. CIHT, London, UK.

CIHT (2015b) *Planning for Walking*. CIHT, London, UK.

CIHT (2018) *Creating Better Streets: Inclusive and Accessible Places. Reviewing Shared Space*. CIHT, London, UK.

Davis CJ (2014) *Street Design for All: An Update of National Advice and Good Practice*. Public Realm Information and Advice Network (PRIAN). London, UK.

DETR (1998) *Places, Streets and Movement: A Companion Guide to Design Bulletin 32 – Residential Roads and Footpaths*. DETR, London, UK.

DfT (Department for Transport) (2004) *Traffic Management Act 2004: Network Management Duty Guidance*. DfT, London, UK.

DfT (2005a) TD 27/05: Road geometry. links, cross-sections and headrooms. In *Design Manual for Roads and Bridges (DMRB)*. DfT, London, UK.

DfT (2005b) TA 90/05: Geometric design of pedestrian, cycle and equestrian routes. In *Design Manual for Roads and Bridges (DMRB)*. DfT, London, UK.

DfT (2005c) TA 91/05: Provision for non-motorised users. In *Design Manual for Roads and Bridges (DMRB)*. DfT, London, UK.

DfT (2005d) HD 42/05: Non-motorised user audits. In *Design Manual for Roads and Bridges (DMRB)*. DfT, London, UK.

DfT (2005e) *Inclusive Mobility: A Guide to Best Practice on Access to Pedestrian and Transport Infrastructure*. DfT, London, UK.

DfT (2005f) *Home Zones: Challenging the Future of Our Streets*. DfT, London, UK.

DfT (2007) *Manual for Streets*. Thomas Telford, London, UK.

DfT (2008) LTN 1/08: *Traffic Management and Streetscape*. The Stationery Office, London, UK.

DfT (2011) LTN 1/11: *Shared Space*. The Stationery Office, London, UK.

DfT (2012) TD 9/12: Road geometry. links. road link design. In *Design Manual for Roads and Bridges (DMRB)*. DfT, London, UK.

DfT (2018) *Design Manual for Roads and Bridges (DMRB)*. DfT, London, UK.

Disability Wales (2014) *Travel and Transport: Retain Kerbs in Shared Space Says Circulated Guidance from DfT*. Disability Wales, Caerphilly, UK. http://www.disabilitywales.org/travel-and-transport-retain-kerbs-in-shared-space-says-circulated-guidance-from-dft/ (accessed 08/10/2018).

European Commission (2004) *Reclaiming City Streets for People: Chaos or Quality of Life?* European Commission, Brussels, Belgium.

HMG (Her Majesty's Government) (1991) New Roads and Street Works Act 1991. The Stationery Office, London, UK.

HMG (2000) The Building Regulations 2000. The Stationery Office, London, UK.

HMG (2000) Transport Act 2000. The Stationery Office, London, UK.

HMG (2004) Fire and Rescue Services Act 2004. The Stationery Office, London, UK.

HMG (2004) Traffic Management Act 2004. The Stationery Office, London, UK.

HMG (2010) Equality Act 2010. The Stationery Office, London, UK.

Holmes, Lord of Richmond (2015) *Accidents by Design: The Holmes Report on 'Shared Space' in the United Kingdom*. House of Lords, London, UK.

Home Zone News (2013) http://www.homezonenews.org.uk/ (accessed 08/10/2018).

IHE (Institution of Highway Engineers) (2016) *Traffic Signals Design Course*. IHE, London, UK.

IHIE (Institute of Highway Incorporated Engineers) (2002) *Home Zone Design Guidelines*. IHIE, London, UK.

IHT (Institution of Highways and Transportation) (1997) *Transport in the Urban Environment*. IHT, London, UK.

IHT (2000) *Guidelines for Providing for Journeys on Foot*. IHT, London, UK.

Jones PM, Boujenko N and Marshall S (2007) *Link and Place: A Guide to Street Planning and Design*. Landor, London, UK.

National Assembly for Wales (2007) The Fire and Rescue National Framework (Wales) 2005 (Revisions) Order 2007. The Stationery Office, London, UK.

National Assembly for Wales (2012) The Fire and Rescue Services (National Framework) (Wales) Order 2012.

Parkin J (2018) *Designing for Cycle Traffic, International Principles and Practice*. Thomas Telford, London, UK.

Pedestrian Facilities, Second edition

Schoon, John G
ISBN 978-0-7277-6309-9
https://doi.org/10.1680/pfse.63099.091

Chapter 6
Crossings at unsignalised locations

Crossings on carriageways, where pedestrians and vehicles interact, are typically the most dangerous, intimidating and inconvenient locations on most pedestrians' journeys. Crossing design is therefore of critical importance. The location and geometry of carriageways and footways are, ideally, related to the stopping sight distances of drivers and vehicles and the crossing characteristics of pedestrians, as described earlier. Particularly at and near junctions and bends, property lines and other features may affect the layout; in many cases, the final design becomes a compromise between theoretical and practical considerations.

This chapter focuses initially on crossings at junctions. It then applies many of the principles underlying their analysis and design to straight sections of roadway and continuous bends where stand-alone or midblock crossings occur.

As indicated earlier, current guidance is divided broadly into the categories of trunk roads and residential streets, as described in *DMRB* (DfT, 2018) and *MfS* (DfT, 2007a), respectively. More recent policies associated with the 'shared-space' concept are also described. Coordination with the design needs of cyclists and equestrians and pedestrian facilities at signalled junctions and roundabouts, are discussed in later chapters.

6.1. Driver and pedestrian tasks at junctions

Before examining current practice in the design of crossings, this section compares the actions that pedestrians must take in comparison with drivers, and then presents some of the related times taken to conduct certain essential manoeuvres during the crossing process. The physical abilities that pedestrians need to safely negotiate a crossing include eye and body movements and mental coordination. The extent of effort and attention required by pedestrians will also depend on the crossing location and its environment and the characteristics of vehicles encountered.

6.1.1 Comparison between pedestrians' and drivers' tasks at junctions

Considerable mental and physical activity is required of all road users in order to negotiate even the simplest priority T-junction safely, and the principles also apply to non-junctions.

Illustrating this, Figure 6.1 shows the most basic scanning requirements for a pedestrian to cross the minor arm at a three-way priority junction. A pedestrian's scanning angle may be described as the angle that must be scanned to adequately identify approaching vehicles and their characteristics so that suitable gaps may be identified in order to start and complete a safe crossing.

It is evident from Figure 6.1 that pedestrians, shown waiting to cross the minor arm of the junction, require the greatest scanning angle – close to 265° if they are to satisfy their need to be aware of vehicles approaching from all possible directions in order to avoid a collision as they cross. During this scan, the pedestrians must also focus and refocus several times to evaluate the speed and intentions of vehicles approaching from at least three different directions.

Although many pedestrians are able to mentally and physically undertake the scanning process quickly, for a large segment of the population, such as children, people accompanied by small children, people in wheelchairs and their helpers, people with upper body and neck impairments and visually impaired people, the task of quickly scanning nearly 270° is often difficult and cannot be done quickly. The presence of heavy traffic obscuring sight lines, or poor visibility due to rain or fog, complicates an already difficult task. Furthermore, it requires a finite amount of time to complete the *Highway Code*'s procedure (a routine considered to afford a safe crossing if followed correctly), as indicated in Chapter 3.

6.1.2 Pedestrians' tasks in crossing

A basic requirement of all road users is a general awareness of the environment, such as the weather conditions, surface conditions, amount of illumination, likelihood of unexpected appearance of obstacles, general traffic density and unpredictability of other road users. An ability to be aware of these conditions, although not necessary to cope with all eventualities, will be 'built in' to road users' attitude and reflected in their conduct on the road and how the carriageway should be crossed.

The observation angles needed by a pedestrian standing on the corner of the minor arm of a junction similar to that shown in

Figure 6.1 Comparison of required scanning angles for a pedestrian and drivers at three-way uncontrolled junctions. Note: angles quoted are approximate and depend on exact dimensions and configuration of junction

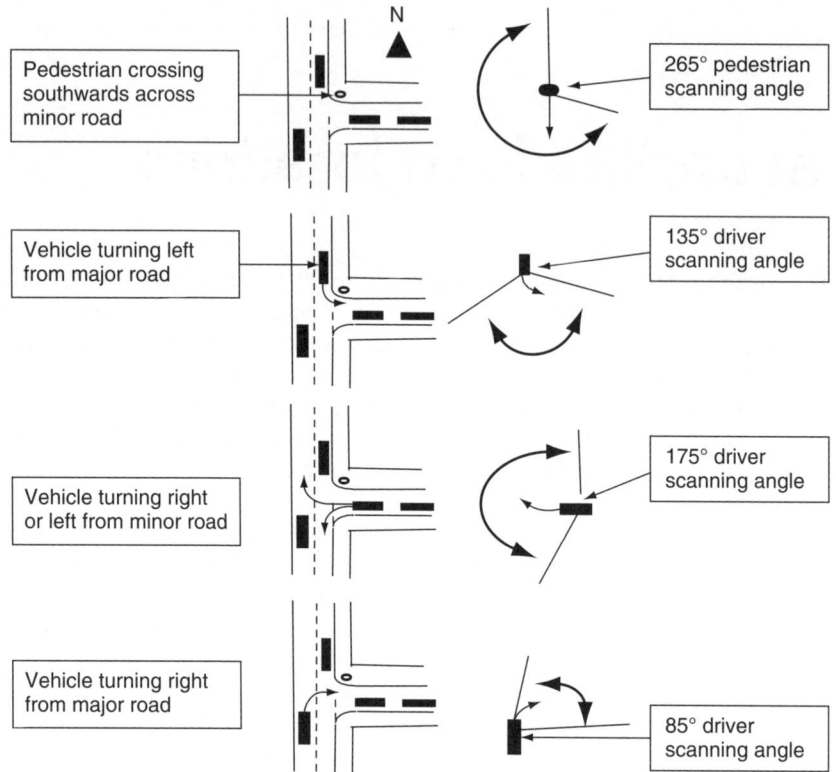

Pedestrian crossing southwards across minor road — 265° pedestrian scanning angle

Vehicle turning left from major road — 135° driver scanning angle

Vehicle turning right or left from minor road — 175° driver scanning angle

Vehicle turning right from major road — 85° driver scanning angle

Figure 6.1 and wishing to cross directly southward are shown in Figure 6.2, which indicates portions of the pedestrian's total scanning angle divided into sectors A to D. Note that the total range of scanned angle is approximately 265°. Each sector of the pedestrian's total scanning angle is briefly described as follows.

- *Sector A*. Vehicles turning left, essentially approaching from behind the pedestrian's position.
- *Sector B*. Vehicles turning right and approaching from between approximately +10° and +90° of the pedestrian's look-ahead direction.

Figure 6.2 Example comparison of pedestrian's scanning angles and sectors related to approaching vehicles during pedestrian's observation–reaction time

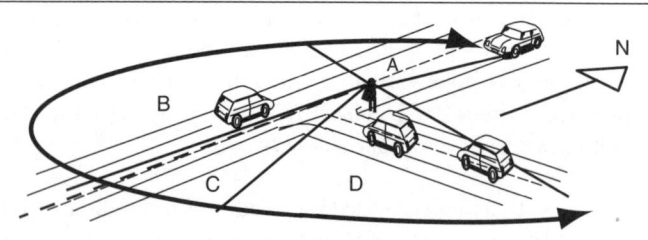

- *Sector C*. Pedestrians approaching on the opposite side of the crossing from ahead of the pedestrian.
- *Sector D*. Vehicles approaching from between the pedestrian's look-ahead direction, and −90° to the left direction, and turning right or left.

An essentially similar sector and time-scanning angle diagram can be constructed for the pedestrian crossing in the opposite direction (northwards across the minor arm), as shown in Figure 6.3.

The scanning angles required of drivers approaching the junction are, in general, considerably less than for pedestrians wishing to cross the minor arm. The implications of this larger scanning angle for pedestrians are important for several reasons.

- The greater angle will require extensive head and some shoulder movement – a movement that some pedestrians, particularly those with physical disabilities, those encumbered and those in wheelchairs, may find difficult or in some cases impossible.
- The greater angle, in and of itself, requires a longer time to scan. That is, the greatest time requirement of all road users negotiating a junction is that of the pedestrian.

Figure 6.3 Required visual scanning angle and sectors by a pedestrian about to cross northwards over the minor road of an uncontrolled junction

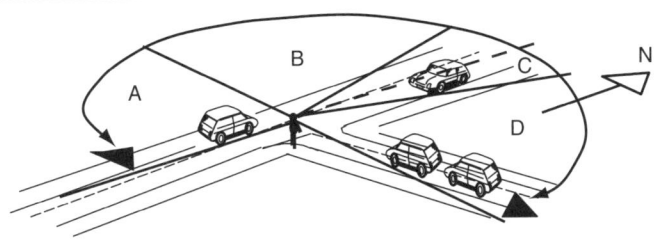

■ The time taken to scan the areas of approaching vehicles will mean that certain areas within the necessary visibility sector will be unobserved during part of the scanning process, as the pedestrian's direction of gaze changes. Therefore, unless the analysis and geometric design of the intersection recognises this (closely related to the pedestrian's observation–reaction time described in Chapter 3), the pedestrian may begin to walk, unaware that a hitherto unseen vehicle is approaching along a collision path.

NOTE
As noted in previous chapters, the concept of the pedestrian's observation–reaction time is not addressed in current design guidance in the UK. The assumption in current design procedures is that the driver will take action to avoid the pedestrian, not that the pedestrians must be given the ability to avoid a vehicle in which the driver does not take avoiding action. The assumption that a pedestrian, when crossing a carriageway, can look both ways and thereby avoid a vehicle that is not stopping (presumably by the pedestrian's stopping and reversing direction) appears tenuous at best, and could not apply for encumbered or disabled pedestrians (particularly wheelchair users) or those accompanied by small children.

6.1.3   Intervisibility: a need for further research

There appears to be no readily available or agreed definition of intervisibility.

The term 'intervisibility' is used in a number of cases in *DMRB* (DfT, 2018) and referred to in other documents. Generally, it is taken to mean that two moving elements on a potential collision course can see each other and each can take action to stop if the other takes no action to do so in order to just avoid a collision. Furthermore, the time or distance for this visibility appears to be assumed as the driver's stopping sight distance (SSD) described earlier, based (as outlined in Chapter 2) on the driver's observation and reaction time plus the time for the vehicle to brake to a stop.

In the case of intervisibility between drivers and pedestrians, a driver, with minimal head turning, can have the pedestrian in view during this whole stopping time and distance. Yet a pedestrian about to cross a carriageway, immediately after looking for an approaching vehicle (i.e. sees no vehicle) in one direction, must then look back and focus across the crossing and begin to cross it. The time it takes to do this may be longer than the time associated with the SSD for an approaching vehicle that appears in the intervisibility line just after the pedestrian looked in that direction. This puts the entire onus of avoiding the collision on the driver, who may be distracted, speeding or otherwise not responsibly in control of the vehicle. It also means that, for the pedestrian to know that he or she can cross safely, particularly at junctions with poor visibility, the 'intervisibility' distance should be based on the pedestrian's observation–reaction time plus walking time to a position beyond that of a potential collision, not the drivers' SSD; otherwise the pedestrian cannot be in control of his or her own safety.

The matter is illustrated in Figure 6.4, which shows a scenario where the pedestrian at point A cannot see the approaching

Figure 6.4 Sketch of potential pedestrian–vehicle collision scenario due to inadequate pedestrian sight line

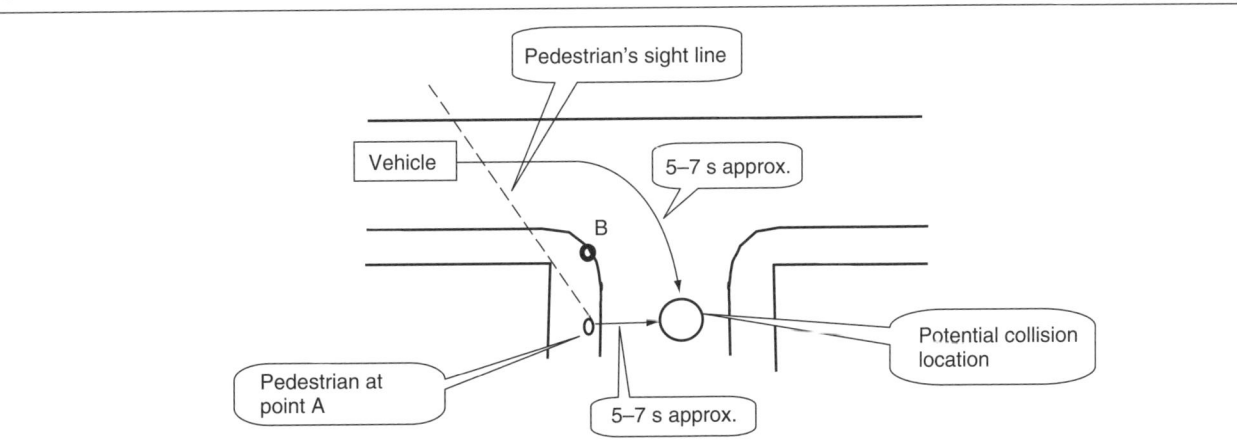

vehicle and walks, while scanning (if able) to his right, while the driver does not stop. Both require approximately the same time to reach the collision point. However, a pedestrian starting to cross at point B can see an approaching vehicle earlier; this will assist in making a decision on whether to cross.

A further consideration concerning intervisibility is the mostly anecdotal – yet important nevertheless – evidence indicating that limited sight distances encourage drivers and pedestrians to show greater caution and that fewer collisions have occurred where sight distances were less than theoretically required. While there is undoubted validity to this assertion, several points should be considered.

- Most drivers behave cautiously under conditions of reduced sight lines. But some will proceed at a speed that does not enable them to stop within the reduced sight distance in time to avoid a pedestrian who may have looked away from his or her sight line immediately before seeing the approaching vehicle. At the very least, this situation can cause intimidation and hence a reluctance by the pedestrian to use the route involved.
- Although greater driver caution due to reduced sight lines may decrease the number of accidents due to the inattention of some pedestrians, this must be weighed against the inconvenience to pedestrians who habitually use caution but are put at risk because of the reduced sight lines.
- A prudent pedestrian venturing into a situation where reliance must be placed on the driver to avoid a potential collision cannot allow him- or herself to rely on the driver and, for safety's sake, will invariably give priority to the vehicle. This is effectively a form of intimidation, which can lead to a reluctance to walk, especially for disabled or encumbered people.
- Often, the exercise of caution relies on eye contact between users. This can be unreliable in such cases as poor visibility, tinted windows on vehicles, distracting reflections from the vehicle's windows; for partially sighted or blind people, it is completely impossible.

With these considerations in mind, the designer must clearly consider the appropriate requirements and guidance, such as those provided in the *DMRB* (DfT, 2018) and *MfS* (DfT, 2007a). This is an area where further research would be beneficial. Such research would address the relative advantages of providing stopping sight distances with appropriate consideration of pedestrians' abilities, and the effects of the risks of potential collision and intimidation on walking, preferably involving psychological and attitudinal variables.

NOTE
In terms of forensic analysis, or the role of accident reconstruction in improving design guidance, it would seem that procedures could

be improved. The listing of contributory factors on the STATS 19 accident report form (form MG NSRF/D) for reporting accidents (DfT, 2011) has no suitable category for indicating that a pedestrian has insufficient sight distance for deciding to start to cross. Whereas drivers and riders may warrant a contributory factor of 'Vision affected by', i.e.:

> 703 Road layout (e.g. bend, winding road, hill crest). *Only use this Code where drivers/riders' vision was affected by the road layout (e.g. failing to see pedestrian crossing at bend, or a vehicle overtaking near crest of hill)*

(DfT, 2011)

there is no equivalent contributory factor that identifies problems that pedestrians might have with road layout, which contribute to their injury and death. County or local jurisdictions may, however, differ in reports affecting pedestrians, and these should be checked for the relevant contributory factors.

6.2. Visibility considerations
6.2.1 General note
This section mainly addresses the design of junctions, primarily where a minor road intersects a major or busier road. Throughout, the basis of geometric layouts as given in current guidance is the need for drivers to see and react to the presence of other vehicles. However, only minor consideration is given in quantitative terms to the needs of pedestrians. Therefore, when designing for the safety of all road users, reference should be made to the description of pedestrians' abilities and crossing characteristics mentioned earlier, especially Chapters 1, 3 and 4.

6.2.2 Visibility for emerging drivers
Drivers emerging from a minor road of a priority junction must have adequate visibility to left and right along a single-carriageway major road. Where the major road is a dual carriageway, with a central reserve of adequate width to shelter turning traffic, the standard visibility splay (the horizontal range of visibility in the viewed direction) in the direction to the left is not required from the minor road but visibility to the left is needed in the central reserve. If the major road is a one-way street then, clearly, visibility is only required towards oncoming traffic. Key dimensions associated with priority junctions typically encountered by pedestrians in urban areas are shown in Figure 6.5, where the lines over which the unobstructed visibility of drivers is specified.

The x distance gives a good field of view for drivers approaching, or stationary at, the give-way line. It also allows oncoming traffic to see emerging side-road traffic. TD 42/95 (DfT, 1995a) advises a desirable x distance of 9 m but acknowledges that, in difficult circumstances, this may be relaxed to 4.5 m for lightly trafficked simple junctions and, exceptionally, to 2.4 m. An x distance of 2.4 m should normally be used in most built-up

Figure 6.5 Visibility standards, based on DfT (1995a). MSSD, minimum stopping sight distance

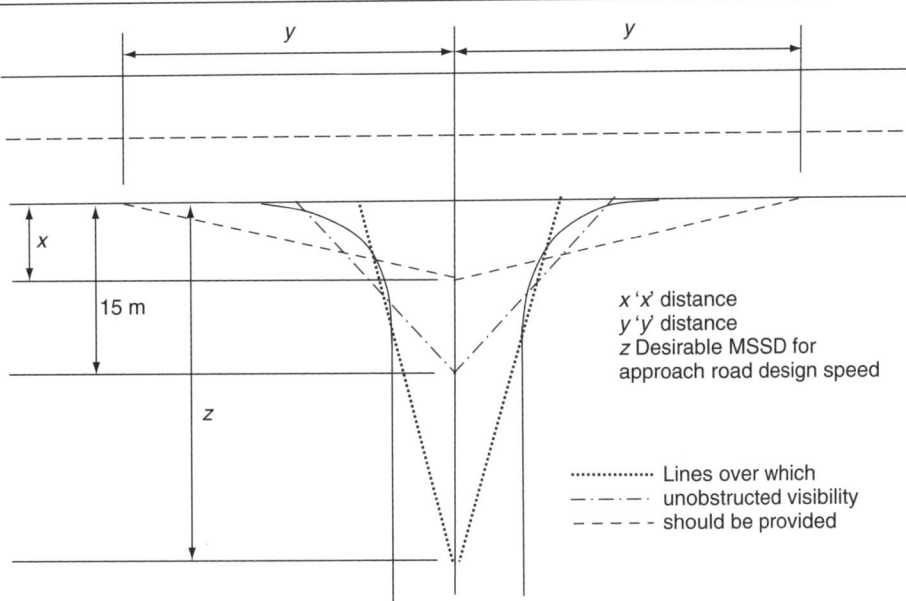

x '*x*' distance
y '*y*' distance
z Desirable MSSD for
approach road design speed

⋯⋯⋯ Lines over which
—·—·— unobstructed visibility
— — — — should be provided

situations and a minimum of 2.0 m may be considered in some low-speed situations where flows on the minor arm are low (CIHT, 2010, Figure 10.2). However, this low value may mean that the front of some vehicles might protrude into the carriageway of the major road, leading to the need for additional caution by drivers, pedestrians and cyclists.

6.2.3 Major-road visibility distance

The *y* distance, measured along the major road, is determined by the speed of the main road traffic. It must be sufficient to allow side-road traffic to emerge safely and to provide forward visibility to allow major-road traffic to stop, if required. The *y* distance is determined by the 85th percentile wet-weather speed or, if this is not known, by the speed limit of the road. Table 6.1 lists the appropriate speeds. For curved major roads, the layout of the appropriate *x* and *y* distances is shown in Figure 6.6.

6.2.4 Forward visibility

The distance a driver needs to see ahead to stop safely (the SSD) on a single bend is illustrated in Figure 6.7. For designing new facilities or checking existing facilities, this distance is measured between points on the curve along the centreline of the inner traffic lane.

6.2.5 Visibility at property frontages and driveways

Drivers emerging from and entering driveways and similar accesses to property, and pedestrians walking along the

frontage footways need to have adequate visibility to ensure their safety and convenience.

Studies referenced in *MfS* (DfT, 2007a) and *MfS2* (CIHT, 2010) were conducted, with the following major conclusions.

It was found that very few accidents occurred involving vehicles turning into and out of driveways, even on heavily-trafficked roads.

Links with direct frontage access can be designed for significantly higher traffic flows than have been used in the past,

Table 6.1 '*y*' visibility distances from the minor road (relaxations not available) (DfT, 1995a)

Design speed of major road: km/h	*y* distance: m
50	70
60	90
70	120
85	160
100	215
120	295

Figure 6.6 Visibility standards with a curved major road (DfT, 1995a)
© Crown Copyright, 1995
This information is licensed under the Open Government Licence v3.0. To view this licence, visit http://www.nationalarchives.gov.uk/doc/open-government-licence/ **OGL**

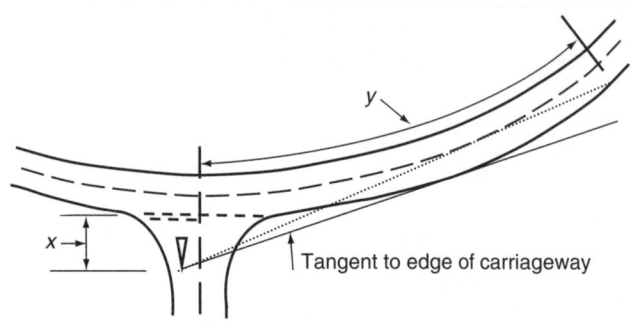

Figure 6.7 Measurement of forward visibility (DfT, 2007a)
© Crown Copyright, 2007
This information is licensed under the Open Government Licence v3.0. To view this licence, visit http://www.nationalarchives.gov.uk/doc/open-government-licence/ **OGL**

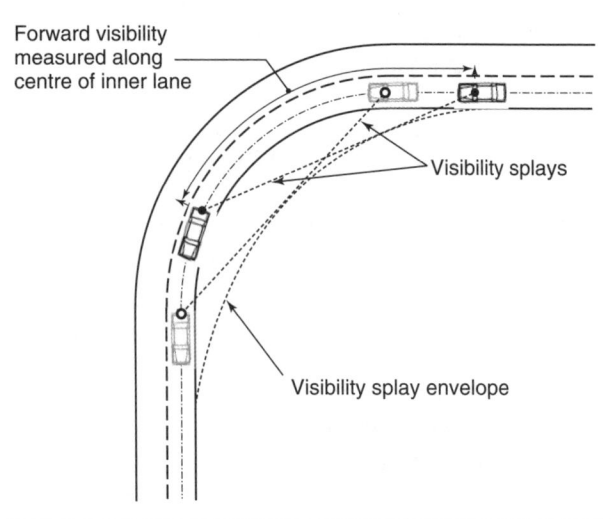

and there is good evidence to raise this figure to 10 000 vehicles per day. It could be increased further, and it is suggested that local authorities review their standards with reference to their own traffic flows and personal injury accident records. The research indicated that a link carrying this volume of traffic, with characteristics similar to those studied, would experience around one driveway-related accident every five years per kilometre. Fewer accidents would be expected on links where the speed of traffic is limited to 20 miles per hour or less, which should be the aim in residential areas.

(DfT, 2007a)

In terms of dimensions, it would seem prudent to ensure that the distance from the driver's viewpoint, when exiting a driveway, to approaching vehicles on the road or street is equal to at least the SSD of the approaching vehicles. The relevant dimensions would have similar characteristics, in some respects, to those for vehicles exiting a minor road to join a major road, but with modifications to suit the dimensions, such as the footway width and the presence of pedestrians.

6.2.6 Other visibility considerations

As well as visibility between drivers, visibility between all combinations of drivers, pedestrians and cyclists using the junction should be considered. Drivers approaching on the minor road should have an unobstructed view of the junction. The dimensions shown in TD 42/95 (DfT, 1995a) ensure that drivers can slow down sufficiently to see the junction form clearly. On curved alignments, displays should be made tangential to the nearside edge of the road; see Figure 6.6.

Vehicles turning right from the major road need good visibility towards oncoming traffic and this should never be less than the desirable minimum stopping sight distance; TD 9/93 (DfT, 2002) states that the minimum stopping sight distance must be achieved on the major approach to a junction. The immediate approach is defined as

- minor road: 1.5× the desirable minimum stopping sight distance upstream of the give-way line
- major road: 1.5× the desirable minimum stopping sight distance from the centre line of the minor road.

Visibility splays should preferably line within the curtilage of the highway to ensure that they are not obstructed. Planning controls can be used to prevent any future obstructions and Section 79 of the Highways Act 1980 provides highway authorities with the powers to pursue measures to improve or safeguard visibility.

As mentioned earlier, the x and y distances are measured along the centreline of the minor road and along the kerb line of the major road, respectively. In reality, these are not the points from which drivers observe oncoming traffic or the paths of approaching vehicles. In some circumstances, where decisions on visibility standards are marginal, it may be beneficial to consider the visibility distances that are achieved in reality. Such an approach can assist pragmatic decision-making.

6.3. Crossings at priority junctions: trunk roads

The requirements detailed in the relevant *DMRB* manuals on highway design, are the key focus of the design procedure for trunk and collector roads.

Table 6.2 Possible junction types for different major-road carriageway types (DfT, 1995a)

Carriageway type		Junction type								
Standard	Location	Simple			Ghost island			Dualling		
		⊤	⊤⌐	⊥⊤	⊤	⊤⌐	⊥⊤	⊤	⊤⌐	⊥⊤
S2	Urban	Yes	Yes	Maybe	Yes	Yes	No	Yes (D1)	Yes (D1)	No
	Rural	Yes	Yes	Maybe	Yes	Yes	No	Yes (D1)	Yes (D1)	No
WS2	Urban	No	No	No	Yes	Yes	No	Yes (D1)	Yes (D1)	No
	Rural	No	No	No	Yes	Yes	No	Yes (D1)	Yes (D1)	No
D2	Urban	No	No	No	No	No	No	Yes (D2)	Yes (D2)	No
	Rural	No	No	No	No	No	No	Yes (D2)	Yes (D2)	No
D3	No	No	No	No	No	No	No	No	No	

Various configurations of priority junction are summarised in Table 6.2; guidance on their applicability, related to traffic flows on the major and minor arms, is provided in Figure 6.8.

6.3.1 Trunk road design process

Table 6.3 lists the major elements described in the TD 42/95 design process (DfT, 1995a). This is shown as a series of steps with decision points that refer the designer through an iterative process to the final design. A major feature of the design process is that, in general, the defining parameters focus mostly on drivers' tasks and motorised traffic flow. Safety and specialised users' needs, including the needs of pedestrians, are brought into the process in Steps 3a (safety) and 3b (users' specific requirements). While this approach, to some extent, addresses

Figure 6.8 Approximate level of provision of T-junctions on new single-carriageway roads for various major and minor road design year traffic flows (DfT, 1995a). AADT, annual average daily traffic flow

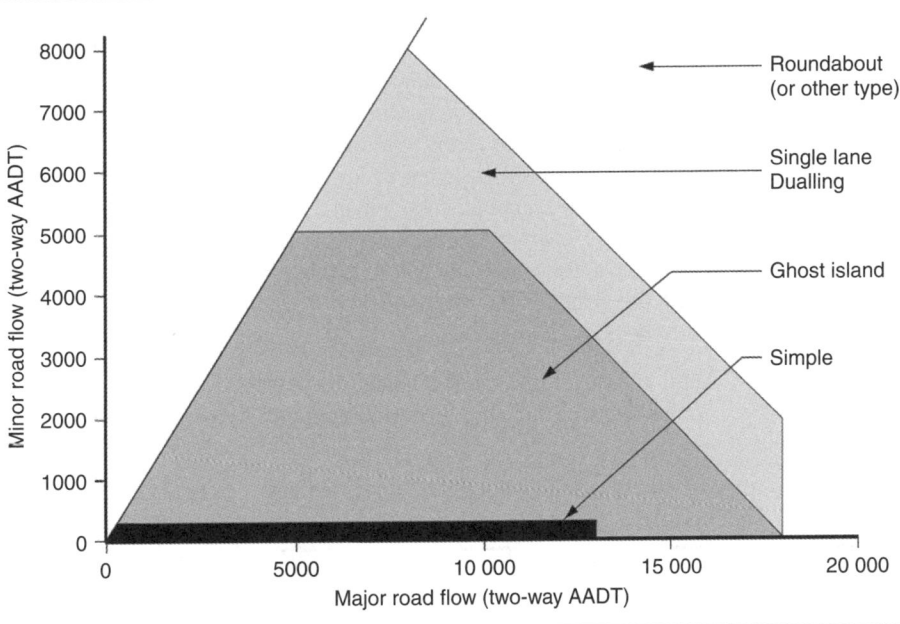

Table 6.3 Key elements of junction design procedure, based on DfT (1995a), Chapter 2, Figure 2/1, with additional considerations relative to pedestrians' needs

Design procedure step[a]	References to user and facility design[b]	Potential additions relevant to pedestrians[c]
Step 1 Choose most appropriate type of junction (DfT, 1981, 1982) leading to major or minor road (DfT, 1995a).	See DfT (1981), Section 5.1 and DfT (1982) Section 6.2. A sequence of too many different junction types along a road would create confusion and uncertainty for drivers and might result in accidents.	Consider the effects of junction types on pedestrians. Account for pedestrian desire lines and neighbourhood cohesion.
Step 2 Choose most appropriate form and size of major or minor priority junction. (DfT, 1995a, Chapter 2)	2.5 'Safe road schemes are usually no surprise for the driver.' 2.6 'Account for design year traffic flow, goods and passenger-carrying vehicles, geometric and traffic delays [. . .] turning stream capacities, site characteristics.'	Safe road schemes should also always be no surprise to pedestrians. Account for pedestrians' presence and characteristics.
Decision point – re-cycle to Step 2 Is junction type appropriate for site characteristics? (DfT, 1995a, Chapter 3)	Drivers' visibility is related to horizontal and vertical geometries.	Pedestrians' visibility related to vertical and horizontal geometries is also important.
Step 3a Address all relevant safety issues. (DfT, 1995a, Chapter 4)	4.3.j Installation of pedestrian guardrails, central refuges and pedestrian crossings in urban areas.	Consider integration of pedestrian needs from Step 1.
Step 3b Take account of road users' specific requirements. (DfT, 1995a, Chapter 5)	Note Sections 5.12 to 5.16. Section 5.12: 'The requirements of pedestrians should be carefully considered in the design and choice of major or minor junctions. Although it is preferable to provide separate pedestrian routes away from the junction, where road widths are less and traffic movements more predictable, this is rarely practical, in which case the following facilities should be considered:' (lists central refuge, zebra crossing, displaced controlled crossing, subway and footbridge.)	Consider including pedestrians' needs earlier in the process in order to be consistent with this requirement.
Step 3c Preliminary landscape recommendations. (DfT, 1995a, Chapter 6)	No specific driver or pedestrian involvement.	Obstruction to visibility for pedestrians by vegetation or other features should be checked.

Table 6.3 Continued

Design procedure step[a]	References to user and facility design[b]	Potential additions relevant to pedestrians[c]
Step 3d Assess key geometric parameters. (DfT, 1995a, Chapter 7)	Driver visibility along major and minor road; *x*, *y*, etc. Minimum radii 10 m with taper of 1 : 5 over 30 m distance in urban areas. Note 7.6.b.	Pedestrian visibility along minor and major roads. Consider effects of radii dimensions on pedestrians in addition to effects on drivers.
Decision point Does the junction still have adequate capacity? (Re-cycle to the extent necessary.)	Assumes that 'capacity' refers to vehicle capacity.	Pedestrian capacity should be considered.
Step 4 Assemble design elements. (DfT, 1995a, Chapter 8)	Comments are directed exclusively at driver capabilities: 8.1 'The layout shall be designed so as to follow the traffic pattern, with the principal movements being given the easiest path. This improves the smoothness of operation and makes it more readily understood by drivers.' 8.2 'It is important that, on entering a junction, drivers should be able to see and understand, both from the layout and advance traffic signs, the path they should follow and the likely actions of crossing, merging and diverging vehicles.'	Consider design elements so that pedestrians may also more readily understand the smoothness of operation and the likely actions of crossing, merging and diverging vehicles.
Decision point Is 'driveability' threshold satisfied? (Re-cycle the process to the extent necessary.)	Driveability is seen predominantly in terms of drivers' convenience and safety by conducting an 'on site' check.	Consider 'walkability' in addition to 'driveability'.
Step 5 Final design.	2.9 'As a whole, the layout should be designed to suit the traffic pattern, with the vehicular movements following smooth vehicular paths. This improves the smoothness of operation and makes it more readily understood by drivers.'	More emphasis on walking patterns, and ease of operations for pedestrians.

[a] Source: Main steps in DfT (1995a), Figure 2/1.
[b] Selected comments in design procedure in item (a) (DfT, 1995a).
[c] Comments by the author regarding inclusion of potential pedestrian concerns in the design process.
© Crown Copyright, 1995
This information is licensed under the Open Government Licence v3.0. To view this licence, visit http://www.nationalarchives.gov.uk/doc/open-government-licence/ **OGL**

the needs of pedestrians, it does not include their needs in an integrated way from the beginning of the design process. The points that it would appear appropriate to consider from a pedestrian facilities design perspective (added by the author) are shown in the third column of the table.

The process shown is clearly based on the needs and characteristics of vehicular traffic. Although other road users are included part of the way through the process, they are not included as a part of the whole design process from the initial steps. Nor is there a parallel consideration of pedestrian characteristics as the design proceeds. Furthermore, when the essential feedback process of viewing the design as built is undertaken, the designer is advised to 'walk through' the built intersection to see if there are any undue problems for drivers. A 'walk through' to see if there are any undue problems for pedestrians would also clearly be beneficial.

Key elements of TD 42/95 (DfT, 1995a) guidance and publications as they relate to pedestrians' characteristics and inclusion in the design process are summarised in Box 6.1.

NOTE
Many of the requirements listed in Box 6.1, particularly for locating the crossing away from the mouth of the junction (a minimum of 15 m back from the give-way line), might be in conflict with measures to encourage walking and generally reduce vehicle speeds. The implied use of large-radii kerbs, necessitating the crossing to be back from the give-way line to reduce the distance to be crossed (and encouraging higher vehicle speeds through the junction) or the use of a central refuge to ease the pedestrian's task of crossing the wider minor road's mouth (but simultaneously increasing the total crossing time and exposure to moving vehicles) are factors that cause further impediments and tasks for the pedestrians in order to cross safely and conveniently.

Box 6.1 Key elements of TD 42/95 (DfT, 1995a) as they relate to pedestrians' characteristics and inclusion in the design process (authors' emphasis)

5.12 The requirements of pedestrians should be carefully considered in the design and choice of major/minor priority junctions. *Although it is preferable to provide separate pedestrian routes away from the junction*, where road widths are less and traffic movements are more predictable, this is rarely practical, in which case the following facilities should be considered:

   a.   a minor road refuge at an unmarked crossing place
   b.   zebra crossing, with or without a central refuge
   c.   displaced controlled pedestrian crossing
   d.   subway or footbridge.

5.13 The type of facility selected will depend upon the volumes and movements expected of both pedestrian and traffic, and shall be designed in accordance with current recommendations and requirements – BD 29 (*DMRB* 2.2); TD 36 (*DMRB* 6.3.11); TD 28, TA 52 (*DMRB* 8.5). The use of different types of pedestrian facility at the same junction is not recommended as this could lead to confusion by pedestrian and drivers.

5.14 *At-grade pedestrian crossing points should not be placed in the mouth of the junction, instead they should be located away from the mouth where the carriageway is relatively narrow.* In urban areas, where pedestrian flows are relatively low, it is possible to provide a central refuge in the hatched area of a ghost island junction. However, where pedestrian flows are high, consideration should be given to a single-lane dualling junction, even in circumstance where the traffic flows may not warrant such a provision, in order to enable pedestrians to make the crossing manoeuvre in two stages, and have a safe central waiting area.

5.15 *Defined at-grade pedestrian crossing points on the minor road should be a minimum of 15 m back from the 'Give Way' line and should be sited so as to reduce to a minimum the width to be crossed by pedestrians provided they are not involved in excessive detours from their desired paths.* Central refuges should be used wherever possible, but not in the major road in a rural situation.

5.16 In urban areas, where large numbers of pedestrians are present, guardrails or other deterrents should be used to prevent indiscriminate crossing of the carriageway. The design of guardrailing should not obstruct drivers' visibility requirements. *Guardrails which are designed to maintain drivers' visibility of pedestrians through them, and vice versa*, are available, but should be checked in case blind spots do occur. TA 57 (*DMRB* 6.3) refers.

Consideration should therefore be given wherever possible to reduce kerb radii or construct build-outs to reduce the distance for pedestrians to cross. Such measures will also necessitate vehicles to travel slower in the major and the minor arms of the junction, enable drivers and pedestrians to see each other earlier, give pedestrians a shorter time to be exposed to moving vehicles and ensure that pedestrians can maintain their journey along their desire lines. Regarding the provision of guardrails, although it is stated that 'the guardrail should not obstruct drivers' visibility requirements' it would also seem prudent for designers to ensure that the guardrail does not impede pedestrians' visibility towards approaching vehicles, which, as indicated at the beginning of this chapter, can be a considerably more complex task, even with a central refuge, than the need for drivers to observe the pedestrians. Of particular importance is the matter of inappropriate positioning or the use of guardrails affecting drivers' visibility of smaller children and people in wheelchairs or on mobility scooters.

The question of drivers' reaction times is not dealt with specifically in the foregoing guidelines; instead, safe stopping distances are related to the speeds of approaching vehicles. The speed of approaching vehicles to be used in the design guidelines is based on the 85th percentile of those observed approaching the junction, such as those described in Box 6.1.

6.4. Junctions at residential streets

A number of recommendations are made in *MfS* for junction design in predominantly residential areas. Many of these recommendations are also applicable to locations of higher vehicular traffic volumes.

6.4.1 Planning and design considerations

Junctions are places of high visibility and provide a balance between place and movement functions; their basic form should be determined at the master planning stage when the perspective of the overall street network and land uses can be viewed.

Junction design should facilitate pedestrians' direct desire lines; this will often mean using small corner radii. Figure 6.9 illustrates these points. The use of swept path analysis will ensure that the junctions are usable by vehicles.

Figure 6.9 Pedestrian considerations at priority junctions (DfT, 2007a)
© Crown Copyright, 2007
This information is licensed under the Open Government Licence v3.0. To view this licence, visit http://www.nationalarchives.gov.uk/doc/open-government-licence/ **OGL**

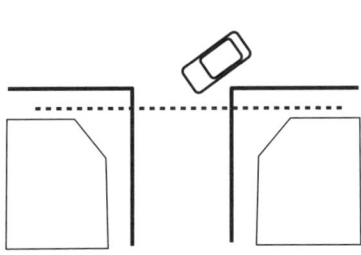

- Pedestrian's desire line is maintained.
- Vehicles turn slowly (10 mph–15 mph).

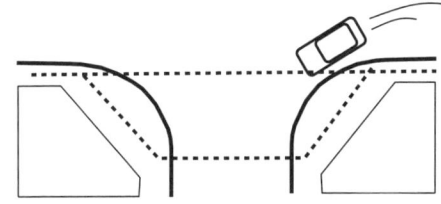

- Pedestrian's desire line is deflected.
- Vehicles turn faster (20 mph–30 mph).
- Pedestrian's detour required to minimise crossing distance.

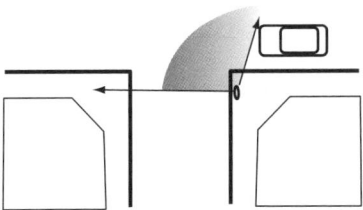

- Pedestrian does not have to look further behind to check for turning vehicles.
- Pedestrian can more easily establish priority because vehicles turn more slowly.

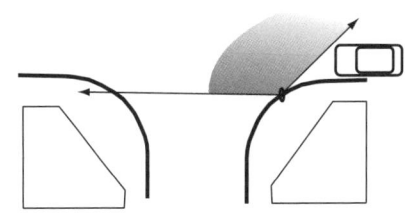

- Pedestrian must look further behind to check for faster turning vehicles.
- Pedestrian cannot normally establish priority against fast turning vehicles.

It should be noted that several other reasons for providing direct access across a junction are evident, apart from the matter of intervisibility addressed earlier in this chapter.

- In being able to cross directly in line with opposite footways, in common with vehicular (including cycle) traffic, the pedestrian can more obviously have priority over turning vehicles moving along the major road.
- Drivers approaching along the main road from either direction can see pedestrians waiting to cross (or starting to cross) as they approach the junction sooner than if the pedestrians are obscured by a building corner or by vehicles stopped or moving in the junction. This ability of drivers to see pedestrians earlier can therefore enable them to better adjust their actions to permit pedestrians to exercise their priority in crossing.
- In being able to see pedestrians earlier and slowing down in good time, the speed of traffic on the major road is likely to be reduced. This can enhance the effects of speed restrictions and traffic calming, as well as attaining a smoother traffic flow, with less likelihood of rear-end collisions.
- As described earlier, pedestrians must scan approximately 270° before making a decision about crossing. This view is often restricted by obstructions and moving vehicles approaching from several directions if the crossing is not direct.
- Pedestrians, when looking ahead and behind them, for approaching vehicles can see a greater distance and are therefore better able to judge if an approaching vehicle is being driven too fast, erratically or in some other way that appears to pose a threat of collision.
- Pedestrians using mobility scooters equipped with rear-view mirrors are able to see behind them for approaching vehicles on the main road; this may be difficult (and therefore time-consuming and uncertain) or impossible if the crossing is not direct.
- Pedestrians are less likely to have to cross behind a vehicle that is waiting for vehicles to clear on the main road in order to exit the minor road. Such a crossing location, behind the waiting vehicle, obscures the pedestrians from vehicles turning from the main road, decreases their and drivers' intervisibility, and can leave the pedestrian in a dangerous position as the vehicular traffic changes. For someone with a wheelchair or mobility aid, or accompanied by small children or a helper, the intimidation, danger and delay in making such a crossing (or the unacceptable delay in not being able to make it) can essentially render the walking journey unacceptable.

Conventional design justification for not providing direct crossing for pedestrians includes the recommendation that vehicles turning from the major arms should not be delayed, as this will tend to reduce the capacity of the major arm. It has also been stated that such a position enables the drivers turning to better see pedestrians waiting to cross or crossing. However, if a pedestrian has been unable to see an approaching vehicle because of the location of the crossing away from the corner, and the driver is distracted, the pedestrian is clearly at risk.

In terms of pedestrian priority, locating the crossing away from the direct desire line path tends to diminish the priority that pedestrians have for continuing across the minor arm in the directions of the major traffic flows. It also tends to negate the general rule that traffic (including motor vehicles, pedestrians and cyclists) moving straight ahead has priority over turning traffic. See also comments regarding the *Highway Code* in Chapter 1.

In terms of general layout, crossroads are convenient for pedestrians, as they minimise diversion from desire lines when crossing the street (DfT, 2007a). Although staggered junctions can reduce vehicle conflict compared with crossroads, they reduce directness for pedestrians. If potential user conflict exists, consideration may be given to replacing the junction with a speed table or closing one of the arms to motor traffic.

6.4.2 Junction dimensions at straight major roads and at bends

Dimensions at junctions are detailed in guidance and depend, to some extent, on the configuration of the junction, as well as the presence of median strips and other features. These dimensions are based on drivers' stopping sight distances. Pedestrians' needs in this respect (described earlier in Chapter 3), in terms of observation–reaction times, walking speeds and a safety margin, are not addressed. The key dimensions are shown in Figure 6.10.

6.5. Other junction characteristics

Although mentioned briefly in the *DMRB* (DfT, 2018) and *MfS* (DfT, 2007a), crossing features in addition to the kerb radii and build-outs, mentioned earlier, that can assist pedestrians and other road users are refuges (islands) and vertical deflections. Key features of the dimensions and applicability of these devices are described in this section.

6.5.1 Overrun areas

These areas may be used at bends and junctions with a surface texture or appearance intended to deter overrunning by cars and other light vehicles. This allows the passage of large vehicles while maintaining 'tight' corner dimensions that deter smaller vehicles from speeding.

In general, overrun areas should be avoided in residential and mixed-use streets because they can

Figure 6.10 Measurement of junction visibility splays on straight roads and bends (DfT, 2007a)
© Crown Copyright, 2007
This information is licensed under the Open Government Licence v3.0. To view this licence, visit http://www.nationalarchives.gov.uk/doc/open-government-licence/ **OGL**

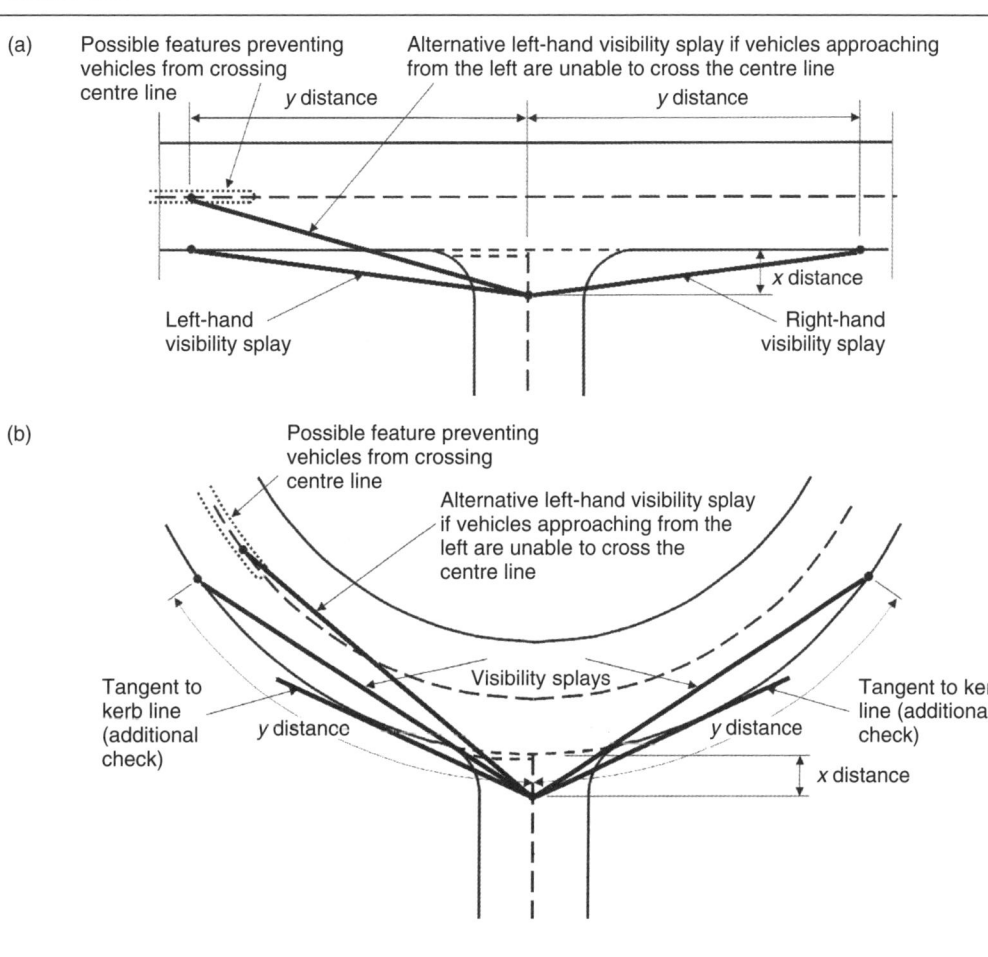

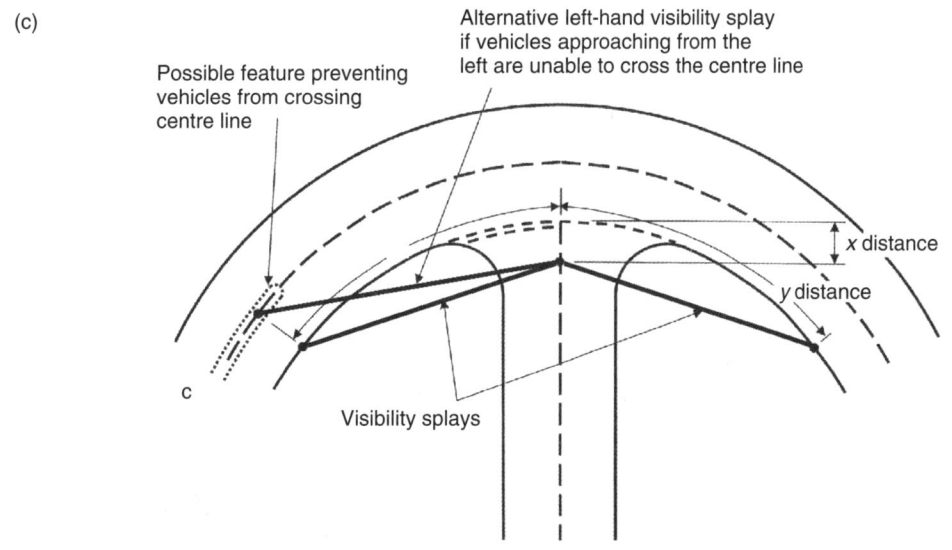

- be visually intrusive
- interfere with pedestrian desire lines
- pose a hazard to cyclists.

They can, however, help to overcome some problems with access for larger vehicles and might be the best solution when the presence of vehicles, such as refuse trucks, is limited.

6.5.2 Refuges

6.5.2.1 General comment

Refuges provide convenience for motor traffic in that the traffic stream is not interrupted by pedestrians who need to cross the road safely and along a convenient route. Given the speed of such traffic and, often, the heavy vehicle flows, a refuge can provide some measure of assistance to pedestrians in maintaining a satisfactory route for their activities. Yet several considerations in terms of the safety and desirability of crossings should be considered, including the following.

- A pedestrian standing at a refuge, especially with young children or in a wheelchair, can feel intimidated by the presence of vehicles, often travelling at speeds of up to 40 miles per hour, very close to where they are standing.
- The refuge can present a potential collision point for vehicles. This is recognised by many authorities, which now install flexible or recoverable bollards to reduce repair and maintenance. The risks of these potential collision areas are referred to by CIHT (2015), and by Belcher *et al.* (2015) regarding night-time collisions. The implications of such a collision for pedestrians at a refuge are clear.

Both of these factors can cause intimidation to pedestrians and, in the potential for collision with the refuge, most pedestrians will feel exposed to a risk that they are likely to avoid, if possible, thereby leading to a reduced number of pedestrian trips. The consequences for these concerns are several. Research and a new approach to pedestrian priority should be investigated to address possible solutions.

6.5.2.2 Guidance

An urban junction with a pedestrian refuge is shown in Figure 6.11 (DfT, 1995a). Minimum dimensions for refuges related to the needs of disabled people (DfT, 2005), particularly related to tactile surfaces, are discussed in Chapter 8. Moreover, the 3 m set-back from the edge of the major-road carriageway may be considered a maximum, subject to detailed analysis. Narrowing of the carriageway to accommodate a refuge can cause problems for cyclists, particularly on high-volume roads or where speeds are high. Regarding access for emergency services, refuges that leave gaps of less than 4 m should be avoided (DfT, 2007b).

Figure 6.11 Typical urban separation island (refuge) (DfT, 1995a)
© Crown Copyright, 1995
This information is licensed under the Open Government Licence v3.0. To view this licence, visit http://www.nationalarchives.gov.uk/doc/open-government-licence/ **OGL**

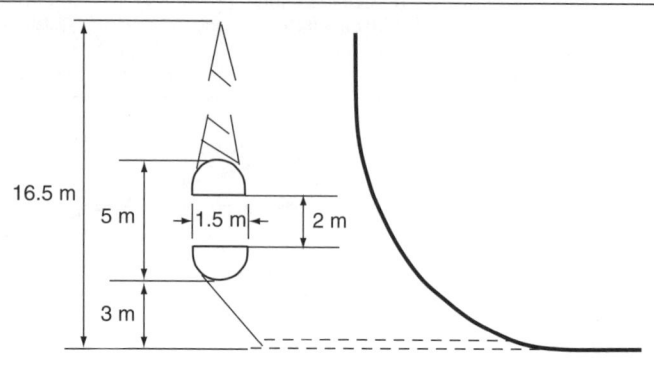

The width of a refuge should be related to the pedestrian flow and composition and should be sufficient to accommodate more than occasional pedestrian use: 2 m width will allow two wheelchairs to pass each other. There is no upper limit but if pedestrian flows are such that wide crossings are needed, a controlled crossing solution would be preferable.

NOTE
The 1.5 m minimum length shown in Figure 6.11 is inadequate for many pedestrians, especially people with small children with a pram or buggy, and, at the least, places people crossing close to passing traffic.

Provision of a central refuge requires several considerations, depending on its location and whether it is at an uncontrolled location, a zebra crossing or a junction. Many of these considerations are summarised in Table 6.4. Other matters related to road users' conduct when negotiating refuges are described in Chapter 1.

6.5.3 Vertical deflections

Coming under the generic term 'road hump' and described in *Traffic Calming* (DfT, 2007b) are several devices that, as well as assisting in reducing vehicle speeds, may also assist pedestrians in crossing at junctions and other locations. The devices are only permitted at locations where there is a 30 miles per hour speed limit. The particular types of device suitable for pedestrians are those that have flat tops and extend the full width of the carriageway, and are

- normal flat-top humps
- speed tables (long flat-top humps)
- junction plateaus.

Table 6.4 Relative advantages and disadvantages of pedestrian central refuges (islands)

Location of refuge or island	Benefits for pedestrians	Benefits for drivers	Disbenefits for pedestrians	Disbenefits for drivers
Midblock, uncontrolled	Pedestrian needs to look only 90° before starting to cross Waiting time for each half is shorter than for the entire crossing, but total waiting time may be greater Shorter distance to be walked in one go	Drivers are less hindered by requirement to stop for pedestrians crossing entire carriageway	Two crossings are necessary Available waiting space may be inadequate and intimidating May require longer total time to cross owing to vehicle traffic on second half of crossing	No apparent disadvantages
Midblock, zebra	Same as above	Drivers need stop only for pedestrians who move onto the relevant half of the crossing	Same as above	Drivers must stop if pedestrian moves into relevant half of crossing
Priority junction, uncontrolled minor road crossing	Pedestrian needs to look only maximum of 180° instead of maximum 270° before starting to cross Waiting time for each half is shorter than for the entire crossing, but total waiting time may be greater Shorter distance to be walked in one go	Drivers are less hindered by requirement to stop for pedestrians crossing entire carriageway	Two crossings are necessary Available waiting space may be inadequate and intimidating May require longer total time to cross, owing to vehicle traffic on second half of crossing Longer-term tendency for pedestrians' priority at junctions to be negated	No apparent disadvantages

Key features of each of these devices as they apply to pedestrians are summarised in Table 6.5.

A particular form of raised table is the side raised entry treatment (SRET). Side entry treatments are applied to road sections at junctions of side roads with major roads. They are placed either at the entrance to the junction or within a short distance of it. Based on a study by Buchanan *et al.* (ca. 1995) and a more detailed investigation by Wood *et al.* (2006), TfL (2007) posed the following research questions.

(*a*) Are SRETs beneficial to pedestrians' safety on the features themselves?
(*b*) Are SRETs beneficial to road safety at junctions and on side roads?

Side raised entry treatments have been installed to achieve a combination of objectives (e.g. TfL streetscape guidance), including

- creating a strong visual threshold for traffic leaving or entering a minor road
- providing easier pedestrian movement by raising the treatment
- assisting in establishing pedestrian priority
- deterring parking close to junctions
- slowing vehicle speeds
- reducing collisions involving vulnerable road users.

The findings suggest that SRETs do provide easier pedestrian movement and assist pedestrians in asserting priority. The findings further indicate that SRETs reduce the number of collisions involving pedal cyclists but not those involving other vulnerable road users.

Table 6.5 Selected vertical deflection traffic calming devices for pedestrian use (DfT, 2007b)

Type of road hump	Brief description	Suitability	Dimensions	Comments
Flat-top hump	Normally full width of the carriageway with a short, flat top, minimum 1 m long	Assisting pedestrians crossing the road, especially if the hump is level with the footway	■ 50–100 mm high (75 mm maximum on a bus route) ■ Spacing 70–100 m ■ 1.5–4 m long ■ Ramp gradient 1 in 10 to 1 in 20	Minimum length for pedestrians would be 2 m to permit adequate cross movement
Speed table	Similar to flat-top humps but 6 m long on top to accommodate buses			Useful pedestrian crossing point
Junction plateau	As flat-top hump but covers the full length and breadth of a junction	Used in junctions in 20 miles per hour zones and as part of gateway features	■ 50–100 mm high (75 mm maximum on a bus route) ■ Plan area to cover the whole junction, including radii ■ Ramp gradient 1 in 10 to 1 in 20	Only good for pedestrians if long enough to include pedestrian crossing route

6.6. Stand-alone crossings

In general, the principles that apply to pedestrians crossing at junctions also apply to crossings on straight sections of road, in that visibility, vehicle speed and stopping sight distances and pedestrians' crossing times are key elements in the crossing design. Guidance on establishing and designing stand-alone crossings is contained in LTN 1/95 (DfT, 1995b), LTN 2/95 (DfT, 1995c) and, more recently, in LTN 1/04 *Policy, Planning and Design for Walking and Cycling* (DfT, 2004). In addition, pedestrian facilities associated with traffic calming measures are relevant to stand-alone crossings; these are described in LTN 1/07 (DfT, 2007b).

6.6.1 Design process

A flow chart of the key steps involved in selecting a crossing, including a stand-alone crossing, is shown in Figure 6.12. The actual geometric design procedure follows a route review and consideration of the hierarchy of provision and cycles through these activities during the audit process.

6.6.2 Site and option assessment

The hierarchy of provisions for pedestrians, outlined earlier (DfT, 2007a), consists of

■ traffic reduction

Figure 6.12 Walking infrastructure design process – main elements, based on DfT (2004)

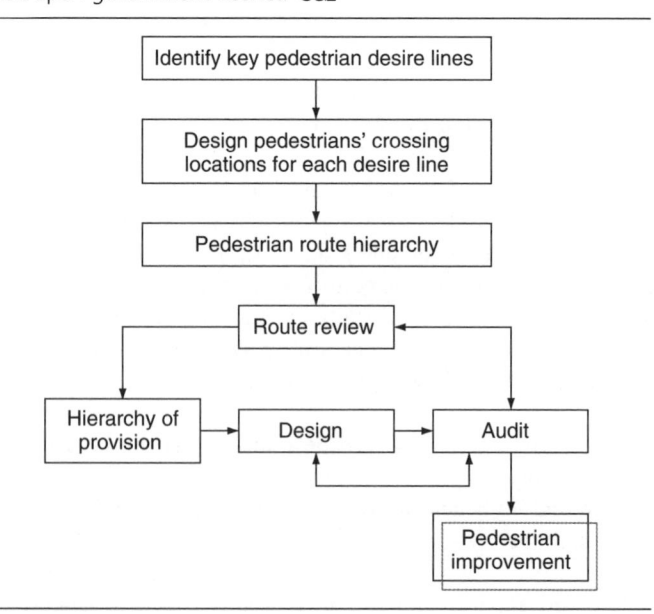

- speed reduction
- reallocation of road space to pedestrians
- improved pedestrian routes on existing desire lines
- improved pedestrian alignment or grade separation.

Each of these factors will affect the location and design of the crossing; an assessment process for evaluating the site and potential design options is outlined in LTN 1/95 (DfT, 1995b). An important determinant of the ease (or difficulty) of crossing for pedestrians is provided by consideration of the acceptable gap in traffic. An acceptable gap in which to cross from kerb to kerb (or from kerb to refuge) varies for each individual; the majority of pedestrians will accept a gap of 4–6 s to cross two lanes of traffic for normal urban traffic speeds and even shorter gaps at slower vehicle approach speeds. Others may require gaps of up to 10–12 s or longer. Locations where it is evident that pedestrians are already crossing, or where there is an obvious need for a crossing, such as between important land uses, require an analysis of traffic volumes and related factors to determine the appropriateness for a crossing. These are listed further in LTN 1/95 (DfT, 1995b). The results of an illustrative example of the main considerations are indicated in Tables 6.6 and 6.7.

Examples of factors most likely to affect the choice of pedestrian crossing type are

- difficulty in crossing
- vehicle delays during peak periods
- carriageway capacity

- local representations
- cost (including maintenance)
- vehicle speeds.

Several options for action when considering the provision of pedestrian crossings include

- do nothing
- provide traffic management (including refuge island)
- provide a zebra crossing
- provide a signal-controlled crossing.

Where a crossing is thought necessary but crossing flows are relatively low and traffic flows are no more than moderate, a zebra crossing may be suitable. Pedestrians establish precedence by stepping onto the crossing, so delays to them are minimal. Vehicle delays are typically 5 s for a single able person crossing but can be much longer where irregular streams of people cross over extended periods. The likely effect of installing a zebra crossing can be tested by checking the availability of sufficient gaps in the traffic flow. Where gaps are few, and waiting times long because people feel it may be hazardous to establish precedence, a zebra crossing is likely to be unsuitable. The number of people at the site will also give an indication of the likely performance of a zebra crossing. Higher flows of pedestrians will cause substantial delay to vehicles and a zebra crossing is less likely to be a satisfactory choice.

Where traffic speeds exceed 30 miles per hour, pedestrians will require longer gaps in the traffic flow or be exposed to the risk

Table 6.6 Illustrative site assessment (DfT, 1995b)

Characteristic	Data and comments at 31 March 1995
Highway facilities	Road lighting is to a recent traffic route standard and no re-arrangement is needed. The road surface gives adequate skid resistance.
Visibility	Desirable visibility standards can be met. There is no need to further restrict parking, on visibility grounds, and the road is not a bus route.
Complexity	There are no road junctions, other pedestrian crossings, public buildings or facilities, other than the local primary school, within 250 m.
Crossing traffic	About 1250 people cross the road daily with an average breakdown into groups. Crossing time and difficulty of crossing are typical for roads of this character in this area.
Vehicles	5600 vehicles a day with 2% of heavy goods. Highest two-way peak hour flow 985. Highest 85th percentile in peak periods is 33 mph. There is a 30 mph speed limit.
Road accidents	There were 3 P.I. accidents in 1994, none in the previous 4 years. None have been recorded this year.

Table 6.7 Illustrative option assessment (DfT, 1995b)

Factor	Do nothing	Refuge island	Zebra	Signalled crossing
Difficulty of crossing, average wait in seconds	20 (able)/120 (elderly) in peak periods	15 (able)/40 (elderly) in peak periods	1 to 3 for all groups	1 to 3 after end of vehicle minimum green period
Vehicle delay in peak periods	None	None	3 stops/minute of 10 s	2 stops/minute of 12 s
Road capacity	Not reduced	Not reduced	50% reduction	40% reduction
Representations	Police suggest consideration of speed reduction measures may be correct course of action	Police do not favour because of uncontrolled bunching of schoolchildren on island	Local elected representatives think best balance between needs and costs	Public petition and individual letters favour to meet safety needs of children, elderly and disabled people. Stimulated by accident to girl on crutches after other incidents in 1994
Installation cost	None at this time	1000	3000	20 000
Operating cost	100	300	2000	5000

of more serious injury if precedence is not conceded for any reason. Zebra crossings should not be installed on roads with an 85th percentile speed of 35 miles per hour or above.

Care should be taken at unusual sites, such as contraflow bus lanes and one-way streets, as uncertainty can be caused. A signal-controlled crossing may then be more suitable.

It is important to ensure that the safety and convenience of pedestrians and cyclists are not compromised by roadworks (DfT, 1995b, Section 4.14.4). Where roadworks limit the carriageway to a single lane, a minimum lane width of 4.0 m is desirable to enable cars to pass cyclists safely, and a minimum width of 4.25 m is necessary for heavy goods vehicles to pass safely. Advice about streetworks on footpaths is given in *Inclusive Mobility* (DfT, 2005).

Selected points of the various site characteristics directly relevant to geometric design (DfT, 2004) are summarised as follows.

■ A site location description and a map reference enable precise definition of the location of the crossing. Carriageway features, such as widths and the number and directions of lanes, are key determinants of the pedestrians' walking distance when crossing.

■ The width of the current footpath or footway may need to be increased to accommodate pedestrians waiting to cross.

■ The presence or lack of a refuge island will be important in determining the directions in which a pedestrian must look, as well as the distance that must be walked without a rest.

■ Minimum visibility of the pedestrian to the vehicle and the vehicle to the crossing location are key elements in the geometric design and are related to the minimum stopping sight distance (but see also the earlier note in this chapter on intervisibility).

■ Restrictions and other regulations affecting the location at which vehicles must stop may significantly affect available disability distances.

■ Public transport stopping points may significantly affect visibility between the pedestrian and approaching vehicles. In addition, public transport stops may generate considerable pedestrian traffic, including people wishing to cross the carriageway adjacent to the stop.

■ Distances to such features as the nearest junction will affect the physical layout, influencing and being influenced by the crossing.

■ Other pedestrian crossings in the vicinity, and the distance to them and the type of crossing, may affect the characteristics of the users and also may affect the placement of the crossing under consideration.

- Skidding risk is a factor that should be considered, although, in the geometric design, a conservative value of skidding, consistent with tables of values of minimum stopping sight distance related to speed, should be employed.

- Entrances listed within 100 m of the proposed crossing are likely to generate considerable pedestrian traffic. The characteristics of such pedestrian traffic will be a matter for individual consideration in terms of volume and characteristics of the pedestrians. Such features as a fire station or hospital, which may generate emergency vehicle traffic, should be given special consideration with regard to adjacent or proposed pedestrian crossings.

- The flow and composition of the crossing pedestrian traffic is vitally important and category characteristics, in which the ability of the various pedestrian groups in particular are addressed, should be clearly noted.

- Latent pedestrian demand for a crossing should be estimated from estimates of future traffic generation, desire lines and assigned pedestrian traffic routes.

6.7. Shared-space schemes

Most of these descriptions of the interaction of pedestrians and motor vehicles have emphasised the role of sight and stopping distances for pedestrians and drivers. This has assumed a demarcation between carriageway and footways consisting of kerbs or other means of vertical surface separation, often accompanied by differing surface colours or tactile surfaces, or both. Certain approaches in continental Europe and in the UK to the 'shared-space' concept are typified by the integration of traffic, pedestrians and other road users to reduce the dominance of vehicles on the roads (JCMBPS, 2005).

6.7.1 Characteristics of shared-space schemes

Application details of shared space vary, but key features are the removal or reduction of traffic signs, markings and other instructions. Such removal is thought to induce drivers to reduce their speeds. The concept has also been extended in several cases to town centres and is flexible in its details. However, its full application de-emphasises the separation between motor vehicles and other road users, often by removal of raised kerbs between the footway and carriageway. Specific methods of reducing vehicle speeds are proposed, including: narrowing the carriageway and increasing the width of the footway; improvement of street furniture; increased lighting; and provision of seating related to walking distances.

A study of public transport in London borough pedestrian-priority areas undertaken by TRL for the Bus Priority Team at Transport for London (TfL) concluded that there is a self-limiting factor on pedestrians sharing space with motorists, of around 100 vehicles per hour. Above this, pedestrians treat the general path taken by motor vehicles as a 'road' to be crossed rather than as a space to occupy. The speed of vehicles also has a strong influence on how pedestrians use the shared area. Although this research project concentrated on pedestrian-priority areas, it is reasonable to assume that these factors are relevant to other shared-space schemes. The relationship between visibility, highway width and driver speed identified at links was also found to apply at junctions. A full description of the research findings is available in *MfS* (DfT, 2007a), *MfS2* (CIHT, 2010) and York *et al.* (2007).

6.7.2 Issues in the implementation of shared-space schemes

Because of the emerging nature of the concept of shared space in the UK and the preliminary nature of evidential data and experience in its operation and level of success, any guidelines for the auditing of schemes featuring the concept are as yet tentative. However, a number of concerns have been raised, particularly by representatives of visually impaired people (JCMBPS, 2005), which may assist in establishing guidelines for road safety auditing procedures. Comments on each of these are as follows.

- *Reducing or removing demarcation between surfaces used by cars and other vehicles and pedestrian areas.* This requires the definition of 'right of way' to be decided by negotiation, significantly by means of eye contact between driver and pedestrian. The difficulties for blind and partially sighted people in undertaking this successfully are obvious. Given the number of driving violations reported daily in the press and other documents, this clearly cannot inspire confidence in pedestrians, particularly visually impaired and other disabled people, whose walking speed and ability to change direction is limited.

- *Removal of controlled crossing points.* Several concerns for visually impaired pedestrians include those again for eye contact and the difficulty imposed when regular crossing locations, possibly equipped with audible signals or tactile surfaces, are no longer available.

- *Absence of way-finding features.* The effect of way-finding for blind and partially sighted people and the absence of a recognised kerb to assist way-finding can lead to disorientation, movement into a vehicle's path and a general lack of directional ability.

- *Mixed use with pedestrians and cyclists.* The possibility of cyclists travelling faster because of slower motorised traffic is mentioned as a possible threat to visually impaired pedestrians and to others also.

As regards evidence of the safety of shared-space schemes, it is noted (JCMBPS, 2005) that:

It is essential that the research into the shared-space experiment includes not just accident figures, which incorporate blind, deaf blind or partially sighted people, but that the number of users before and after the implementation of shared space, is considered. There is a strong possibility that people may use the area less due to safety concerns, therefore the figures would not be reliable. It is also requested that safety figures include a range of accidents from minor to major to enable assessment of the full impact on safety. From what records of reported accidents can be assessed, it is more difficult to take account of the near misses and minor accidents which were not reported, because these will affect perceptions of safety.

(JCMBPS, 2005)

Work has been initiated to determine whether some minimum differentiation in height between surfaces will be adequate to enable visually impaired people to distinguish between footways and carriageways. However, as yet, no definitive values have been determined.

A further concern is the matter of negotiation of priority at crossings between non-motorised and motorised users. Shared space and surface schemes require eye contact between users in order to 'negotiate' priority. However, as mentioned earlier when considering intervisibility, eye contact can often be difficult, owing to lighting and weather conditions, the use of tinted windscreens, obstructions due to vehicles and other obstructions. Thus, pedestrians cannot be certain of receiving priority and crossing in safety until a vehicle actually stops in order to accord that priority. Under these circumstances of uncertainty, it would seem that pedestrians who are in some way encumbered, physically impaired, accompanied by small children, or children themselves, will be intimidated by the presence of shared space and surfaces. Even though most vehicles will travel slower and drivers behave responsibly, a prudent pedestrian knows that he or she cannot rely on every driver acting in this way. The result may well be a reluctance of many pedestrians to use the shared-space areas, and to engage in walking in general, if a shared-space area is a part of their route between other places.

Undoubtedly, the implementation of shared-space schemes can have benefits in terms of urban form and social cohesiveness. However, as regards the safety and accessibility for disabled people, particularly those with visual impairments, considerably more investigation and development of responsive safety auditing would seem desirable. Shared-space schemes feature extensively in considering general street layout; related issues in this respect are also addressed in Chapter 5.

6.8. Review of options in improving crossings for pedestrians

Devices related to the engineering and geometry of junctions and midblock locations that can improve pedestrians' safety and convenience have been briefly described in the foregoing sections. Key points in the decision process and potential options are shown in conceptual form in this section.

6.8.1 Objectives

In general, based on the advantages described earlier of direct movement of pedestrian traffic across junctions and along desire lines, it is possible to identify the following objectives of improving crossings at junctions and at stand-alone crossings.

- Design geometric physical features, such that they reinforce safety and convenience for pedestrians and other road users and foster compliance with the *Highway Code*.
- Enable drivers to better see pedestrians crossing and about to cross at junctions and midblock locations and, accordingly, modify their speed or prepare to stop if intending to cross the pedestrians' routes. An added benefit of achieving this is to assist drivers in avoiding sudden changes in speed – a major cause of rear-end collisions and reduction in service volume.
- Enable pedestrians to see vehicles approaching from the several directions, described earlier, along the major and minor arms of a junction in order to evaluate the driver's likely actions and speed before crossing.
- Enable disabled people using mobility scooters to see vehicles approaching from behind at a junction by use of their rear-view mirrors.
- Enable pedestrians to cross the carriageway in as short a time as they are able and comfortable in achieving.
- On particularly wide or busy carriageways, enable pedestrians to interrupt their crossing by means of a refuge island, when approaching drivers do not observe the priority of pedestrians who have started to cross.

In addition to these objectives of the geometric design, community concerns, aesthetics and other local issues must be addressed in order to achieve a satisfactory outcome.

Several design devices and their key dimensions, which have evolved to assist pedestrians in crossing, have been mentioned earlier in this chapter. They include

- reduced corner radii
- build-outs, the width of which would accommodate a raised table and the length of which, in the direction of crossing, would generally not exceed the typical width of a parallel kerbside parking lane
- vertical deflections (flat-top humps, speed tables and junction plateaus)
- central refuges (sometimes referred to as islands or separators).

All of these devices, either singly or in combinations, have an effect on traffic in general; an example of their combined use is given in the following example.

6.8.2 Example of priority junction design

For illustrative purposes, consider a priority junction, such as is typically found along an urban trunk road, with a distributor road leading to a residential area. This form and location is one of the most often encountered forms of junction in urban areas as people travel between town centres and residential areas. It is therefore a commuter route and a route that people need to take to access major services and shops. Consequently, its potential for use by people who might switch from car use if facilities for walking are improved can be considerable. It is assumed in the example that the conditions do not warrant a zebra crossing or traffic signals.

6.8.2.1 Approach

With no central refuge (see also the discussion about this in Chapter 1), pedestrians have priority over all turning vehicles (left and right turns between the major and minor arms) for the entire width of the minor arm of the carriageway when crossing (*Highway Code*, Rule 206 (DfT, 2017)). Two factors affect the functioning of the crossing, as follows.

- Drivers making turns from the major arms or approaching the junction along the minor arm often do not observe the pedestrian-priority rule and do not stop for pedestrians who have started to cross.
- In UK practice, the triangular surface marking superimposed over the pedestrians' route, combined with a stop or yield sign beyond where a pedestrian would cross, would seem to offer a visual deterrent to pedestrians who wish to cross and would seem to give

drivers no clues that pedestrians who have started to cross have priority.

Given these conditions, one approach to modify the existing junction or design a new scheme can be as follows.

- Encourage lower vehicle speeds – by reducing kerb radii and installing a raised table at the crossing location, in addition to general lowering of speeds by the methods described in the *MfS* (DfT, 2007a).
- Alert drivers to the need to stop before the crossing location in order to give priority to crossing pedestrians – by installation of a raised table, including relevant surface markings and studs if appropriate.
- Minimise the crossing distance for pedestrians by providing build-outs. This also has the benefit of reducing delays to waiting drivers because pedestrians can clear the crossing faster.
- Provide a central refuge to reduce the total walking distance to two smaller parts, if sufficient width exists to provide a sufficiently wide refuge to accommodate wheelchair and other multiple-component crossing units.

The conceptual layouts, shown in Figure 6.13, resulting from these considerations will depend not only on the physical dimensions of the available space and existing layouts but also on pedestrian characteristics, vehicular traffic flow, roadway width, parking needs, location of property lines, sight lines, prevailing speed and the use of other traffic calming measures.

NOTE
Practice in some other countries, notably the United States and Canada, requires a 'stop' sign and surface markings for vehicles before reaching the pedestrian's crossing point. The crossing itself

Figure 6.13 Examples of options in junction design

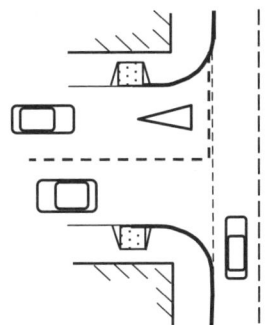

(a) Original priority three-way junction

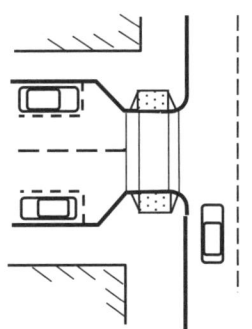

(b) Option for maximum narrowing of minor road: raised table and reduced radii and build-outs and parking on each side

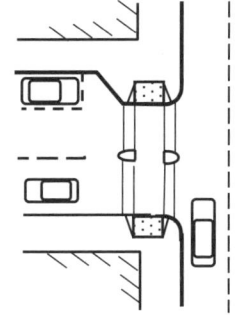

(c) Option for partial narrowing of minor road: raised table and reduced radii and build-out and parking on one side with refuge to reduce individual crossing distances

is marked on the carriageway surface, sometimes with bold cross-markings, otherwise with lines delineating the width of the crossing. This contrasts considerably with the markings in UK practice which, with the triangular marking superimposed over the pedestrians' route, would seem to pedestrians to offer a visual deterrence to crossing.

6.9. Summary

This review of the pedestrian's tasks in crossing a carriageway has provided a basis for examining, in outline fashion, the design approaches and guidance offered by *DMRB* (DfT, 2018), the CIHT (2010, 2015), and *MfS* (DfT, 2007a). In addition, an introduction to the concept of shared space was made – typically applicable in schemes introduced so far to locations for vehicular volumes of 100 vehicles per hour or less. It will be evident that the variety of current assumptions about driver and pedestrian behaviour, the priority ascribed to the different users and the variation in physical environments and locations in which junctions are located will indicate the level of care and detail needed in the analysis and conduct of the design. In addition, owing to the frequency of pedestrian–vehicle collisions at crossings of all kinds, much more research would be appropriate to improve the safety and mobility of pedestrians and the performance of the traffic system in general.

REFERENCES

Belcher M, Proctor S and Cook C (2015) *Practical Road Safety Auditing*, 3rd edn. Thomas Telford, London, UK.

Buchanan C *et al.* (ca. 1995) *Justification and Design of Entry Treatments. Assessment Study for the London Transport Director.* TfL, London, UK.

CIHT (Chartered Institution for Highways and Transportation) (2010) *Manual for Streets 2 – Wider Application of the Principles.* CIHT, London, UK.

CIHT (2015) *Designing for Walking.* CIHT, London, UK.

DfT (Department for Transport) (1981) TA 23/81: Junctions and accesses: determination of size of roundabouts and major/minor junctions. In *Design Manual for Roads and Bridges (DMRB).* DfT, London, UK.

DfT (1982) TA 30/82: Choice between options for trunk road schemes. In *Design Manual for Roads and Bridges (DMRB).* DfT, London, UK.

DfT (1995a) TD 42/95: Geometric design of major/minor priority junctions. In *Design Manual for Roads and Bridges (DMRB).* DfT, London, UK.

DfT (1995b) LTN 1/95: *The Assessment of Pedestrian Crossings.* The Stationery Office, London, UK.

DfT (1995c) LTN 2/95: *The Design of Pedestrian Crossings.* The Stationery Office, London, UK.

DfT (2002) TD 9/93: Highway link design, Amendment no. 1. In *Design Manual for Roads and Bridges (DMRB).* DfT, London, UK.

DfT (2004) LTN 1/04: *Policy, Planning and Design for Walking and Cycling.* DfT, London, UK.

DfT (2005) *Inclusive Mobility: A Guide to Best Practice on Access to Pedestrian and Transport Infrastructure.* DfT, London, UK.

DfT (2007a) *Manual for Streets.* Thomas Telford, London, UK.

DfT (2007b) LTN 1/07: *Traffic Calming.* The Stationery Office, London, UK.

DfT (2011) *STATS 20: Instructions for the Completion of Road Accident Reports from non-CRASH Sources.* DfT, London, UK.

DfT (2017) *The Highway Code.* DfT, London, UK.

DfT (2018) *Design Manual for Roads and Bridges (DMRB).* DfT, London, UK.

HMG (Her Majesty's Government) (1980) Highways Act 1980. The Stationery Office, London, UK.

JCMBPS (Joint Committee on the Mobility of Blind and Partially Sighted People) (2005) *Shared Space in the Public Realm – Policy Statement.* JCMBPS, Reading, UK.

TfL (Transport for London) (2007) *Effect of Side Raised Entry Treatments on Road Safety in London. London Road Safety Unit Research Summary No. 9.* TfL, London, UK.

Wood K, Summersgill I, Crinson LF and Castle JA (2006) *Effect of side raised entry treatments on road safety in London.* TRL, London, UK.

York I, Bradbury A, Reid S, Ewings T and Paradise R (2007) *The Manual for Streets: Evidence and Research.* TRL Report 661. Transport Research Laboratory, Crowthorne, UK.

Schoon, John G
ISBN 978-0-7277-6309-9
https://doi.org/10.1680/pfse.63099.113

Chapter 7
Signalised pedestrian crossings – geometric design

Signalised crossings for pedestrians are appropriate for several reasons, including where signals are needed primarily for vehicular traffic and where high volumes of pedestrian and vehicular traffic interact.

The analysis and design of the geometric elements of signalled crossings relevant to pedestrian requirements is extensively influenced by the needs of vehicular traffic, for which signals were originally introduced. Together with consideration of the safety and convenience of bicycle and equestrian traffic, capacity and encouragement of walking are key concerns.

Following an introduction and background to the needs for, and characteristics of, signalised crossings, geometric design layouts, largely based on LTN 2/95 *The Design of Pedestrian Crossings* (DfT, 1995); TAL 5/05 *Pedestrian Facilities at Signal-controlled Junctions* (DfT, 2005); TD 50/04 *The Geometric Layout of Signal-Controlled Junctions and Signalised Roundabouts* (DfT, 2004) and related publications.

7.1.   Sources of information
Extensive documentation exists on signalised pedestrian crossings. A summary of key documents is provided here; these will be referred to selectively later when discussing specific analysis and design matters.

The need for pedestrian traffic signals is typically established from traffic volumes, accident records, encouragement of walking policies, or local plans or strategies (DfT, 2004, 2003a). Detailed codes of practice exist for safety considerations; there is documentation related to traffic control and information systems for all-purpose roads (DfT, 2001a). Information on cyclists and equestrians is contained in separate documents (TAL 4/98 *Toucan Crossing Development* (DfT, 1998) and TAL 3/03 *Equestrian Crossings* (DfT, 2003b), respectively). Some advice on signalled stand-alone pedestrian crossings is contained in LTN 2/95 *The Design of Pedestrian Crossings* (DfT, 1995). This document is also relevant to signal-controlled junctions. In addition, there are common references in TD 50/04 *The Geometric Layout of Signal-Controlled Junctions and*

Signalised Roundabouts (DfT, 2004). Useful information is also given in TAL 1/01 *Puffin Pedestrian Crossing* (DfT, 2001b), TAL 1/02 *The Installation of Puffin Pedestrians Crossings* (DfT, 2002), *Puffin Crossings: Good Practice Guide* (DfT, 2006) and TAL 2/03 *Signal-control at Junctions on High-speed Roads* (DfT, 2003c). Also relevant are the Traffic Sign Regulations and General Directions 2016, which supersede the Traffic Signs Regulations and General Directions 2002 and the Zebra, Pelican and Puffin Pedestrian Crossings Regulations and General Directions 1997.

Site options depend on pedestrian flow patterns, degree of saturation, topographical layout and nearside or farside signal location, as described in TAL 1/01 (DfT, 2001b) and TAL 1/02 (DfT, 2002).

7.2.   Geometric design and signalisation
The geometry of signalised crossings is closely connected to the signal design and timing, the nature of the vehicular traffic and several related factors.

■ Stopping sight distances for drivers associated with prevailing speeds, visibility and timing of the signals; this affects the location of the crossing.
■ Signal timing associated with the observation–reaction, 'start-up' or 'comfort' time, walking speed and safety margin time of pedestrians; this affects the total time required for pedestrians to cross the carriageway and safely gain the opposite footway.
■ Visibility of and for drivers and pedestrians (intervisibility) associated with the geometry and proximity of junctions, driveways and street furniture; this affects the location of the crossing and the location and dimensions of surrounding features.
■ Footway width and capacity associated with pedestrians' ability to arrive and assemble at crossings before walking, and with the crossing's pedestrian capacity; this affects the dimensions of the footway and the width of crossing needed for pedestrians to cross.

■ Categories of pedestrians and their abilities in a traffic environment; this affects several of the aforementioned parameters mentioned and, hence, the location and dimensions of the crossing and its environment.

7.3. Criteria for provision of signalised crossings

7.3.1 Criteria for provision of signalised pedestrian facilities (DfT, 2005)

The need for pedestrian traffic signals is typically established from traffic volumes, accident records, encouragement of walking policies, or local plans or strategies (DfT, 2004, 2003a).

Numerical criteria for the provision of signalised pedestrian facilities were given in TA 15/81 *Pedestrian Facilities at Traffic Signal Installations* (DfT, 1981a), which is now superseded. Justification could be achieved if either the number of pedestrians crossing was high or the headway of vehicles turning into the section was short and there were at least a specified minimum number of pedestrians crossing. More recently, the key points, based on TAL 5/05 (DfT, 2005), can be summarised as follows.

■ Justification for installation – either a high number of pedestrians crossing or short vehicle headways with a specified minimum number of pedestrians crossing; otherwise, the assumption was that pedestrians would choose to cross during intergreen periods or when vehicles turned into a section at lower speeds.
■ New assumptions are necessary, owing to improved control methods, more unexpected movements, complex layouts and higher vehicle flows.
■ Formerly, design focused on vehicular movement, delay and congestion.
■ Current design emphasises all road users and specifically encourages walking.
■ Compliance with 'do not walk' (or red pedestrian figure) signals is generally poor.

7.4. Signal-controlled crossings: types and components

Four types of independent signal-controlled pedestrian crossing are currently in use: pelicans, puffins, toucans and PedEXes, with the following features.

■ *Pelican* (pedestrian light-controlled) crossings use farside pedestrian signal heads with a green 'walking pedestrian' aspect demanded by a pedestrian push-button. They have a fixed duration of flashing amber to traffic, concurrent with flashing green to pedestrians.
■ *Puffin* (pedestrian user-friendly intelligent) crossings use nearside pedestrian signal heads with an extendable all-red crossing period, which is controlled by both kerbside and on-crossing pedestrian detectors to cancel demands that are no longer required. The kerbside detectors confirm the pedestrian demand; the 'on-crossing' detectors extend the all-red crossing period.
■ *Toucan* ('two can') crossings use farside pedestrian and cycle signal heads and the same 'on-crossing' detection as puffins and are used by both pedestrians and cyclists. The farside signal heads are likely to be replaced as standard by nearside heads, similar to puffins.
■ *PedEX* crossings, for pedestrians only, feature red and green pedestrian signals, a junction push-button and a blackout period (IHE, 2016). The PedEX crossing has the same markings as a pelican crossing and either countdown or detection – but not both. PedEX signals will replace pelicans as the latter gradually reach the ends of their useful lives. Note that 'countdown' is an optional addition to the PedEX. It has an extra aspect to count down the blackout period – the time left to cross the road following the green figure. The fixed period of the blackout is not compatible with sensor-operated variable countdowns.

More recently, 'countdown' indicators giving the number of seconds left for a pedestrian to begin crossing have been introduced for PedEXes and toucans. It should be noted that equestrians are the only non-pedestrians who may legally use a pedestrian crossing. If cyclists are to be permitted to ride across the crossing, a toucan crossing should be installed for joint use by pedestrians and cyclists.

7.4.1 Signal equipment

The design of an individual signal installation, as well as that of the geometric features of the footway and adjacent features, must accommodate the necessary equipment. This typically includes a traffic controller to interpret incoming information about the relevant traffic and to operate the signals in some predetermined pattern; signal heads with the required lights, poles and signs, vehicle detectors and possibly pedestrian detectors on the relevant approaches, push-buttons for pedestrians', cyclists' and equestrians' use; and ancillary equipment and markings to guide traffic directions and waiting areas (IHT, 1997, p. 507). In addition, signalled crossings may include audible indicators to assist deaf or hearing-impaired users.

Controller programmes can provide essentially 'fixed-time' or 'vehicle-actuated' operation. They can also provide predetermined timings in order to coordinate with other signals (including some pedestrian crossings) at adjacent junctions and can reset to a 'hurry call' for emergency or special public transport services or 'manual' operation for the police or traffic wardens. Typically, the right of way between the various traffic demands is allocated by provision of a green signal, while red and amber signals denote a need to stop a particular movement.

Figure 7.1 Example of a phasing and staging diagram, based on IHT (1997). Note that phasing is denoted by letters and staging is denoted by numbers

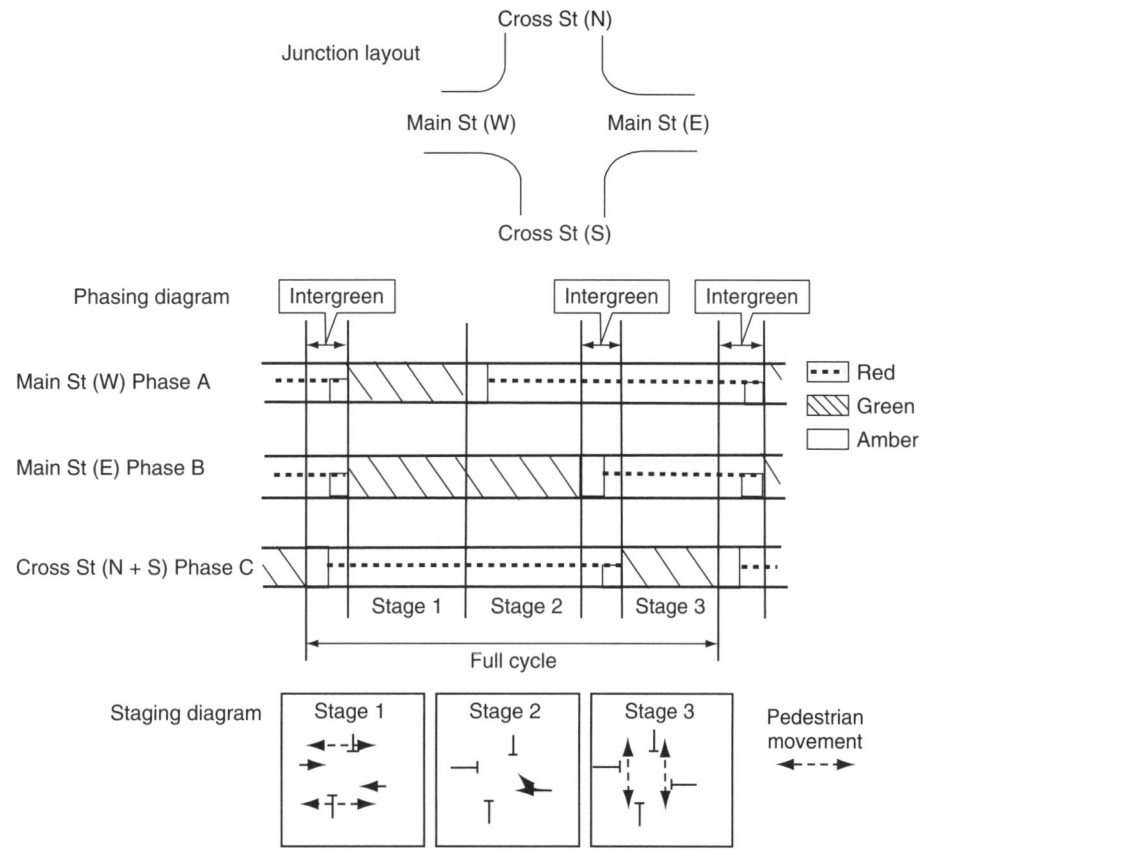

7.4.2 Traffic control sequence

The various movements for a simple four-way signalised junction with a separate pedestrian stage for all pedestrians crossing simultaneously, during which time all vehicular movements are stopped, is shown in Figure 7.1, for a signal-controlled junction. The diagrams, for illustrative purposes, show the three phases, each of which describes a set of movements taking place simultaneously or the sequence of signal indications received by such movements. In each controller, there is a normal sequence in which the various phases receive green. Each repetition of this sequence is called a 'cycle'. A 'stage' is that part of the cycle during which a particular set of phases receives green. In particular, the period between the end of the green for one phase and the start of the green for another conflicting phase, gaining right of way at the same change of stage, is known as the intergreen period.

7.5. Geometric design of signalised crossings

The range of geometric layouts of signalised crossings reflects the location, signal type, vehicle and pedestrian demands, visibility, proximity to specific land uses and related features.

Of immediate relevance to the geometric design are the general configuration, the length of the crossing and its width. The width must accommodate the volume of pedestrians and their movement characteristics. In addition, signing should be provided for signalised pedestrian crossings in accordance with the relevant Traffic Signs Regulations and General Directions 2002.

7.5.1 Siting and key geometric features of signalised pedestrian crossings

A wide spectrum of safety, capacity and convenience issues must be addressed in the siting of signalised crossings. These items range from proximity to nearby junctions, installation at stand-alone locations, contiguous land uses and bus stops to detailed geometric design matters, such as visibility and vehicle speeds. A summary of selected siting issues and related features and dimensions is provided in Table 7.1.

7.5.2 Options in pedestrians' crossing timing and geometric configuration at junctions

Signalled pedestrian crossing options may be divided into a combination of timing and geometric configurations, typified as (DfT, 2005)

Table 7.1 Siting and key features of signalled crossings, based on LTN 2/95 (DfT, 1995)

Item	Crossing siting considerations	Rationale and dimensions
Proximity of junctions	Distance along major-road approach to the minor road	Suggested 20 m minimum from signal-controlled crossing for uncontrolled priority junction 5 m minimum for zebra crossing; safe distance depends on geometry and type of minor road
	Junction with yellow box marking	Zig-zag marking not to interfere with yellow markings; length of zig-zag marking can be varied
	Signal-controlled junctions	See section on linking with other signalling systems
School crossing patrols	Existing school crossing within 100 m of possible site	Select mutually acceptable site for patrol and other pedestrians; more guardrail sections may be required Operator should be given instructions for appropriate use of signals
Visibility	All approaches to crossings	Minimum visibility distances for drivers – see Table 7.2. Pedestrians must be in a position to see and be seen by approaching traffic. Visibility must not be obstructed by fixed or moving objects
	Where alignment problem exists	Queue lengths must be accounted for in the design
Crossing width	Zebra, pelican, puffin Toucan 600 pedestrians per hour or greater Refuge island width	2.4 m minimum; regulations may allow greater widths 4 m ideal Use wider crossing 2 m reasonable minimum
Guardrailing	All locations	Intervisibility important Avoid gaps in railings Should start at signal post but not encroach past push-button position
Crossing approaches – footways	Approach ramp from dropped kerbs	Maximum slope 1 : 12; minimum 1 : 20. Provide adequate drainage Install tactile paving (see facilities for disabled pedestrians) No surface obstructions permitted
Crossing approaches – carriageway	Surfacing	Provide high skid resistance Surfacing length to depend on approach speed and accident record
Facilities for disabled pedestrians	Footway facilities	Provide dropped kerbs and ramps with tactile surfaces See separate section on disabled pedestrian facilities
	Audible signals	Pulsed tone or tactile signals to be provided
Bus stops	Locate to avoid buses obscuring vision of pedestrians or drivers	Generally, a bus stop is better sited on the exit side of a junction

- no pedestrian phase or stage at junction
- full pedestrian stage – all vehicular approaches stopped during pedestrians' walk phase
- parallel pedestrian facility
- staggered pedestrian facility
- displaced pedestrian facility.

The key characteristics of these configurations are summarised in Table 7.3; further considerations are outlined next (DfT, 2005).

7.5.3 No pedestrian phase or stage at junction

This installation tends to be the least popular with pedestrians because the lack of a separate pedestrian-only 'green' period means that pedestrians have to rely on vehicles giving way in order to cross safely – this is often perceived as stressful and uncertain. The presence of a refuge in order to more safely interrupt the crossing while vehicles continue in an opposite lane or lanes can help to reduce the walking distance and permit pedestrians to make a more ordered crossing (but note that this

Table 7.2 Visibility requirements (LTN 2/95, DfT, 1995)

85th percentile approach speed: miles per hour	25	30	35	40	45	50
Desirable minimum visibility: m	50	65	80	100	125	150
Absolute minimum visibility: m	40	50	65	80	95	115

Table 7.3 Types of pedestrian facility at signal-controlled junctions, based on IHT (1997)

Type of facility	Characteristics
No pedestrian signal	Traffic signals, even without pedestrians signals, can help pedestrians to cross by creating gaps in traffic Especially applicable where there are refuges, enabling the conflict at each crossing movement to be with only traffic in one direction, and on one-way streets
Full pedestrian stage	All traffic is stopped Demanded from push-buttons More delay to vehicles than combined vehicle–pedestrian stages
Parallel pedestrian facility	Combined vehicle–pedestrian stage, sometimes accompanied by banned vehicle movements Useful for crossing one-way streets
Staggered pedestrian facility	Pedestrians cross half of the carriageway at a time Large waiting area at the centre of the carriageway is required
Displaced pedestrian facility	For junctions close to capacity The crossing point is situated away from the junction but within 50 m Normal staging arrangements apply

statement may contradict the *Highway Code* requirement that drivers give priority to pedestrians over the whole pedestrian crossing, even if a refuge is present). The presence of metal studs or, more recently, painted lines, to delineate the crossing may also assist.

Most pedestrians cross during the intergreen period (defined earlier) (DfT, 2005), although an extended intergreen phase to assist pedestrians is not recommended because of delays to vehicles and resulting driver disobedience (DfT, 1981b). An alternative may be a key switch for use by an authorised person for special crossing occasions (e.g. a school children's crossing). One problem with this approach might be that if pedestrians become familiar with an increased intergreen time, they may inadvertently cross during a normal green time, with resulting inadequate crossing time and associated danger.

7.5.4 Full pedestrian stage – all vehicle approaches stopped

The major advantage of this kind of installation is that it is relatively easily understood by all road users – pedestrians and drivers alike – and the separation between vehicle and pedestrians during the intergreen and crossing times, respectively,

is the most obvious. This simplicity, however, has the worst effect on overall vehicular capacity of a junction, in that signal cycle times are longer (because of the intergreen plus crossing time) and pedestrians arriving at the end of the 'invitation to walk' period might have longer to wait. Additionally, two pedestrian stages per cycle will increase vehicular delay. Other considerations include the following.

- The facility should be called by a 'demand button' – at nearside signals – to encourage pedestrians to look at both pedestrian signals and oncoming vehicles.
- Permanent pedestrian demand, often by time of day, may be used where pedestrian demand is high and minimal safety problems exist.
- Refuges, if used, should be straight-across type.
- Diagonal crossings can be difficult for disabled people; special design considerations are needed for locations where disabled people need to cross.

7.5.5 Parallel pedestrian crossing

In this type of installation, shown diagrammatically in Figure 7.2, the pedestrian 'walk' signal is activated so that pedestrians walk parallel to the moving traffic stream. This can

Figure 7.2 Parallel pedestrian crossing, based on TAL 5/05 (DfT, 2005)

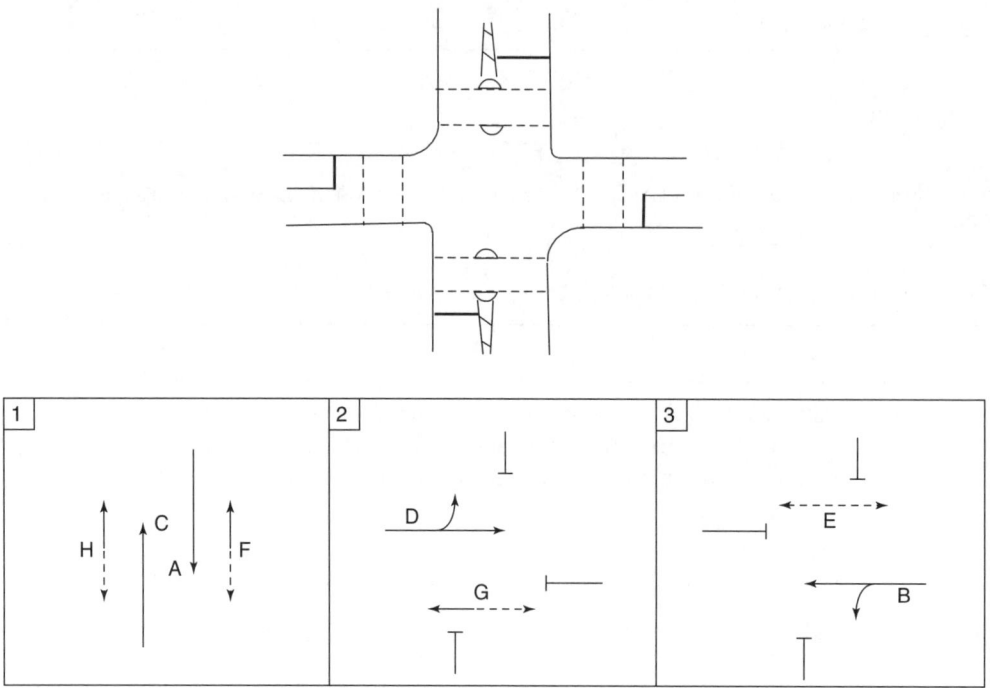

improve the overall efficiency of the junction and, from a pedestrian's point of view, may also reduce delay due to long red periods. Refinements to the overall principle can include prohibited vehicle turns, to ensure that pedestrian priority is reinforced. This can be emphasised by using squared kerb corners and appropriate signage.

Splitter islands may be useful at T-junctions with one-way streets to facilitate pedestrian crossing from either side of the minor road and across the major road during appropriate 'walk' periods. A typical example is shown diagrammatically in Figure 7.3.

7.5.6   Staggered pedestrian crossing

Where carriageway widths permit, it is possible to economise on cycle time by the provision of a larger refuge. The pedestrian movement, which is normally staggered, can then be integrated with vehicular staging. A minimum size of 10 m × 3 m for the central refuge is recommended, although widths over 3 m may be required to meet the needs of those crossing. At some refuges, where there may be a number of pedestrian routes to cater for, the size of the waiting area must be carefully designed accordingly.

NOTE

1. Staggered crossings may require a considerable amount of manoeuvring and inconvenience to wheelchair and other disabled users.

2. Total waiting time and overall crossing time for pedestrians using a staggered crossing can be considerable; it is possible that replacement with a straight-across crossing will save overall pedestrian time, although increasing the time for vehicular traffic. See the example in Chapter 15.

The recommended stagger at stand-alone crossings is left–right, as shown in TAL 1/02 (DfT, 2002). However, a right–left stagger, as shown in Figure 7.4, is probably more common at junctions. Table 7.4 lists the advantages and disadvantages of each. It should be remembered that the staggers should meet the pedestrian desire lines as closely as possible. If staggers are dividing two flows of vehicles travelling in the same direction, such as at a bus gate, appropriate signs should be considered. The guidance in TD 50/04 (DfT, 2004) on intervisibility between drivers and pedestrians should always be part of any assessment.

Sites located close together should have the same layout to save confusion of vulnerable users. Pedestrians can negotiate one half of the carriageway at the entry stop line when traffic on that approach is held on red.

NOTE

Staggered crossings therefore tend to reduce the motorists' obligation to give priority to pedestrians crossing the entire road because the stagger and holding area renders each half of the total carriageway a separate crossing.

Figure 7.3 Parallel pedestrian crossing – one-way street arrangement, based on TAL 5/05 (DfT, 2005)

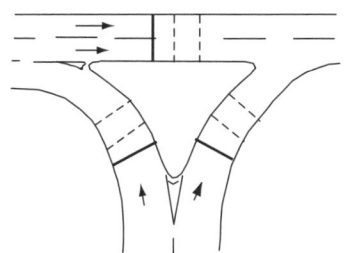

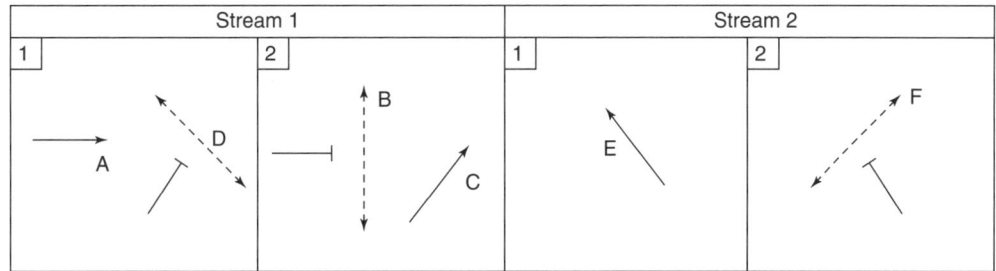

Normal pedestrian signals are shown during this period. The other pedestrian phase can use a parallel-stage stream, as shown in Figure 7.4. This type of arrangement can work well when the route follows a natural pedestrian desire line. The facility would be demanded by push-buttons associated with two phases.

Another type of staggered arrangement includes an on-locking right-turn stage and may also have a left-turn parallel-stage stream with a slip road. With this arrangement the locations of studs, give-way markings and intervisibility are important factors. There is also more scope for positioning the crossing nearer to the pedestrians' desire line.

7.5.7 Displaced pedestrian crossing
In this type of crossing, an example of which is shown diagrammatically in Figure 7.5, the crossing is displaced a

Figure 7.4 Staggered pedestrian crossing, based on TAL 5/05 (DfT, 2005)

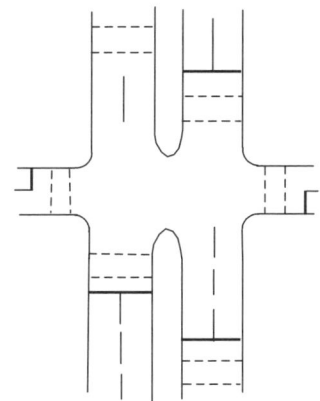

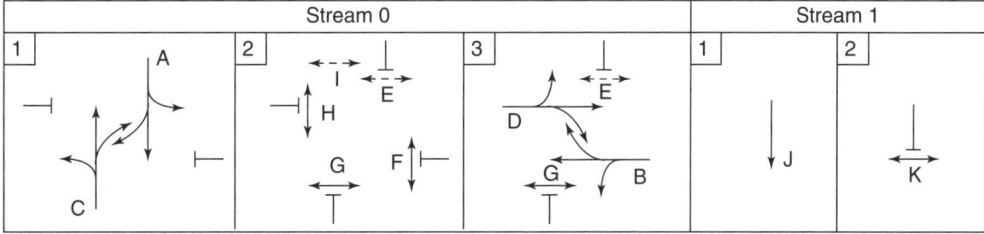

Table 7.4 Summary of advantages and disadvantages of alternative orientation at signalled crossings, based on TAL 5/05 (DfT, 2005)

Stagger	Advantages	Disadvantages
Left–right	Consistent with stand-alone crossings. Encourages pedestrians to face oncoming vehicles. Pedestrians on exit of junction are nearest to the junction, improving intervisibility.	Moves stop line and queue further from junction. May increase intergreen periods and therefore lost time if crossing points, on all approaches, are not the same. If a stop line were needed for the crossing on exit, it would be very close to the junction.
Right–left	Brings the stop line nearer to the junction. Moves the exit crossing away from the side road and allows drivers of turning vehicles more time to assess possible dangers. Allows possible stop line for exit crossing to be a reasonable distance from the junction.	Not consistent with stand-alone crossings. Pedestrians are not encouraged to face oncoming vehicles while walking between crossings. May cause problems with intervisibility between side road and pedestrians.

recommended maximum of 50 m from the junction in order to improve intervisibility and satisfy pedestrian demand. The placement may result in vehicular traffic capacity improvements, therefore resulting in shorter waiting times for pedestrians. However, a possible disadvantage is that pedestrians may be diverted somewhat from their preferred desire line. Conditions for the parallel stream should be specified to ensure that main vehicular flow is not interrupted and that vehicles turning from the side road are not impeded by a queue.

Visibility distances should be checked carefully, owing to the displacement; separate detection will be needed if the installation operates under vehicle actuation.

7.5.8 Stand-alone signalised pedestrian crossings
In most respects, stand-alone signalised pedestrian crossings will have many of the characteristics of crossings at junctions. However, because the signals control for either vehicle or pedestrian movements only (rather than vehicle movements in

Figure 7.5 Displaced pedestrian facility, based on TAL 5/05 (DfT, 2005)

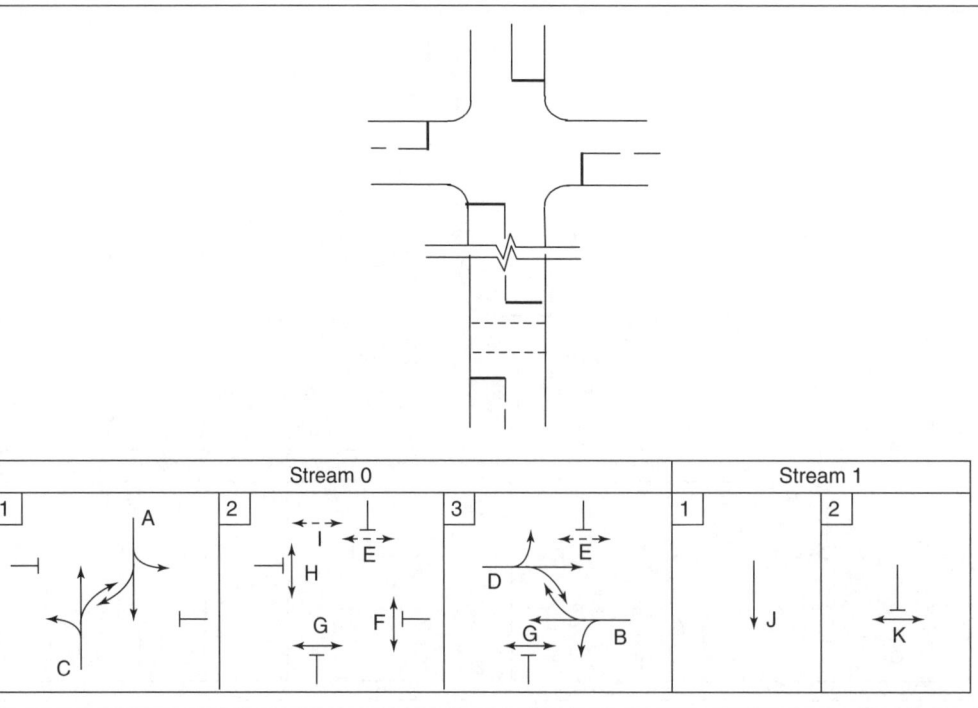

different directions), certain specific precautions should be taken in their siting. In particular, envelopes of visibility and stopping sight distances should be observed. These are discussed in greater detail in Section 7.8.

General guidelines are presented in TAL 1/02 *The Installation of Puffin Pedestrian Crossings* (DfT, 2002), which indicates that, for all signalled crossings, a sufficient area should be provided on the footway where pedestrians can gather. This should be large enough to accommodate the number of pedestrians expected under normal conditions. It should not normally be less than 2 m from the kerb to the back of the footway for a 2.4 m wide crossing. If large volumes of pedestrians are anticipated, a wider crossing should be considered. Care should be exercised when converting pelican sites to puffin sites, to ensure that the site layout is suitable for proper operation.

Tactile paving should be provided at all crossing points between the crossing studs, in accordance with the advice contained in *Guidance on the Use of Tactile Paving Surfaces* (DETR, 1998).

Additionally, adequate footway widths should be available for pedestrians to pass who are not intending to cross.

The gradient at all crossing points should not exceed 1 : 12 (8%); where space allows, a gradient of 1 : 20 (5%) should be achieved. On very narrow footways, it may not be possible to drop the kerb at this angle while keeping the majority of the pavement level. In this situation, an alternative is to lower the whole pavement for the length of the dropped kerb.

7.6. Influence of urban traffic control systems on geometric design of signalised crossings

Coordinating and improving crossing opportunities at junctions and stand-alone signalised crossings for pedestrians, where urban traffic control systems using the split cycle and offset optimisation technique (SCOOT) are in operation, are addressed in TAL 2/09 (DfT, 2009).

In general, the presence or absence of urban traffic control systems mostly affects the timing of the signals and, hence, the frequency and duration of the pedestrians' green periods. Minimising vehicular traffic delays tends to increase delays at pedestrian signals if they are incorporated in the system. Several options related to the geometric design for improving pedestrians' crossing opportunities are, however, possible, including

■ widening a crossing where possible to increase its capacity and therefore allow for a reduced pedestrians' green period

■ removal or relocation of obstructions, such as signal posts, from the footway, to enable a maximum pedestrian flow to and from a crossing for any given situation

■ the use of diagonal crossings, where this is possible within the urban traffic control system, can be shown to be acceptable in the overall street and land use environment

■ the use of pedestrian volume detectors in order to provide a greater time period to leave the kerb when there are many pedestrians.

Along with the development of split cycle and offset optimisation technique (SCOOT) systems, research is continuing into its application and impacts, and investigations by Kirkham *et al.* (2006) and Bretherton (2007) address several of the key issues.

7.7. Puffin pedestrian crossings: nearside pedestrian signals (DfT, 1995, 2005)

It is intended that the puffin operational cycle will become the standard form of pedestrian crossing at stand-alone crossings and junctions. For this reason, and to provide an example of the timing calculations for a signalled pedestrian crossing, the procedures for puffin crossings are outlined next. The method of calculating timings shown here has been tried at a number of junction installations and is recommended as advice. Tables 7.5 and 7.6 refer to cycle timings.

Of particular importance to the geometric design of the crossing are periods 4 to 8, summarised as follows.

Period 4
The timing for the pedestrian green walking figure period, with the option of the audible and/or tactile signal, should normally be set to 4 or 5 s at crossings with light to moderate pedestrian flows. Where one or more of the following conditions occur the length of this period should be increased to 6–9 s, as appropriate

■ the crossing is in an area where heavy pedestrian flows are generated
■ the distance between kerbs is greater than 11 m
■ a central refuge is provided
■ space in the pedestrian waiting area is limited
■ areas where there is a higher proportion of disabled or elderly people.

Period 5
The all-red period of 1–5 s.

Period 6
The all-red is extended by the on-crossing detectors up to 25 s.

Table 7.5 Puffin crossing – operational cycle, use and variations (DfT, 1995)

Period	Use	Variation for
1	Vehicle running time	Traffic volumes
2	Standard amber to vehicles	None
3	Vehicle clearance period	Vehicle actuation
4	Pedestrian invitation to cross	Road width, disabled pedestrians, crossing with central refuge
5	Pedestrians must not start to cross	Type of detector
6	Completion of pedestrian crossing time	Road width
7	Additional pedestrian clearance time	Pedestrian detection
8	Additional pedestrian clearance time	Pedestrian gap change
9	Standard red and amber to vehicles	None

The extension period for the pedestrian on-crossing detector should normally be set within the range 1.6 to 2.2 s.

Period 7
If the normal maximum of the clearance period is reached when pedestrians are still being detected on the crossing, this operates to permit the pedestrians to clear before Period 9

commences. The duration of this period is normally 3 s but can be adjusted between 0 and 3 s.

The maximum duration of the pedestrian extendable clearance period (Periods 6 and 7 together), in seconds, should normally be set to 5 + 1.67 (the length of crossing − 3 m).

Table 7.6 Puffin crossing – operational cycle and timings (DfT, 1995)

Period	Signals shown		Timings: s
	To pedestrians	To vehicles	
1	Red standing figure (wait)	Green (proceed if way is clear)	20–60 (fixed) 6–60 (vehicle actuation)
2	Red standing figure	Amber (stop unless not safe to do so)	3
3	Red standing figure	Red (stop, wait behind stop line on carriageway)	1–3
4	Green walking figure with audible signal if provided (cross with care)	Red	4–9
5	Red standing figure (do not start to cross)	Red	1–5
6	Red standing figure	Red	0–22 (pedestrian extendable period)
7	Red standing figure	Red	0–3 (only appears on a maximum change if pedestrians are still being detected)
8	Red standing figure	Red	0–3 (only appears at a pedestrian gap change)
9	Red standing figure	Red with amber (stop)	2

Table 7.7 Timing chart (DfT, 2002)

		Carriageway width (m)											
		4	5	6	7	8	9	10	11	12	13	14	15
Pedestrian signal period		4	5	6	7	8	9	10	11	12	13	14	15
All-red period		1	Or up to 3 s for a forced change										
Green-man period		4	4	4	4	5	5	5	6	6	7	7	7
Clearance period	Minimum	3	3	3	3	3	3	3	3	3	3	3	3
	Maximum	4	5	7	9	10	12	14	15	17	19	20	22
	Forced	7	8	10	12	13	15	17	18	20	22	23	25
Total period	Minimum	9	9	9	9	10	10	10	11	11	12	12	12
	Maximum	13	14	16	18	20	22	24	26	28	31	32	34

Note: The minimum period can be adjusted to between 1 and 3 s. The value chosen will depend on the detector's response time to pedestrians when they step onto the carriageway. Care should be taken to choose a value which is appropriate for the detection system, because this period is a fixed time burden on the overall pedestrian stage.

Period 8
If the normal maximum of the clearance period is not reached Period 7 will be followed by this period. Normally set to 0 s but can be adjusted in steps of 1 s to a maximum of 3 s.

(DfT, 1995)

7.7.1 Using a timing chart for carriageway width
Although timing for the signals may be computed by use of formulae, for illustrative purposes and to indicate typical values, Table 7.7 shows the various timing periods related to the carriageway width, and can be used to check or augment calculation, or as an alternative to calculation. The carriageway width is a key geometric feature on which the timing depends. Neither the use of the formulae nor the tabulated values provide the width of the crossing itself; moreover, these would have to be adjusted beyond the minimum requirement based on the expected volume and characteristics of the pedestrians using the crossing.

7.8. Geometric design considerations for trunk and high-speed roads (DfT, 2004, Section 2)
Traffic signals with associated pedestrian signals may be provided on trunk and other high-volume and high-speed roads. For this reason, and from consideration of the need for intervisibility between drivers, pedestrians and other road users, sight distances related to vehicle approach speeds and location of potential obstructions to sight lines are particularly important.

7.8.1 Design speed (DfT, 2004)
Some design standards are dependent on the approach speed of vehicles; reference should be made to TD 9/93 (DfT, 1993) to

determine appropriate design speeds for each entry arm. Where these design speed-related standards cannot be achieved, traffic management measures should be introduced to reduce the approach speed to an appropriate value for the available stopping sight distance. Table 7.8 lists the values of these speeds related to the various visibility distances.

7.8.2 Visibility on the approach to junctions
Stopping sight distances and visibilities of drivers to signals are as follows.

- The stopping sight distance on the immediate approach to the junction [See DfT (1993), para 1.26] shall be in accordance with the standards contained in [(DfT, 1993)] and relaxations below [desirable minimum stopping sight distance] shall not be permitted on the immediate approaches to the junction.
- Each traffic lane shall have clear vision of at least one primary signal associated with its particular movement, from a distance equivalent to the [desirable minimum stopping sight distance]. The visibility envelope [see DfT (1993), Figure 3] shall be increased to include the height of the signal head as indicated in [Figure 7.6].

(DfT, 2004)

7.8.3 Junction intervisibility zone (DfT, 2004)
Depending on the configuration of the crossing, the intersecting streets and the immediate surroundings, the intervisibility between all categories of road user will be critical to the safety and effectiveness offered by the crossing and the total junction. This section presents examples of conditions where

Table 7.8 All types of crossing – visibility requirements (DfT, 1988)

Item	Vehicle speed					
85th percentile approach speed: miles per hour	25	30	35	40	45	50
85th percentile approach speed: km/h	40	48	56	64	70	80
Desirable minimum visibility: m	50	65	80	100	125	150
Absolute minimum visibility: m	40	50	65	80	95	115

LTN 2/95 *The Design of Pedestrian Crossings* (DfT, 1995) provides essentially the same information.
© Crown Copyright, 1988
This information is licensed under the Open Government Licence v3.0. To view this licence, visit http://www.nationalarchives.gov.uk/doc/open-government-licence/ **OGL**

intervisibility zones at various pedestrian layouts interact and how some may be adjusted to reflect improved designs (DfT, 2004, Section 2).

The junction intervisibility zone is the area identified for the purposes of assessing intervisibility within the junction and the drivers at each stop line, or between drivers and pedestrians. Delineation of each zone facilitates the identification of measures to mitigate the effect of obstruction to sight lines.

NOTE
See also notes on intervisibility in earlier chapters.

The junction intervisibility zone is defined as the area bounded by measurements from a distance 2.5 m behind the stop line extending across the full carriageway width for each approach arm, as indicated in Figure 7.7, which shows a case where pedestrian crossings are also present.

At new signalled junctions, major obstructions to intervisibility within the junction intervisibility zone, such as that caused by buildings, should be avoided. Under these conditions, each obstruction to visibility shall be considered as a departure from standard and measures shall be taken to mitigate the effects on intervisibility.

7.8.4 Junction intervisibility

An example of a reduction in intervisibility for a vehicle at a stop line on the left-turn entry lane (Arm C) and the exit lane and pedestrians crossing (Arm D) caused by a major obstruction is shown in Figure 7.8. Figure 7.9 illustrates how the intervisibility between Arm C and Arm D in this example can be mitigated, by moving the stop line forward and providing an in-line pedestrian crossing on Arm C. When adjusting the positions of the stop line and the pedestrian crossing, the effect of the position of any pedestrian refuge on vehicle swept paths should be determined, and its effects on pedestrians, particularly visually impaired people, should be considered.

Minor obstructions to visibility caused by slim projections, such as lighting columns, sign supports, signal posts, controller cabinets and guardrails placed within the intervisibility zone might be unavoidable in the optimum design. When placing signs, street furniture and planting, consideration should be given to ensure that their obstructive effect is minimised. Minor obstructions are not considered as either departures from or relaxations within standard.

Depending on the geometry of the junction and the class of vehicles using it, several other factors should be considered in the process of establishing a pedestrian crossing. These include

Figure 7.6 Visibility requirements on approach to junction (DfT, 2004)
© Crown Copyright, 2004
This information is licensed under the Open Government Licence v3.0. To view this licence, visit http://www.nationalarchives.gov.uk/doc/open-government-licence/ **OGL**

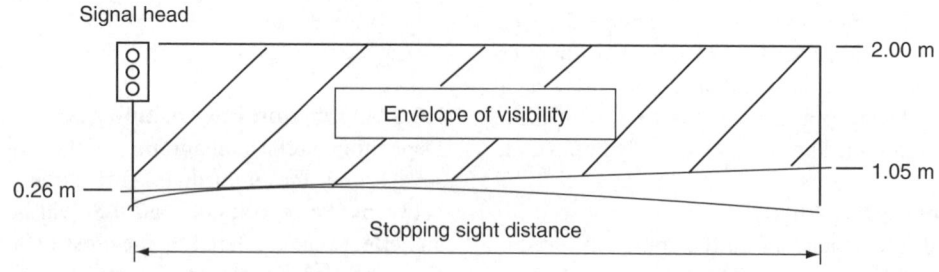

Figure 7.7 Junction intervisibility zone and pedestrian crossings (signal head and actuation points not shown, for clarity) (DfT, 2004) © Crown Copyright, 2004
This information is licensed under the Open Government Licence v3.0. To view this licence, visit http://www.nationalarchives.gov.uk/doc/open-government-licence/ **OGL**

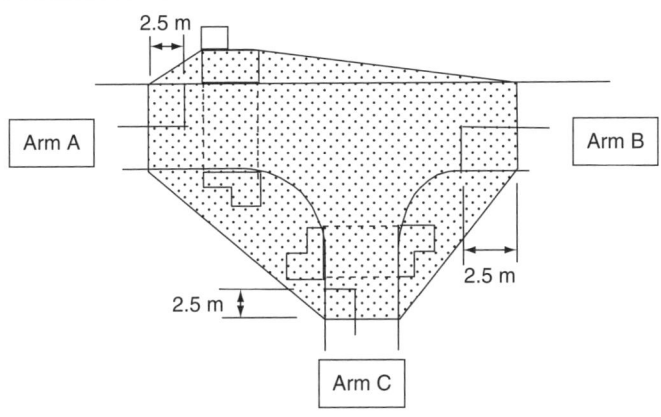

Figure 7.9 Mitigation of major obstruction to intervisibility, based on DfT (2004)

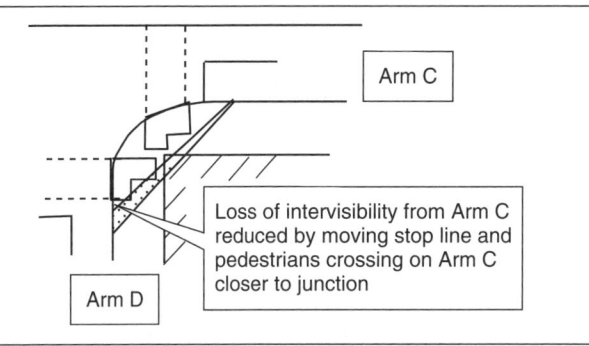

kerb radii, vehicle holding on the junction approach arms in order to avoid vehicle queues, lane conformity and alignment, and carriageway width around curves. Guidelines for these are addressed in TD 50/04 (DfT, 2004).

The location of pedestrian crossings might also be affected by the swept path requirement of the design vehicle. An example of this requirement is shown in Figure 7.10.

Other concerns affecting, primarily, vehicular traffic at junctions should be considered in the geometric design. Depending on the particular situation, some of these matters may be relevant to the provision and location of a pedestrian crossing. These include left-turn slip lanes, separation islands, corner radii and tapers, effects of right-turn movements, signs and markings, high-friction surfaces on the approaches to crossings

Figure 7.8 Mitigation of major obstruction to intervisibility, based on DfT (2004)

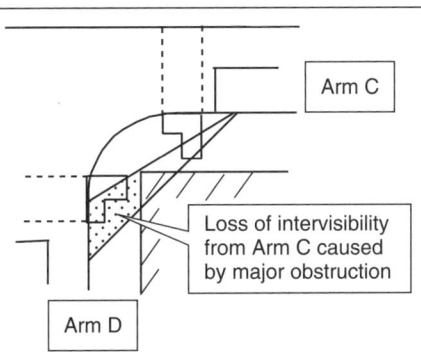

and the location of signals for drivers' visibility. Key features of facilities at separation islands, advanced stop signs for cyclists and combined toucan and equestrian crossings are shown in Figures 7.11 to 7.13. They are addressed in TD 50/04 (DfT, 2004).

7.8.5 Other design details
As well as the overall intervisibility requirements detailed previously, it will be necessary to address other matters, such as geometric design details and junction arrangements and operations. Key points associated with these include the following.

7.8.5.1 Geometric design details
These include carriageway widths, entry lanes, holding lanes, lane continuity, carriageway widths around curves, swept path requirements, left-turn slip lanes, separation islands, right-turning traffic movements and non-hooking arrangements.

7.8.5.2 Junction arrangements and operations
Complex layouts should be avoided; the size and complexity of the junction depend on the combination of geometric features and the number and timing of phases, stages and separately

Figure 7.10 Example of swept path affecting location of pedestrian crossing, based on DfT (2004)

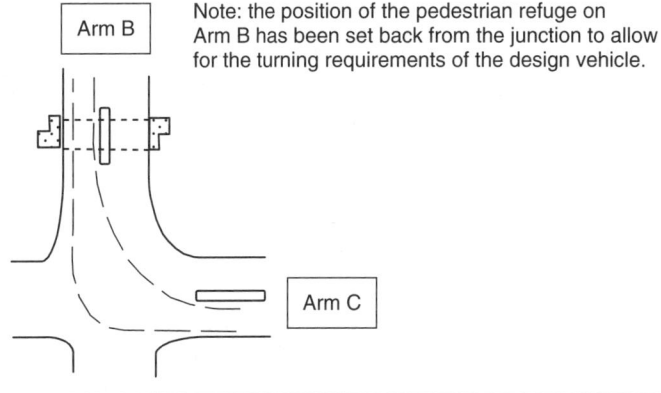

Figure 7.11 Pedestrian crossing facilities on separation island, based on DfT (2004); signal details and guardrail not shown for clarity

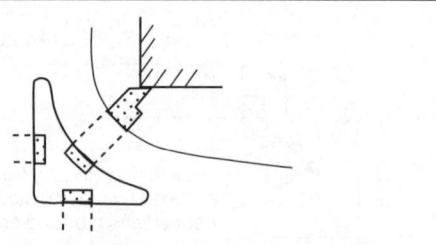

Figure 7.13 Combined toucan and equestrian crossing, based on DfT (2004); signal details and guardrail not shown for clarity
© Crown Copyright, 2004
This information is licensed under the Open Government Licence v3.0. To view this licence, visit http://www.nationalarchives.gov.uk/doc/open-government-licence/ **OGL**

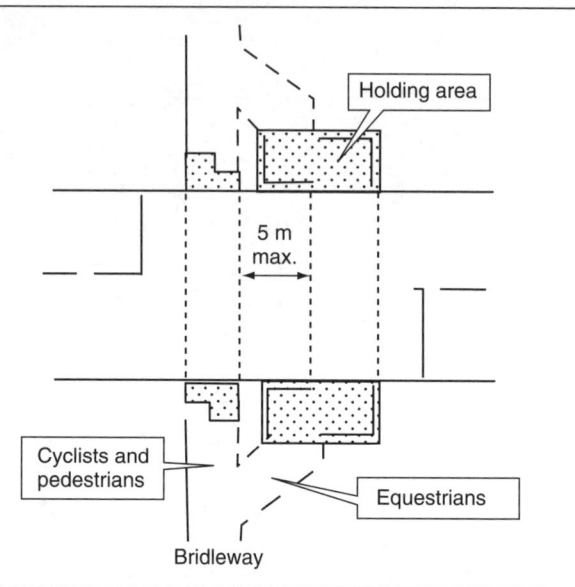

signalled movements. It has been stated that pedestrians' facilities might not be appropriate or possible in some circumstances. TD 50/04 (DfT, 2004) provides a selection of examples of various sizes and complexities of T-junctions, four-way junctions, staggered junctions, skews and one-way roads.

NOTE
Most of the junction configurations of these examples illustrate current preference in design policy for motor vehicle use over pedestrians, often requiring pedestrians to walk farther, divert from their desire lines and rely on drivers to ensure their safety. This is evident in the lack of in-line crossings, the frequent use of displaced and staggered crossings and the lack of inclusion of pedestrians' observation–reaction times in the consideration of intervisibility.

7.9.   Summary
Many pedestrian crossings assist vehicular and pedestrian traffic in safe movement throughout the transport system. In general, signal installations have favoured vehicular traffic over non-motorised users. Attempts have been made, notably with the puffin crossing, to overcome some of the deficiencies from a pedestrian's perspective. However, the geometric layouts and configurations of many junctions continue to require pedestrians to undertake often complex, onerous and unsafe

Figure 7.12 Advanced stop line for cyclists and pedestrian crossing, based on DfT (2004); signal details not shown for clarity
© Crown Copyright, 2004
This information is licensed under the Open Government Licence v3.0. To view this licence, visit http://www.nationalarchives.gov.uk/doc/open-government-licence/ **OGL**

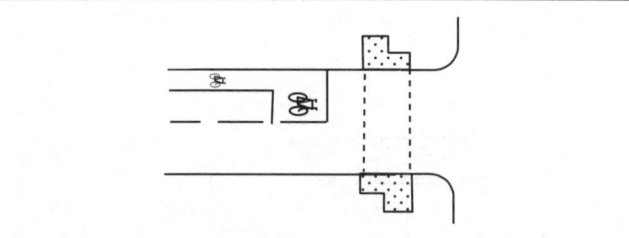

actions in order to complete their journeys by foot. The problems encountered by disabled people and those accompanied by children in crossing conveniently, safely and with confidence, indicate that considerably more exploration and improvements are needed to encourage walking.

REFERENCES

Bretherton D (2007) SCOOT MC3 and current developments. *Transport Research Annual meeting.* TRL, Washington, DC, USA.

DETR (Department of the Environment, Transport and the Regions) (1998) *Guidance on the Use of Tactile Paving Surfaces.* DETR, London, UK.

DfT (Department for Transport) (1981a) TA 15/81: Pedestrian facilities at traffic signal installations. In *Design Manual for Roads and Bridges (DMRB).* DfT, London, UK.

DfT (1981b) TA 16/81: General principles for control by traffic signals. In *Design Manual for Roads and Bridges (DMRB).* DfT, London, UK.

DfT (1988) TA 12/81: Traffic signs and lighting. In *Design Manual for Roads and Bridges (DMRB).* DfT, London, UK.

DfT (1993) TD 9/93: Highway link design. In *Design Manual for Roads and Bridges (DMRB).* DfT, London, UK.

DfT (1995) LTN 2/95: *The Design of Pedestrians Crossings.* The Stationery Office, London, UK.

DfT (1998) TAL 4/98: *Toucan Crossing Development*. DfT, London, UK.

DfT (2001a) TA 84/01: Code of practice for traffic control and information systems. In *Design Manual for Roads and Bridges (DMRB)*. DfT, London, UK.

DfT (2001b) TAL 1/01: *Puffin Pedestrian Crossing*. DfT, London, UK.

DfT (2002) TAL 1/02: *The Installation of Puffin Pedestrian Crossings*. DfT, London, UK.

DfT (2003a) TAL 5/03: *Walking Bibliography*. DfT, London, UK.

DfT (2003b) TAL 3/03: *Equestrian Crossings*. DfT, London, UK.

DfT (2003c) TAL 2/03: *Signal-control at Junctions on High-speed Roads*. DfT, London, UK.

DfT (2004) TD 50/04: The geometric layout of signal-controlled junctions and signalised roundabouts. In *Design Manual for Roads and Bridges (DMRB)*. DfT, London, UK.

DfT (2005) TAL 5/05: *Pedestrian Facilities at Signal-controlled Junctions*. DfT, London, UK.

DfT (2006) *Puffin Crossings: Good Practice Guide*. DfT, London, UK.

DfT (2009) TAL 2/09: *Integration of Pedestrian Traffic Signal Control Within SCOOT-UTC Systems*. DfT, London, UK.

HMG (Her Majesty's Government) (1997) The Zebra, Pelican and Puffin Pedestrian Crossings Regulations and General Directions 1997. The Stationery Office, London, UK.

HMG (2002) Traffic Signs Regulations and General Directions 2002. The Stationery Office, London, UK.

HMG (2016) The Traffic Signs Regulations and General Directions 2016. The Stationery Office, London, UK.

IHE (Institution of Highway Engineers) (2016) *Traffic Signals Design Course*. IHE, London, UK.

IHT (Institution of Highways and Transportation) (1997) *Transport in the Urban Environment*. IHT, London, UK.

Kirkham A, Bretherton D and Wood K (2006) Pedestrian priority within SCOOT. *World Congress, London*. TRL, Washington, DC, USA.

Schoon, John G
ISBN 978-0-7277-6309-9
https://doi.org/10.1680/pfse.63099.129

Chapter 8
Facilities for inclusive mobility: geometric design guidelines

To ensure an inclusive transport system, devices ranging from a white cane to motorised wheelchairs and scooters and fixed infrastructure ranging from dropped kerbs to extensive ramp systems have evolved. A selection of the more common devices and their dimensions, especially as they relate to the geometric design of the facilities that must accommodate them, are presented in this chapter, which should be read in conjunction with Chapter 4.

As with all instances of design, and particularly with respect to the needs of disabled people, as described in Chapter 4, the dimensions and layouts should be regarded as maximum or minimum requirements, depending on the context. Designs at specific locations should take account of the unique circumstances that obtain there, as well as the total trip, from origin to destination. As with all pedestrian trips, and particularly with regard to access for disabled people, the presence of a single unacceptable design at one place along a route can render that entire route unusable by a disabled person, and is also often dangerous for non-disabled people.

8.1. Legislative background (Highways Agency, 2010)

8.1.1   History
The Disability Discrimination Act 1995 was introduced in 1996; Part III gave disabled people a right of access to goods, facilities, services and premises. The original 1995 Act was modified and extended by the introduction of the Disability Discrimination Act 2005 in 2006. This was replaced by the Equality Act 2010, which specifies the public sector's specific duties.

The Acts are also supported by codes of practice issued by the former Disability Rights Commission. The Disability Rights Commission was closed in 2007 and replaced by the Equality and Human Rights Commission (EHRC, 2016). In addition, the Disabled Persons Transport Advisory Committee (DPTAC) advises the government on transport legislation, regulations and guidance and on the transport needs of disabled people, ensuring that disabled people have the same access to transport as everyone else (DPTAC, 2007).

The Acts, codes of practice and guidance should be consulted for detailed definitions and information, but a general indication for the purpose of this guide is given in this section.

8.1.2   Definition of disability
Under the Acts, a disabled person is defined as someone who has a physical or mental impairment that has a substantial and long-term adverse effect on his or her ability to conduct normal day-to-day activities. Persons diagnosed with human immuno-deficiency virus, cancer or multiple sclerosis are also deemed to be disabled under the terms of the Disability Discrimination Act.

8.1.3   Implications
As a result of this legislation, it is unlawful for service providers to discriminate against a disabled person by

- treating them less favourably – for a reason relating to their disability – than it treats others
- failing in a duty to make *reasonable adjustments* in relation to a service by
  - changing practices, policies and procedures
  - providing auxiliary aids and services
  - overcoming a physical feature by removing the feature or altering it.

8.1.4   Reasonable adjustments
With regard to the extent of any adjustments required, it is worth noting the following extract from the *Equality Act 2010 Code of Practice*:

The policy of the Act is not a minimalist policy of simply ensuring that some access is available to disabled people; it is, so far as is reasonably practicable, to approximate the access enjoyed by disabled people to that enjoyed by the rest of the public. The purpose of the duty to make reasonable adjustments is to provide access to a service as close as it is reasonably possible to get to the standard normally offered to the public at large.

(EHRC, 2011)

8.1.5 Guidance to enhancing compliance – chapter emphasis

In addition to the *Equality Act 2010 Code of Practice* (EHRC, 2011), the Department for Transport (DfT) has produced non-statutory guidance outlining best practice on access to pedestrian and transport infrastructure. The document *Inclusive Mobility* (DfT, 2005), on which much of this chapter is based, is designed to help service providers ensure a barrier-free pedestrian environment. This guide includes information from further sources for other assets.

Items directly related to the geometric design of infrastructure that are addressed in this chapter include

- basic information on human factors (wheelchair, mobility scooter and related characteristics and dimensions)
- tactile paving surfaces
- car parking
- bus stops
- taxi ranks
- access to and within transport-related buildings
- transport buildings: facilities.

Matters related to signage and information, lighting, access in the countryside, consultation and training and management, although important in overall geometric design considerations, are not addressed here; the relevant reports should be consulted for further details.

8.2. Human factors

Approximately 9% of the adult population of the UK have some form of impairment that adversely affects their transport (DfT, 2017a). Many other younger people may also be impaired. In addition are those 'encumbered' (with packages, buggies, small children, etc.), all of whom will benefit from improved facilities design. Models of disability include locomotion, seeing, hearing, reaching, stretching, dexterity and learning disability.

8.2.1 Footway and related dimensions

Footways and footpaths have been identified by external stakeholders as the main reason that access across the network is sometimes difficult for disabled users. It is therefore important that footways and footpaths are constructed and maintained or modified correctly.

The distinction between a footway (see Figure 8.1) and a footpath is that a footway (usually called the pavement in the UK) is the part of a highway adjacent to, or contiguous with, the carriageway, on which there is a public right of way on foot. A footpath has no contiguous carriageway. Where reference is made to one, it can generally be regarded as applying to the

Figure 8.1 Example of a footway (HA, 2010)
© Crown Copyright, 2010
This information is licensed under the Open Government Licence v3.0. To view this licence, visit http://www.nationalarchives.gov.uk/doc/open-government-licence/ **OGL**

other for design. Footway dimensions to accommodate selected mobility needs are shown in Figure 8.2.

8.2.2 Basic human factors (wheelchair and related characteristics and dimensions)

Wheelchair design has developed over many years from wicker basket and steel wheels to more recent lightweight and folding models constructed of aluminium and plastics. In configuration, wheelchairs may now be classified as

- attendant propelled
- electric wheelchair
- new-style manual
- older-style manual
- electric mobility scooter.

The basic dimensions and turning space needed for wheelchairs have stabilised somewhat. The dimensions shown in Figure 8.3, based on the dimensions shown in Table 8.1, are considered to accommodate the majority of wheelchairs.

8.2.3 Reach distances

Reach distances are important where people must touch switches, handles and buttons, such as those at signalled pedestrian crossings. Reach varies with personal physical and mental characteristics and direction of reaching, and can be measured as an arc from shoulder level, sideways and forwards and categorised as comfortable, easy or extended. The

Figure 8.2 Footway dimensions to accommodate selected mobility needs (DfT, 2005)
Crown Copyright, 2005
This information is licensed under the Open Government Licence v3.0. To view this licence, visit http://www.nationalarchives.gov.uk/doc/ open-government-licence/ **OGL**

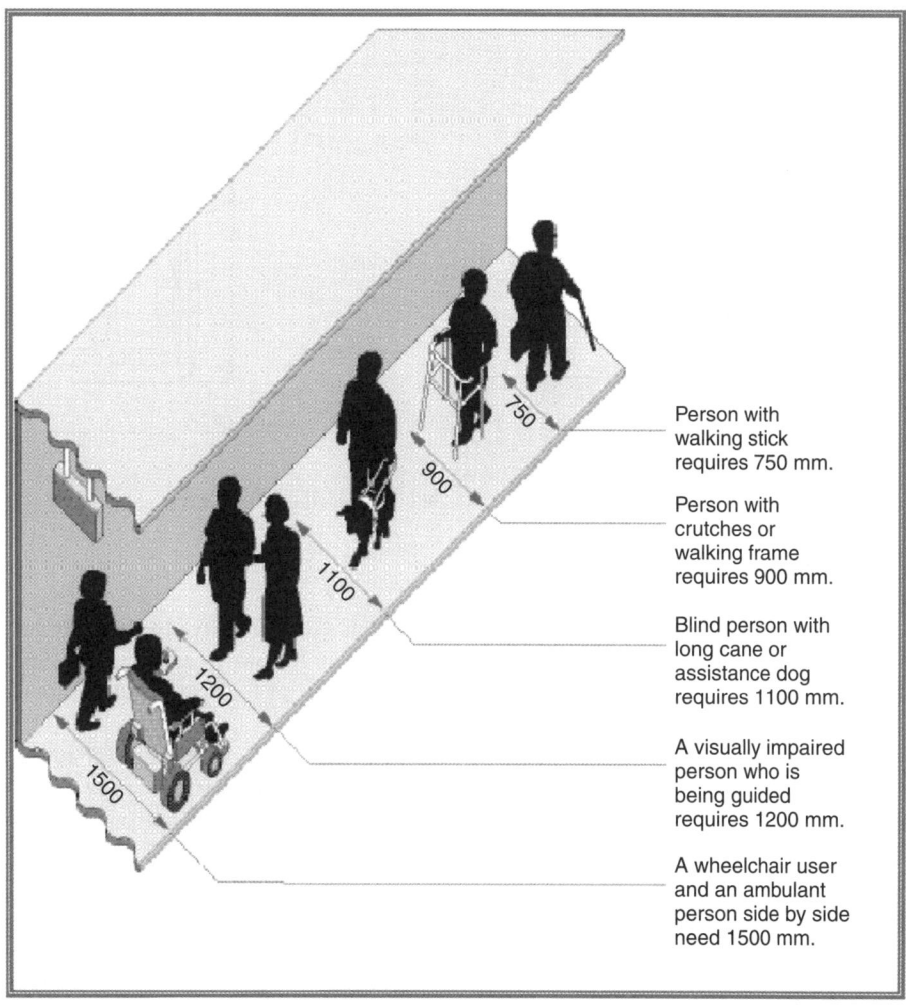

Person with walking stick requires 750 mm.

Person with crutches or walking frame requires 900 mm.

Blind person with long cane or assistance dog requires 1100 mm.

A visually impaired person who is being guided requires 1200 mm.

A wheelchair user and an ambulant person side by side need 1500 mm.

convention used for measuring sideways and upwards reaching is shown conceptually in Figure 8.4; 'comfortable' and 'extended' distances for sideways and frontal movements are listed in Table 8.2.

Reach angles and distances are particularly important when considering the ability to activate pedestrian signal buttons.

8.2.4 Effects of gradient
Generally, an 8% gradient is the permissible maximum for wheelchair use. Most guidelines also agree that 5% is preferred. Steeper than 2.5% is impossible for many manual wheelchair users. The effects of different gradients are shown in Table 8.3.

8.2.5 Walking distances and standing times
Walking distances and related footway features are important considerations and may affect the acceptable distance between crossing points and access to buildings, thereby affecting the geometry of the adjacent infrastructure. Walking distances are recommended as

- wheelchair users, 150 m
- visually impaired, 150 m
- mobility impaired using a stick, 50 m
- mobility impaired without walking aid, 100 m

Limited acceptable standing times may be less than 1 min for some disabled people; nearly 55% could stand for 10 min or

131

Figure 8.3 Wheelchair dimensions, based on DfT (2005) (dimensions in millimetres)

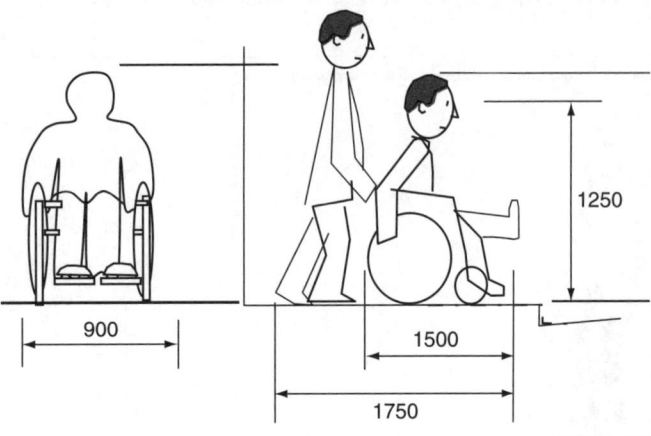

Figure 8.4 Convention for reach distance arc measurement for sideways and upwards reaching, based on DfT (2005)

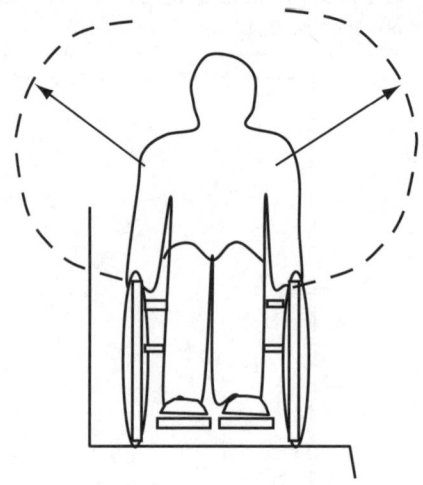

less. Therefore, there is a need to provide frequent seating along pedestrian routes.

8.3. Footways, footpaths and pedestrian areas

Key dimensions for pedestrian facilities that are acceptable in providing inclusive mobility are as follows.

- Widths of footways
  - 2000 mm footway for two wheelchairs to pass comfortably
  - 3000 mm at bus stops
  - 4500 mm at shops
  - 150 mm separation from cycle track (see also LTN 2/04 *Adjacent and Shared Use Facilities for Pedestrians and Cyclists* (DfT, 2004)).
- Gradients
  - 2% can be managed by most people
  - 8% maximum as a general rule, and only over short distances.

- Crossfall on footways
  - 1–2%
  - 2.5% maximum.

Note that crossfall (often the result of installing parking in front gardens) on footways can affect the steering of wheelchairs and that drainage of footways is also important.

In addition to the geometric configuration, appropriate fences, guardrails, seating, barriers, ramps and steps, and street furniture are important in the effective functioning of the completed scheme.

8.4. Electric scooters

Powered mobility vehicles are being used to assist in achieving inclusive mobility, and their dimensions and operating features are relevant to the geometry of pedestrian infrastructure. The two broad categories of these vehicles are powered wheelchairs and mobility scooters.

Table 8.1 Dimensions of wheelchairs and user, based on Stait *et al.* (2001)

Chair type	Length of wheelchair and user		Width of wheelchair and user	
	95th percentile	Maximum	95th percentile	Maximum
Attendant propelled	1197	1318	658	674
Electric wheelchair	1328	11549	706	755
New-style manual chair	1183	1256	702	741
Older-style manual chair	1267	1357	686	722
Electric scooter[a]	1402	1500	658	659

[a] See also Section 8.4.

Table 8.2 Dimensions associated with comfortable and extended reach ranges (BSI, 2009)

Person	Access	Reach angle: degrees	Height (H): mm		Depth (D): mm	
			Comfortable	Extended	Comfortable	Extended
Wheelchair user	Front	+70	1000	1150	90	120
		Horizontal	(750)	97 500	180	230
		−24	650	650	120	200
	Side	+70	1060	1170	100	135
		Horizontal	(750)	(750)	220	310
		−24	665	630	165	230
Ambulant disabled	Front	+70	1500	1625	200	250
		Horizontal	(850)	(850)	280	400
		−24	750	700	180	310

Notes:
1. Dimensions are rounded to nearest 5 mm.
2. Dimensions in brackets are for the horizontal reference plane.
3. It is assumed that any keyhole allows full reach capabilities.
4. Maximum heights are measured from the + 70 line; minimum heights from the −24 line.
5. For some activities, the recommended dimensions in the standard are extended beyond those resulting from the research trials on the basis of accepted practice.
Crown Copyright, 2009
This information is licensed under the Open Government Licence v3.0. To view this licence, visit http://www.nationalarchives.gov.uk/doc/open-government-licence/ **OGL**

8.4.1   Powered wheelchairs

Powered wheelchair configurations are similar in many respects to manual wheelchairs but differ in that they accommodate a battery and electric motor. Control of steering is generally by a small joystick. The greater power of many models over manual wheelchairs may enable other features to be added, such as kerb climbers and adjustable seats.

Table 8.3 Effects of gradient (HA, 2010)

Gradient	Effect
1% (1 in 100)	Never an obstacle
2% (1 in 50)	Can be managed by most people and also provides good drainage
2.5% (1 in 40)	Can be managed by many people
≥2.5%	Impossible for many manual wheelchair users
8% (1 in 12)	Absolute maximum; not only is the physical effort of getting up a steeper gradient beyond many wheelchair users, but there is also a risk of the wheelchair toppling over

Crown Copyright, 2010
This information is licensed under the Open Government Licence v3.0. To view this licence, visit http://www.nationalarchives.gov.uk/doc/open-government-licence/ **OGL**

8.4.2   Mobility scooters (DPTAC, 2007)

Scooters have three or four wheels and are steered by a handlebar on which are mounted most of the necessary controls and warning devices. A battery pack drives two or all of the wheels. Shopping or other bags or containers may be attached to the front or the rear. Larger models may be equipped with all-weather protection, such as plastic frames and canopies.

Legislation separates scooters and powered wheelchairs into categories: Class 2 vehicles and Class 3 vehicles. From a technical viewpoint, they are classified as 'invalid carriages'.

Class 2 vehicles, sometimes referred to as 'pavement vehicles', are designed for use on the footway. They are not allowed on roads other than to cross them and are limited to a maximum speed of 4 miles per hour.

Class 3 vehicles tend to be larger than Class 2 vehicles and are capable of exceeding 4 miles per hour but cannot exceed 8 miles per hour. These vehicles are not permitted to exceed 4 miles per hour on footways. Class 3 vehicle speeds can be switched between 4 and 8 miles per hour. Although allowed on roads, they are not allowed on motorways, cycle lanes or bus lanes. They are required to have lights, indicators, a horn, a rear-view mirror and rear reflectors. Drivers must be disabled and aged 14 or over, but do not require a driving licence to operate the scooter.

Dimensions of mobility scooters are specified in ISO 7193 (ISO, 1985). However, it should be noted that scooter dimensions have tended to increase over the years. The report *Carriage of Mobility Scooters on Public Transport* (DfT, 2006a) lists dimensions that are the same as the reference wheelchair; these give an indication of current trends. The dimensions are

- height, 1350 mm
- length, 1200 mm
- width, 700 mm
- weight (including occupant), less than or equal to 300 kg.

In addition, it is stated that a scooter should have a maximum turning circle of 1500 mm in order to fit into the designated space of a public transport vehicle.

8.5. Fences and guardrails

Steep slopes or drops at the rear of footways require precautions to prevent wheelchair users running over the edge or blind or partially sighted people walking over it. Guardrails and barriers at the side of or across footways should be at least 1100 mm high and preferably 1200 mm, measured from ground level. Colour contrast of guardrails with their surroundings is also necessary.

Guardrails should also be designed to prevent guide dogs from walking under the rails, but there should be sufficient openings between vertical members to ensure that children and wheelchair users can see, and be seen, through the railings. The top rail should have a smooth profile and, if intended to provide support, should be circular with a diameter of between 40 and 50 mm.

There should also be an upstand, a minimum of 150 mm in height, at the rear of the paved area, which can then act as both a tapping rail for long cane users, and a safeguard for wheelchair users.

BS 7818 (BSI, 1995) includes more detailed information on this subject; Figure 8.5 illustrates key features of guardrails and fencing.

8.6. Other features of walking areas

As well as the footways, barriers and fences, most footways will have other features, including seating, staggered barriers, ramps and steps, signs and street furniture, including bollards. All these must be positioned and dimensioned consistently with the geometric design of the footway and related facilities to enable their identification and avoidance by disabled and other people.

8.6.1 Streetworks

Although not usually a permanent feature of the geometric design of pedestrian facilities, guidance has been formulated on key features of temporary works that affect footways and related facilities. Such features include: white marking on scaffolding; continuous, colour-contrasted barriers; barriers and tapping rails at a minimum height; a minimum footway width; and a minimum scaffolding clearance. Major features and dimensions are shown in Figure 8.6.

8.6.2 Road crossings

The recommended layout of a location for pedestrians to cross is shown diagrammatically in Figure 8.7. The location of the tactile surface and the gradients of the ramp on the footway side slopes are shown.

NOTE

The location of this crossing some distance from the corner, thereby possibly adversely affecting the pedestrian's ability to see oncoming vehicles and also interrupting the pedestrian's desire line, may not be ideal. Refer to Chapter 6 for further information.

Information sources providing further details include

- TAL 4/91 *Audible and Tactile Signals at Pelican Crossings* (DfT, 1991a)
- TAL 5/91 *Audible and Tactile Signals at Signal-controlled Crossings* (DfT, 1991b)
- LTN 2/95 *The Design of Pedestrian Crossings* (DfT, 1995a)
- *Interim Changes to the Guidance on the Use of Tactile Paving Surfaces* (DfT, 2015).

8.6.3 Tactile paving surfaces

Whenever a footway or other area used by pedestrians is being constructed, repaired or renewed, consideration should be given to incorporating any appropriate tactile surfaces.

An outline of the main features, the dimensions, colours and configuration of which are detailed in the relevant specifications, is as follows (DfT, 2005).

- At pedestrian crossing points
  - parallel rows of flat-topped blisters – note variations in layout, dimensions and colour for controlled and uncontrolled crossings.
- Hazard warning surface
  - corduroy, *not red* and bars should run transversely to direction of pedestrian travel.
- Off-street platform-edge warning
  - offset rows of flat-topped domes, not red, laid parallel to edge of platform.
- On-street platform-edge warning
  - rows of lozenge shapes with rounded edges to avoid trip hazard, laid at light rail transport platform edges immediately behind edge coping stone.

Figure 8.5 Key features of guardrails and fencing (DfT, 2005)
Crown Copyright, 2005
This information is licensed under the Open Government Licence v3.0. To view this licence, visit http://www.nationalarchives.gov.uk/doc/open-government-licence/ **OGL**

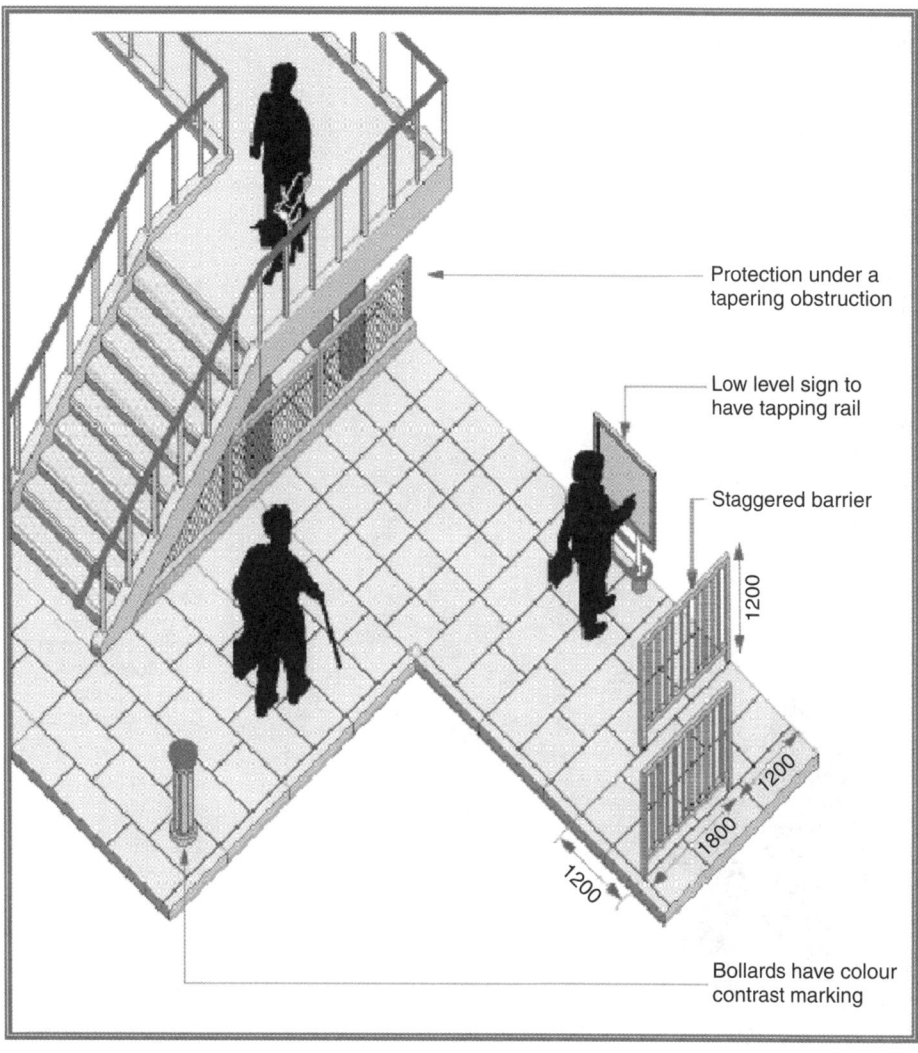

- Segregated shared cycle track and footway surface centre delineator strip
  - start of pedestrian surface – ladder pattern of flat-topped bars
  - start of cycle surface – same but laid parallel to direction of travel
  - centre delineator strip 150 mm wide, white, full length of segregation.
- Guidance path surface
  - raised flat-topped bars, not red, in direction of pedestrian travel
  - used for guidance between property line and carriageway, around obstacles and possibly in transport terminals.

- Information surface
  - not raised but consists of softer material such as neoprene or rubber
  - used to draw attention to bus stops, information boards, ticket offices, toilet entrances, waiting rooms.

Where pedestrians cross at uncontrolled crossing points and where geometric design must be related to pedestrians' and drivers' sight lines, examples of the configuration and dimensions of the blister surfaces are provided (HA, 2010), as shown for

- a typical concrete crossover (Figure 8.8)
- an inset crossing point (Figure 8.9(a))

Figure 8.6 Clearance and related dimensions of streetworks (DfT, 2005)
Crown Copyright, 2005
This information is licensed under the Open Government Licence v3.0. To view this licence, visit http://www.nationalarchives.gov.uk/doc/open-government-licence/ **OGL**

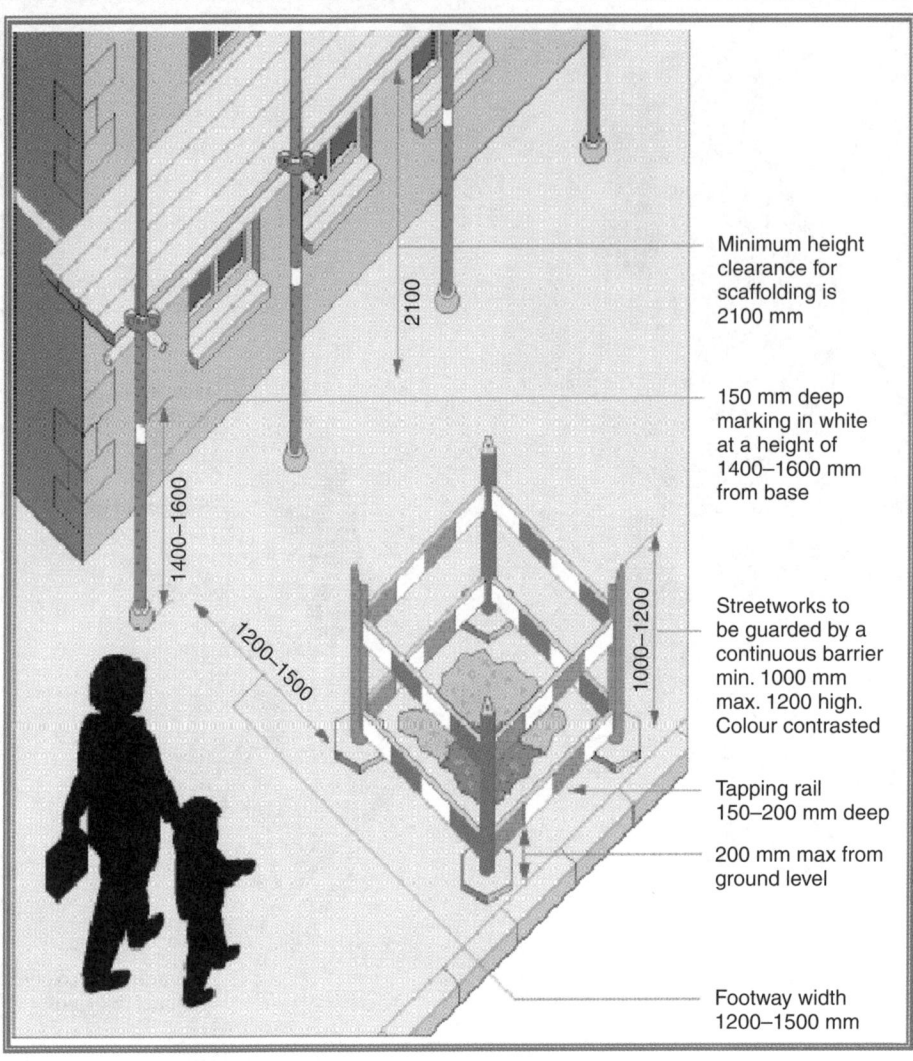

- an in-line crossing point (Figure 8.9(b))
- the layout of blister surfaces at an inset uncontrolled crossing at an acute angle junction (Figure 8.9(c))
- a standard refuge less than 2 m wide (Figure 8.10(a))
- a standard refuge more than 2 m wide (Figure 8.10(b)).

NOTE
Regarding Figure 8.9(a), the location of this crossing some distance from the corner, thereby possibly adversely affecting the pedestrian's ability to see oncoming vehicles and also interrupting the pedestrian's desire line, may not be ideal. Refer to Chapter 6 for further information. Moreover, the placement of the crossing some distance back from the footway in the major road can cause

difficulties for people using wheelchairs or motor scooters because of the additional manoeuvring required at the corner. Figure 8.9(c) shows a placement of crossing points that could be extremely dangerous for pedestrians if sight lines to approaching vehicles are inadequate.

Surface quality is important and affects the ability of disabled and other pedestrians to conveniently and safely move through the entire transport system. Some key concerns related to surface quality include the following.

- Joints between flags should be flush in order to avoid tripping and uneven progress of wheelchairs.

Figure 8.7 Layout of a location for pedestrians to cross (DfT, 2005)
Crown Copyright, 2005
This information is licensed under the Open Government Licence v3.0. To view this licence, visit http://www.nationalarchives.gov.uk/doc/open-government-licence/ **OGL**

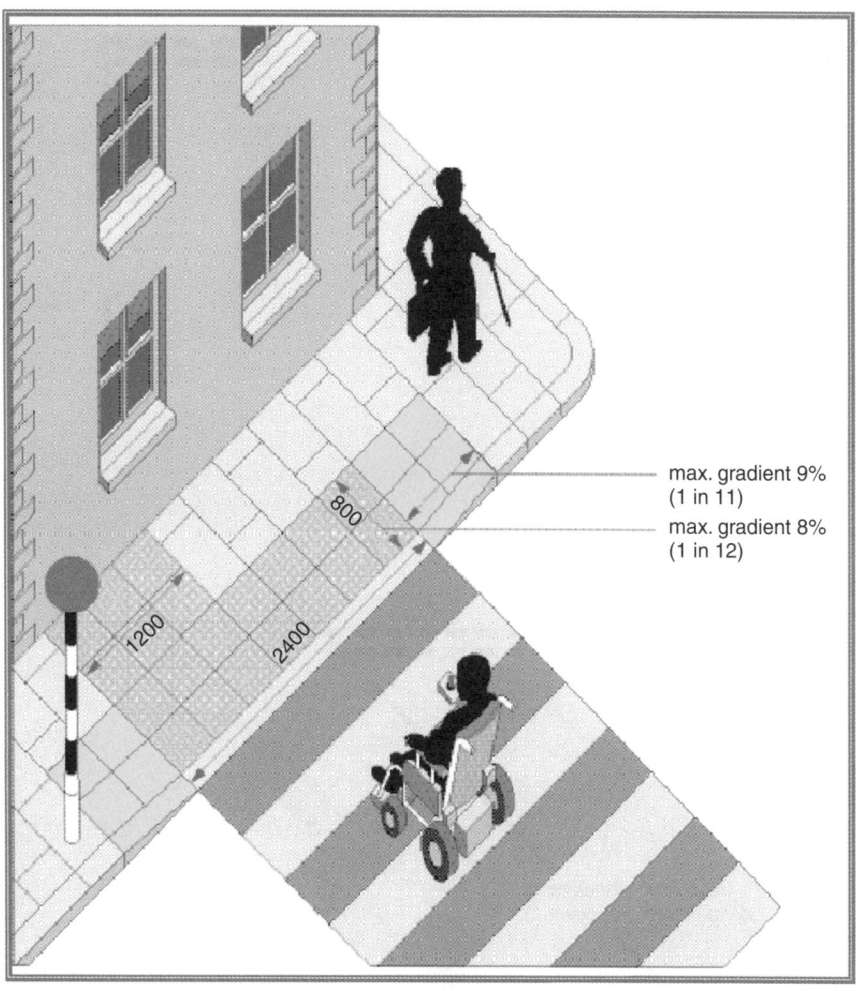

Figure 8.8 Plan of a typical concrete crossover (HA, 2010)
Crown Copyright, 2010
This information is licensed under the Open Government Licence v3.0. To view this licence, visit http://www.nationalarchives.gov.uk/doc/open-government-licence/ **OGL**

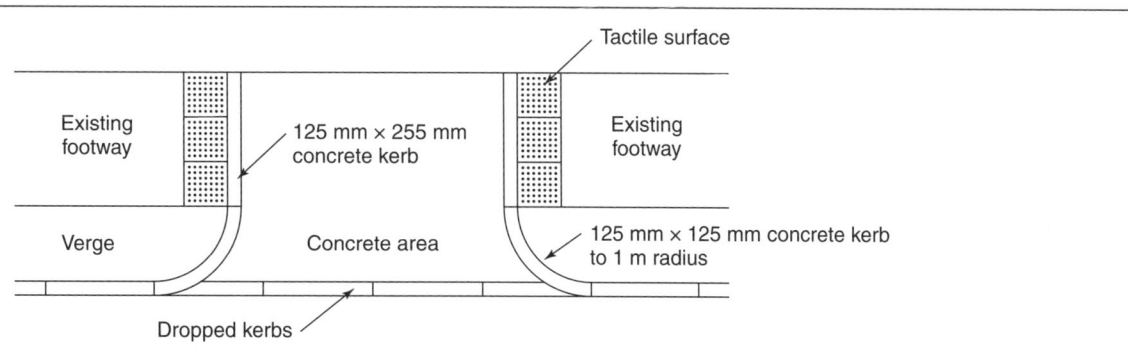

Figure 8.9 (a) Inset crossing point; (b) in-line crossing point; (c) layout of blister surfaces at inset uncontrolled crossing at acute angle junction (HA, 2010) (dimensions in millimetres)
Crown Copyright, 2010
This information is licensed under the Open Government Licence v3.0. To view this licence, visit http://www.nationalarchives.gov.uk/doc/open-government-licence/ **OGL**

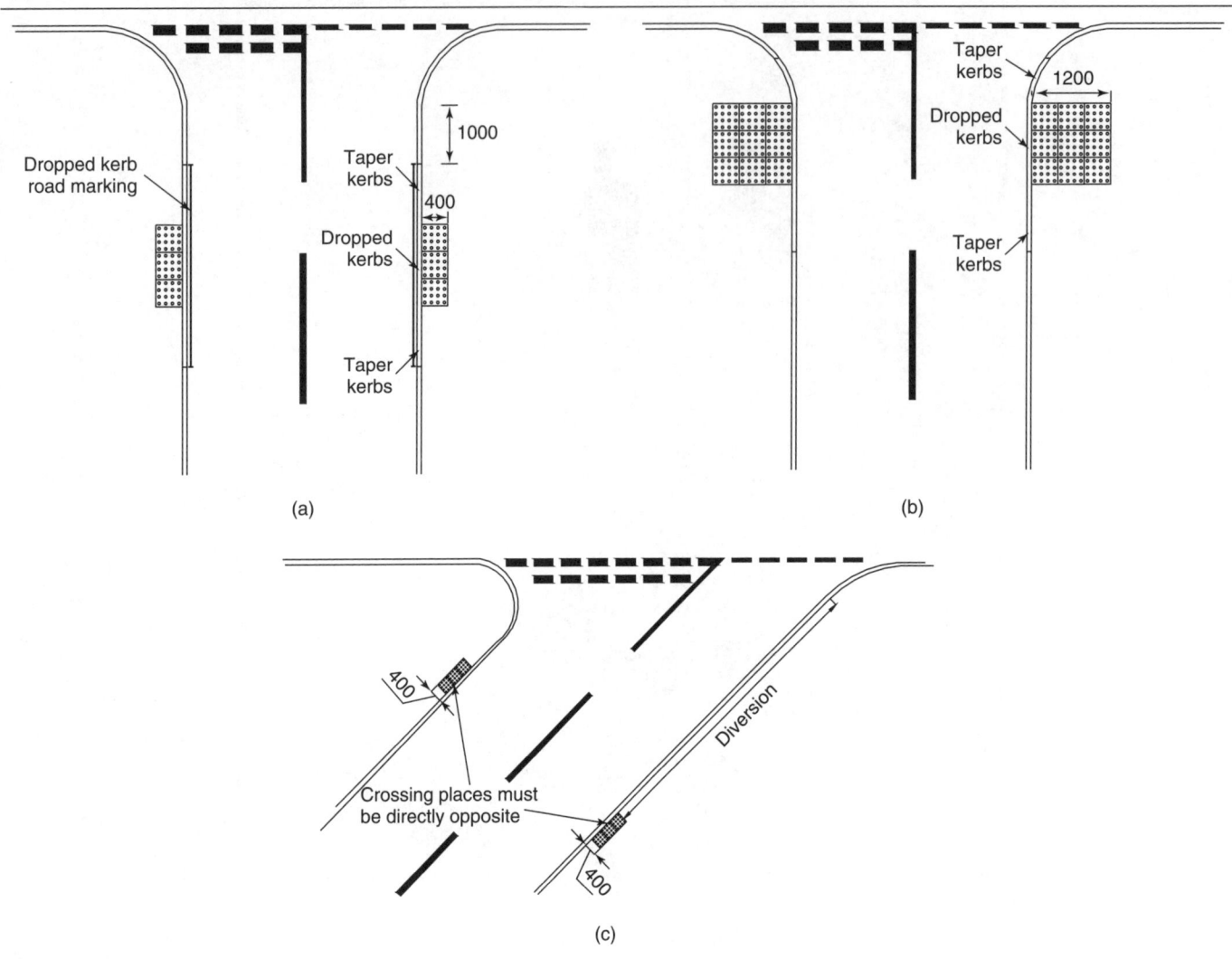

- New cobbled surfaces are undesirable, as they are difficult to negotiate by pedestrians and wheelchairs.
- Covers and gratings should not be of excessive size, in order to minimise areas of possible slipping or differing surface texture.
- Surfaces should be slip resistant.
- Dished drainage channels should not be used.
- Inspections chamber covers should be flush with the surface.

8.6.4   Controlled crossings
The layout and dimensions of zebra and signalled crossings must be adequate, both in the approach configurations and in the kerb details and widths of marked crossings for the passage of disabled people, importantly enabling people with wheelchairs or mobility scooters to pass when travelling in opposite directions. Details of such layouts and dimensions are provided (HA, 2010); an example of a puffin crossing showing the main features and dimensions is shown in Figure 8.11.

8.7.   Transfer to and from other modes of transport
Because of the need for access by disabled as well as non-disabled pedestrians to other modes of transport, including cars, buses, trains and light rail services (trams), the required layout of transfer space must be consistent with these needs.

Figure 8.10 (a) Standard refuge less than 2 m wide; (b) standard refuge more than 2 m wide (HA, 2010)
Crown Copyright, 2010
This information is licensed under the Open Government Licence v3.0. To view this licence, visit http://www.nationalarchives.gov.uk/doc/open-government-licence/ **OGL**

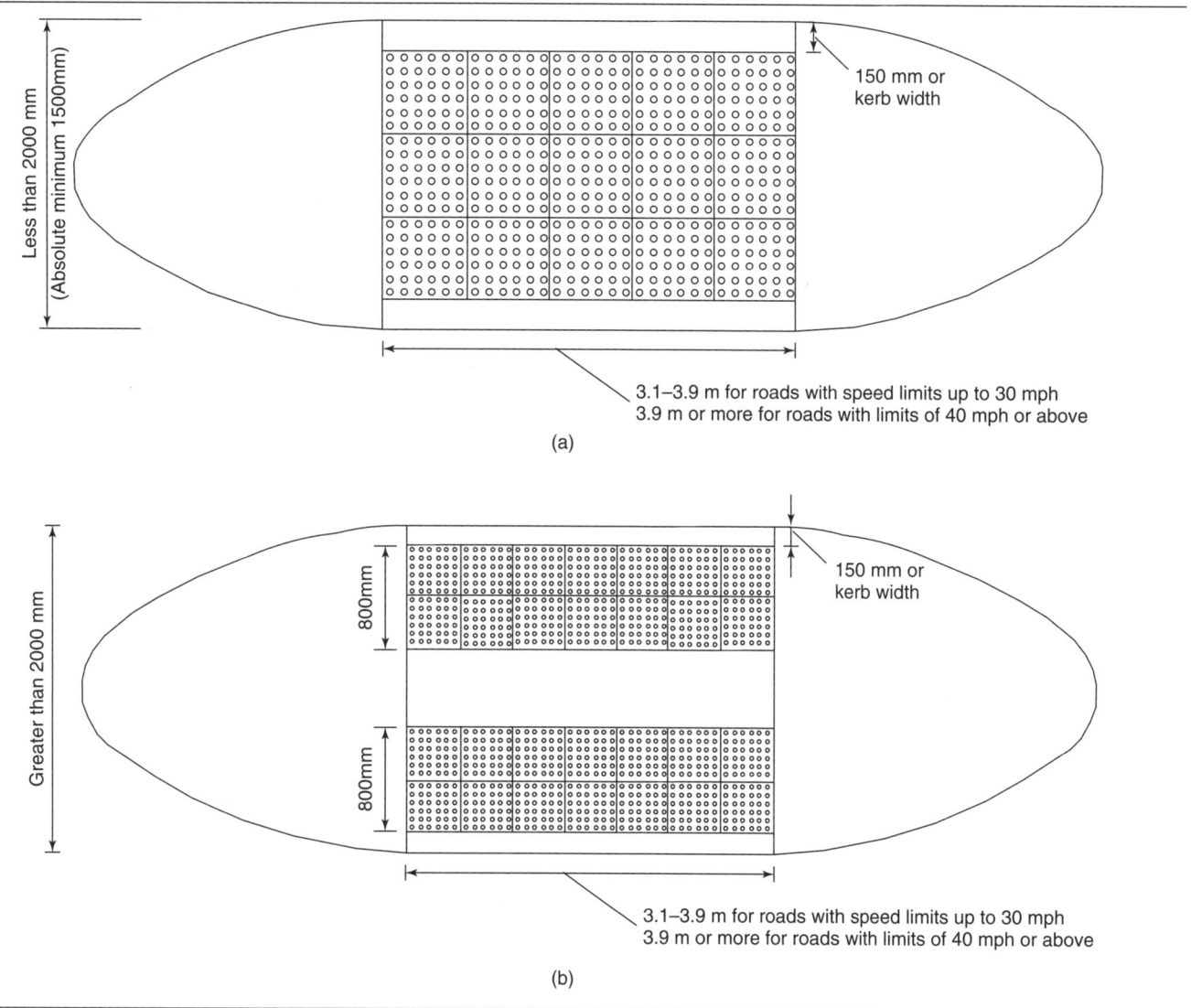

8.7.1 Car parking space

Parking spaces should be provided for disabled people who hold the 'blue badge', preferably a maximum of 50 m from the destination, with a 5% maximum gradient. Tactile surfaces should be used to distinguish between carriageway and footway. The minimum number of disabled spaces should be 2% of the total number of spaces for existing employment premises with an additional minimum of one space for each disabled employee. For new employment premises, disabled parking spaces should make up 5% of the total, while in areas for the general public there should be one disabled space for each disabled employee plus 6% of the total number of spaces for

disabled visitors. At places where groups of disabled people are expected, a greater number of spaces should be provided.

TAL 5/95 *Parking for Disabled People* (DfT, 1995b) gives detailed advice on the provision and design of parking for disabled car users, as does BS 8300 (BSI, 2009), which should be consulted for further details. Bays should conform to the layout and dimensions shown in Figure 8.12 for off-street angle parking, on-street parking at an angle to the kerb and on-street parking parallel to the kerb. For car parks associated with shopping areas, leisure or recreational facilities and other places open to the general public, a minimum of one parking space for

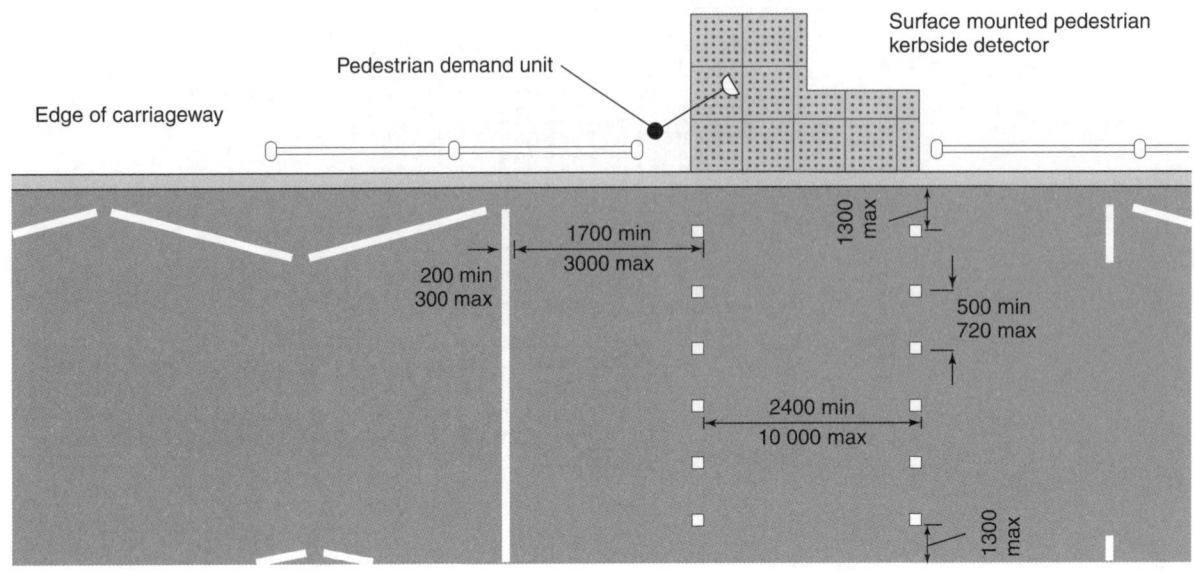

each employee who is a disabled motorist plus 6% of the total capacity for visiting disabled motorists must be allocated for disabled parking.

It may be necessary to provide greater numbers of designated spaces at hotels and sports stadia that specialise in accommodating groups of disabled people.

At railway stations, the Strategic Rail Authority (DfT, 2006b) recommends the following.

- Fewer than 20 spaces, a minimum of *one* reserved space.
- 20 to 60 spaces, a minimum of *two* reserved spaces.
- 61 to 200 spaces, *6%* of capacity, with a minimum of *three* reserved spaces.
- More than 200 spaces, *4%* of capacity, plus *four* reserved spaces.
- An additional space should also be provided for any railway employee who is a disabled motorist.

Parking bays should be marked in accordance with the Traffic Signs Regulations and General Directions 2016 for signs and road surfaces, yellow lines and disabled symbols. Directional signs and distances to key points must be indicated. Control equipment should feature pay-and-display meters conforming to BS 8300 (BSI, 2009). Sufficient clear space for users must be provided at meter locations and maximum allowable vehicle heights should also be indicated.

8.7.2 Bus boarders and shelters (TfL, 2006)
8.7.2.1 General requirements
Many disabled pedestrians, including people in wheelchairs and those employing other devices, rely on bus services for business, shopping and social trips. Of primary importance in enabling buses to be used effectively and safely are the waiting and boarding arrangements.

All buses (but not coaches until 2020) must have accessibility standards, including a wheelchair bay, a wheelchair ramp or other boarding device and priority seating. It will be a criminal offence under the Equalities Act 2010 for a bus not to comply with these access regulations (DVSA, 2016). To comply with these regulations, the geometric design of the bus stop area must facilitate any necessary movements and manoeuvres.

8.7.2.2 Boarders
3.3.2.1 Bus boarders are generally built out from the existing kerb line and provide a convenient platform for boarding and alighting passengers. A raised bus boarding area assists passengers boarding/leaving the vehicle and may enable some wheelchair users to board directly without using a ramp.

3.3.2.4 There are two conventional types of bus boarder: full width and half width. A full-width boarder juts out into the carriageway far enough for the bus to avoid parked vehicles, approximately 1800 mm [see Figure 8.13].

Figure 8.12 Parking bay layout dimensions (DfT, 2005) (dimensions in millimetres)
Crown Copyright, 2005
This information is licensed under the Open Government Licence v3.0. To view this licence, visit http://www.nationalarchives.gov.uk/doc/open-government-licence/ **OGL**

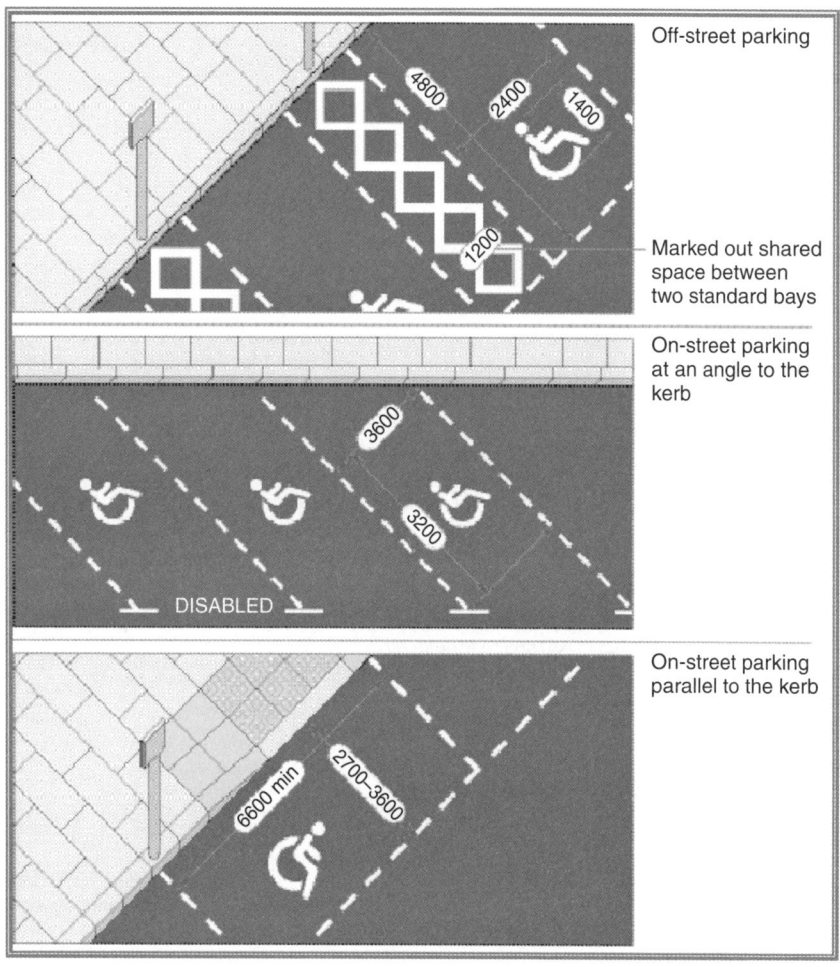

Off-street parking

Marked out shared space between two standard bays

On-street parking at an angle to the kerb

On-street parking parallel to the kerb

Figure 8.13 Full-width boarder bus stop (HA, 2010)
Crown Copyright, 2010
This information is licensed under the Open Government Licence v3.0. To view this licence, visit http://www.nationalarchives.gov.uk/doc/open-government-licence/ **OGL**

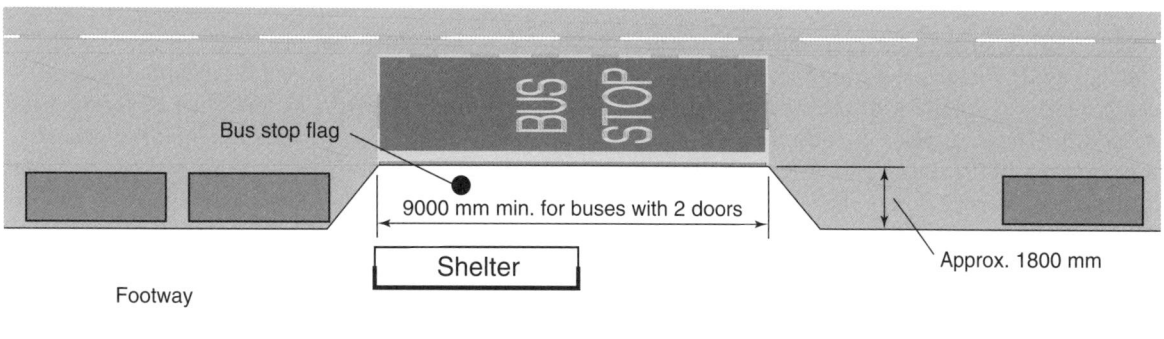

Bus stop flag

9000 mm min. for buses with 2 doors

Shelter

Approx. 1800 mm

Footway

Figure 8.14 Standard bus shelter design (HA, 2010)
Crown Copyright, 2010
This information is licensed under the Open Government Licence v3.0. To view this licence, visit http://www.nationalarchives.gov.uk/doc/open-government-licence/ **OGL**

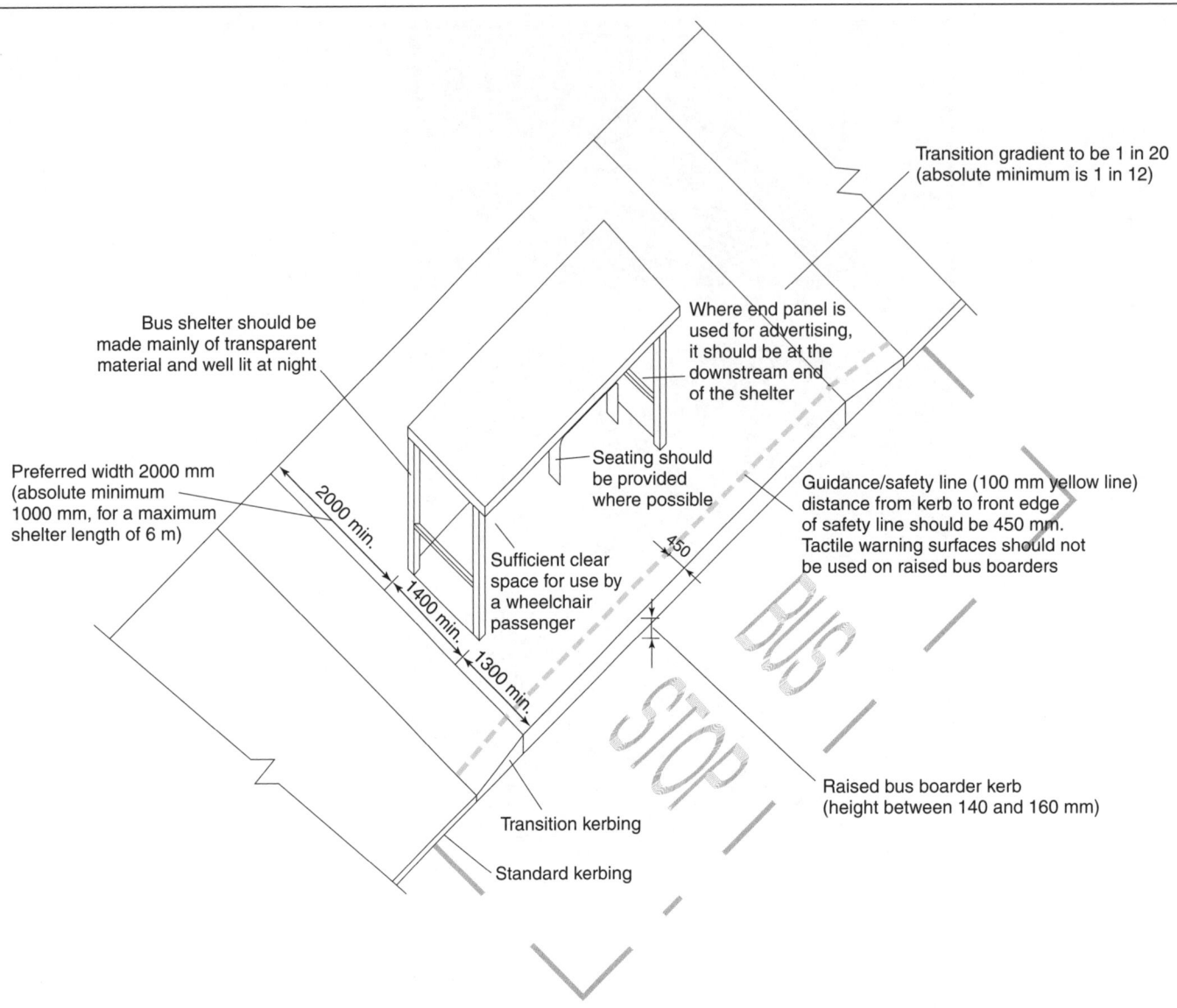

3.3.2.5 A half width boarder, which juts out by between 500 mm and 1500 mm, is a compromise design that can be used where a full-width boarder would unduly delay other traffic or place the bus in or too close to the opposing traffic stream [...] A further alternative is an angled boarder: wedge shaped from up to 2000 mm into the carriageway and tapering back to the original kerb line over the length of the bus stop cage [...] This design is similar to the shallow saw tooth layout used in some bus stations.

(HA, 2010)

8.7.2.3 Bus stop shelters

From the point of view of disabled passengers, particularly wheelchair users, the best location for a shelter is opposite the boarding point. If space constraints mean that this is not possible, an alternative is to place the shelter downstream, leaving 2000 mm length of clear boarding/alighting area. In locations not exposed to severe weather, a cantilever bus shelter with one end panel offers good accessibility and some weather protection.

Figure 8.15 Rail station features (DfT, 2005)
Crown Copyright, 2005
This information is licensed under the Open Government Licence v3.0. To view this licence, visit http://www.nationalarchives.gov.uk/doc/ open-government-licence/ **OGL**

[As shown in Figure 8.14] there should be sufficient space either to the rear of the shelter, or in front of it if the shelter has to be placed at the back of the pavement, to allow easy pedestrian movement. Where shelters are provided in newly built areas, there should be a clear obstacle free footway [width] of at least 2000 mm, [preferably] 3000 mm. Where there are physical constraints, a clear footway width of 1500 mm is acceptable, with an absolute minimum of 1000 mm over a maximum distance of 6 m.

(HA, 2010)

8.7.3 Rail station features

The intent of rail station features for disabled people is to ensure safe movement along platforms and to indicate the proximity and presence of the edge of the platform. A tapping rail is therefore provided at the rear of the platform and tactile paving a minimum of 500 mm from the platform's edge. A white line, at least 100 mm wide, is provided at the platform's edge. These details are shown in Figure 8.15.

8.7.4 Taxi ranks (DfT, 2005)

Key points associated with the provision of taxi services are as follows.

■ The Disability Discrimination Act 2005 required further wheelchair accessible taxis to be provided.
■ Provide ranks at railway, bus, coach stations, major attractions and retail areas.

Figure 8.16 General design of a ramp (HA, 2010)
Crown Copyright, 2010
This information is licensed under the Open Government Licence v3.0. To view this licence, visit http://www.nationalarchives.gov.uk/doc/open-government-licence/ **OGL**

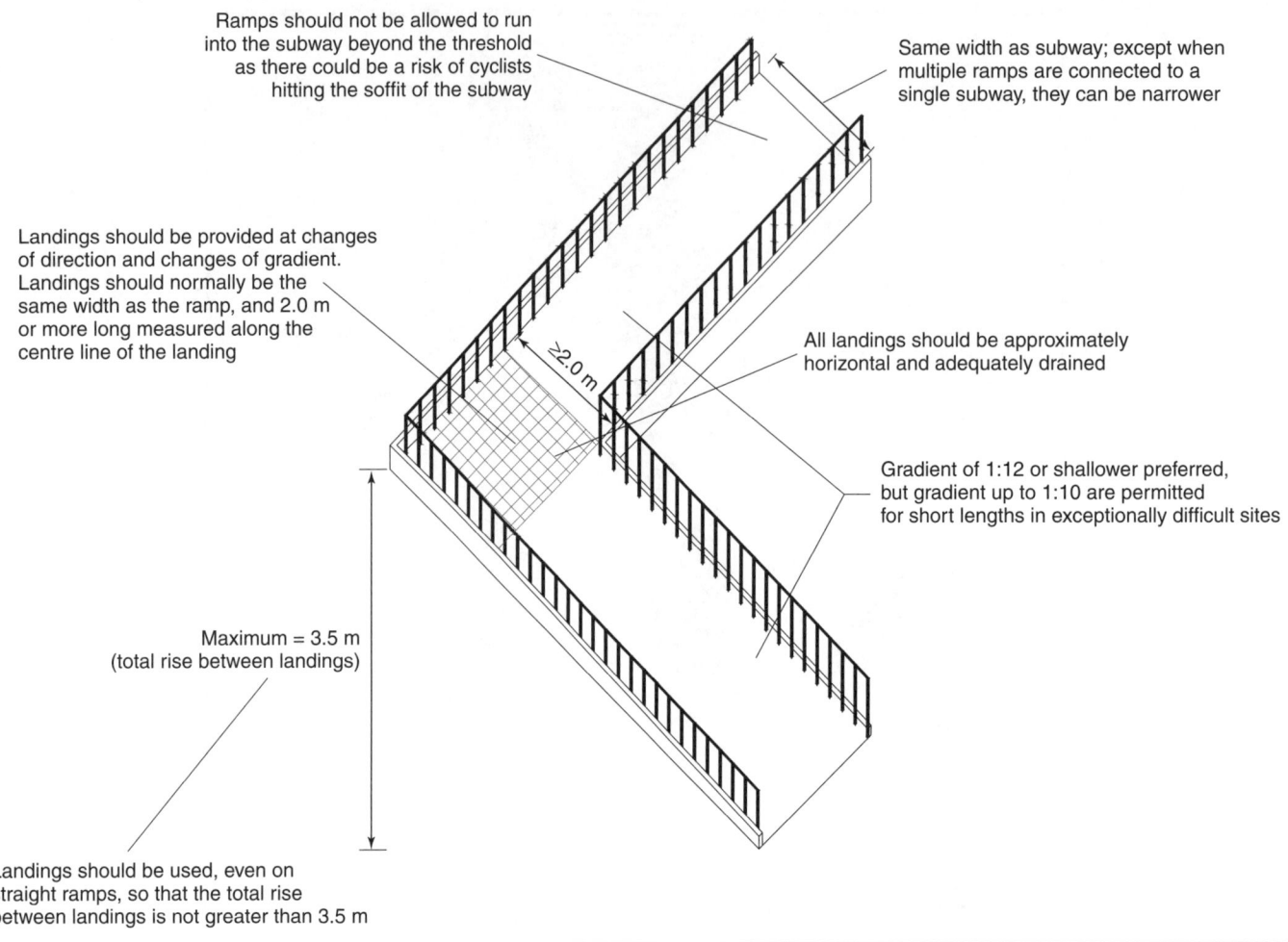

Ramps should not be allowed to run into the subway beyond the threshold as there could be a risk of cyclists hitting the soffit of the subway

Same width as subway; except when multiple ramps are connected to a single subway, they can be narrower

Landings should be provided at changes of direction and changes of gradient. Landings should normally be the same width as the ramp, and 2.0 m or more long measured along the centre line of the landing

≥2.0 m

All landings should be approximately horizontal and adequately drained

Gradient of 1:12 or shallower preferred, but gradient up to 1:10 are permitted for short lengths in exceptionally difficult sites

Maximum = 3.5 m (total rise between landings)

Landings should be used, even on straight ramps, so that the total rise between landings is not greater than 3.5 m

- Ranks are to be located to enable access from the footway to the nearside of taxis.
- Wheelchair ramps up to 1620 mm and manoeuvring space, to total 4040 mm should be provided.
- Dropped kerbs or raised road crossings should be provided adjacent to the rank.
- Ranks should be clearly signed and have seating close by.
- Signs should state times of taxi services, if not continuous.
- Telephone numbers for taxi services are useful, especially when embossed, particularly to partially sighted users.

8.7.5 Ramps, footbridges and stairs

Many railway stations and other pedestrian transfer facilities incorporate a ramp or stairs, or both, between levels. The associated dimensions, including those for rest areas and gradients, are important in ensuring their acceptability for use by

disabled people. Key features of ramps, which afford adequate access to subways or underpasses, as shown in Figure 8.16, include the following.

- All landings should be approximately horizontal, and adequately drained.
- Gradients of 1 : 20 or shallower are preferred for access ramps where significant numbers of disabled persons or heavily laden shoppers are expected to use the subway. In other situations gradients shallower than 1 : 12 are preferred, but gradients up to 1 : 10 are permitted for short lengths in exceptionally difficult sites. Stepped ramps may also be considered at exceptionally difficult sites although wheelchair users find stepped ramps difficult to negotiate.

(HA, 2010)

Figure 8.17 General view of a stair (HA, 2010)
Crown Copyright, 2010
This information is licensed under the Open Government Licence v3.0. To view this licence, visit http://www.nationalarchives.gov.uk/doc/open-government-licence/ **OGL**

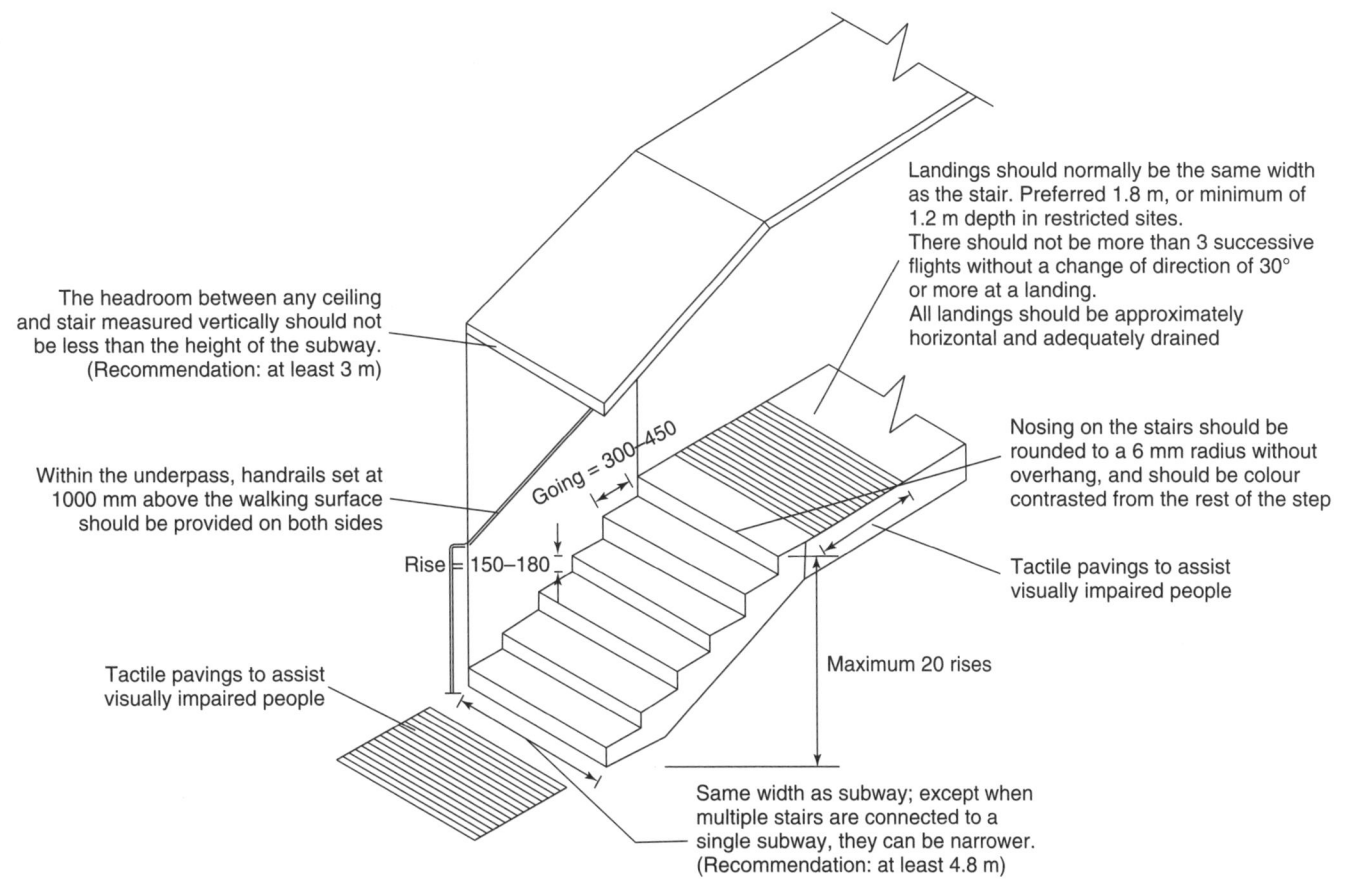

The headroom between any ceiling and stair measured vertically should not be less than the height of the subway. (Recommendation: at least 3 m)

Within the underpass, handrails set at 1000 mm above the walking surface should be provided on both sides

Tactile pavings to assist visually impaired people

Going = 300–450

Rise = 150–180

Landings should normally be the same width as the stair. Preferred 1.8 m, or minimum of 1.2 m depth in restricted sites.
There should not be more than 3 successive flights without a change of direction of 30° or more at a landing.
All landings should be approximately horizontal and adequately drained

Nosing on the stairs should be rounded to a 6 mm radius without overhang, and should be colour contrasted from the rest of the step

Tactile pavings to assist visually impaired people

Maximum 20 rises

Same width as subway; except when multiple stairs are connected to a single subway, they can be narrower. (Recommendation: at least 4.8 m)

For footbridges:

The minimum clear width of the bridge footway, ramps and stairs, which shall be not less than 2 m, shall be derived on the following basis to meet the peak pedestrian flows:

(a) on the level or up to 1 in 20 gradient: 300 mm of width per 20 persons per minute
(b) on steps or ramps steeper than 1 in 20 gradient: 300 mm of width per 14 persons per minute.

(DfT, 2017b)

Stairs must conform to required configurations and dimensions, as shown in Figure 8.17.

8.7.6 Access to and within transport-related buildings

The geometric design of footways must be coordinated with and be consistent with access to and from adjacent buildings and other facilities, such as bus and rail interchanges. Points associated with providing access to and within transport-related buildings, as shown in Figure 8.18, include the notion that layouts should be as compact as possible and the need to ensure that interchanges between modes are fully accessible. Conflicts between pedestrians and motorised vehicles should be minimised. This can often be achieved by a 'horseshoe' layout of footways where pedestrians do not have to cross vehicles' paths, associated with an 'island' type of layout.

As with all interchange facilities, clear signage must be provided, indicating walkways, platforms, assembly locations and directions between modes, platforms and all other points of access and service. Within buildings, the applicable dimensions and guidance methods must be adhered to in the provision of passageways and moving walkways and at changes of level, including steps, stairs, ramps, handrails, escalators, lifts, footbridges, tunnels and underpasses, and at platforms for bus and rail facilities, both off- and on-street (DfT, 2005).

Figure 8.18 Key features of doorways, passageways and counter spaces, based on DfT (2005) (dimensions in mm)
Crown Copyright, 2005
This information is licensed under the Open Government Licence v3.0. To view this licence, visit http://www.nationalarchives.gov.uk/doc/open-government-licence/ **OGL**

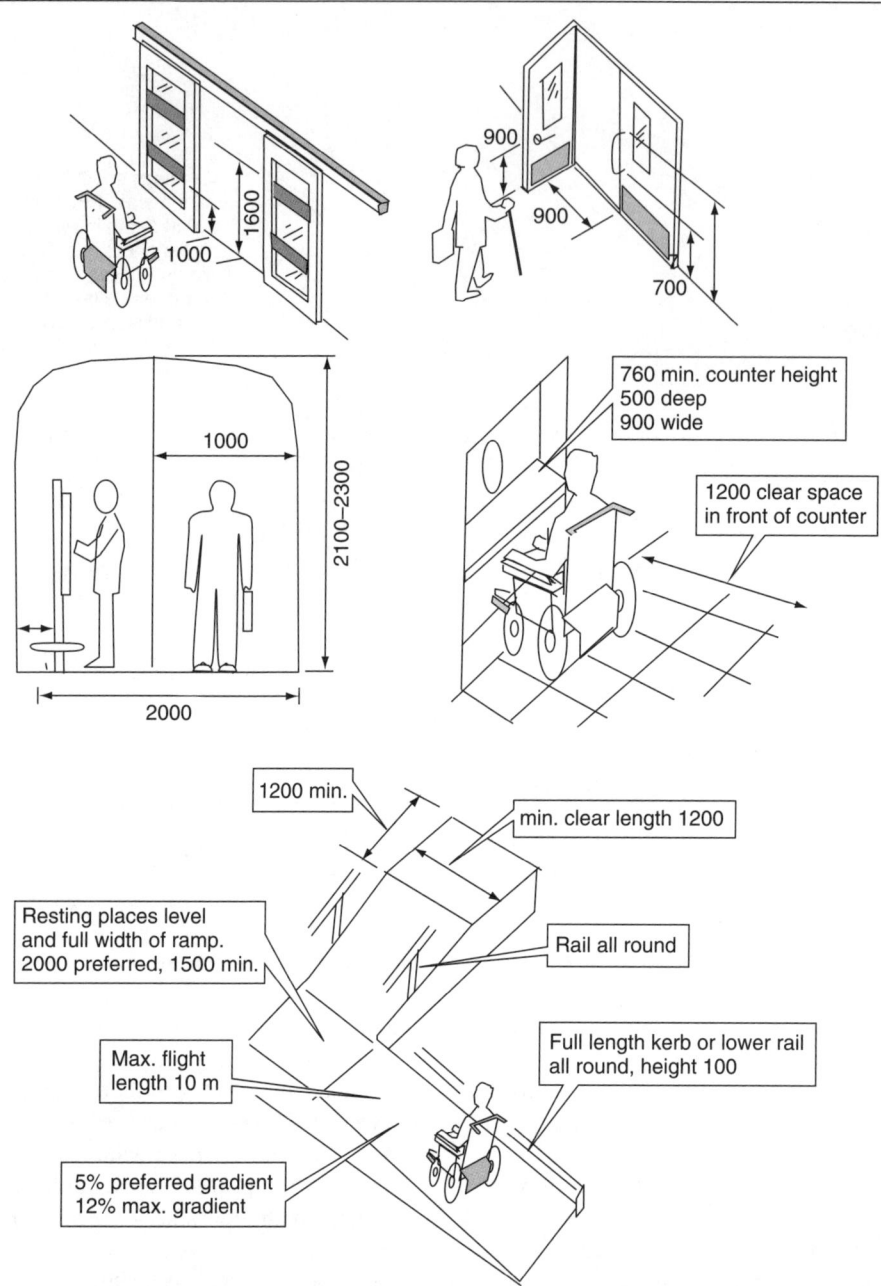

Other areas that should be consistent with the geometric design of larger areas include: locations of telephones (note also height requirements); seating and waiting areas; waiting and refreshment rooms; luggage storage facilities; toilets (see also BS 8300 (BSI, 2009)); and provision of facilities for assistance dogs.

8.8.   Shared-space streets

Shared-space streets related to roadway layouts are described in Chapter 5. Particular aspects affecting their use by disabled people are addressed in this section. A definition of a shared-space street is 'A street or place designed to improve pedestrian movement and comfort by reducing the dominance of motor

Figure 8.19 Example of shared space: High Street, Stonehouse, Gloucestershire (DfT, 2011)

vehicles and enabling all users to share the space rather than follow the clearly defined rules implied by more conventional designs'. An example is shown in Figure 8.19 (DfT, 2011).

It should be noted that the term 'shared surface' means that there is no or negligible kerb height between pedestrian surfaces and those for motorised vehicles. A shared-surface area, therefore, can be extremely difficult for blind or visually impaired people to negotiate because the absence of a kerb means that the use of a cane to indicate where pedestrian facilities begin and end is impossible, and guide dogs are unable to identify safe crossing places.

Concerns have been expressed by organisations for disabled people over the implementation of shared-surface projects, including Holmes (2015). Much effort has been made by these organisations to acquaint design agencies of these concerns. Boxes 8.1 and 8.2 summarise some of the key issues (JCMBPS, 2005).

8.9. Kerb height investigations

8.9.1 Recommendations from research

A key concern for many visually impaired people and people who use wheelchairs is the matter of kerb height. Essentially, for wheelchair users and users of similar wheeled devices, the kerb should be low enough to mount comfortably and with an acceptable level of effort, while, for visually impaired people, it should be high enough for it to be identified visually or by use of a cane.

As a result of investigations into acceptable kerb heights (Childs *et al.*, 2009), it was found that participants could detect kerb heights of 60 mm, 80 mm and 120 mm when stepping up or stepping down from the kerb. There was no difference if the participant approached the kerb at a right angle or obliquely. All kerb heights were easier to detect when stepping up. It was concluded that:

> For confidence that a kerb is detectable by blind and partially sighted people, it is recommended to install a kerb of 60 mm or greater. This applies to kerb profiles approaching vertical and any profile that is significantly different from this would need to be tested. The effects of lighting, weather, additional cognitive loading, and the practicality of using the kerb to assist navigation were not considered in this trial. All these and issues relating to other groups, such as children and people with learning difficulties, would need to be the subject of further research.
>
> (Childs *et al.*, 2009)

8.9.2 Guidance updates concerning LTN 1/11 (DfT, 2011)

The characteristic of 'no level difference' has caused much concern, particularly to people with visual impairments. Pertinent comments relating to LTN 1/11 (DfT, 2011) by Disability Wales (2014) address kerb locations and heights, making the following key points:

> The Department for Transport (DfT) has circulated new guidance that encourages local authorities not to remove kerbs as part of shared-space schemes.
>
> The guidance, entitled *Access for Blind People in Towns*, has been published by the National Federation of Blind of the UK. [...]
>
> 3.1 An ideal shared space accessible to everyone can often be provided by a perimeter footway along the building frontages, protected by a standard height kerb and linked across the access streets by puffin or pelican crossings. Blind pedestrians can then circulate safely around the perimeter of the shared space alongside the building frontages, while non-disabled and sighted people can walk across the central shared space area at any point.
>
> 3.2 Pedestrian-controlled crossings with suitable signage at the road entrances to this shared space will mark the start of the 20 miles per hour speed limit, the change in road surface colour, and the removal of the normal driver priority over pedestrians. Such a design may not need a raised road surface, re-aligned gullies and drains or reinforced footways to support vehicles, all of which create the major part of the usual cost of a normal shared space design.
>
> (Disability Wales, 2014)

Box 8.1 Summary of concerns over shared-space and shared-surface projects based on JCMBPS (2005)

At the heart of the issues is the need to distinguish between shared space and shared surfaces. The former can be successful in meeting everyone's needs, provided that physical clues including kerbs and tactile surfaces are retained. The latter is generally taken to mean the removal of all delineation between areas traditionally used by vehicles and pedestrians. This results in an environment that is both frightening and dangerous, not only for people with reduced vision – and about one million people in the UK are registered as eligible to be registered as blind and partially sighted – but also for many other disabled and older people.

The joint committee on mobility for disabled people will only support shared-surface schemes that make suitable and safe provision for people with mobility, visual and hearing impairments.

The British Council of Disabled People does not believe that schemes where there is no difference in the surface between the road and pavement, and which rely on drivers and pedestrians making eye contact, can possibly meet the needs of blind and partially disabled people.

The Royal National Institute for Deaf People is concerned about people who may be unable to hear vehicles approaching and need to focus on companions rather than their environment in order to be able to communicate.

Arthritis care: without the physical ability to navigate such spaces deftly, people with arthritis are at a level of risk that, as with people with sensory impairments, may reduce their confidence in travelling to such an extent that they will, in effect, be excluded.

For all disabled people, inclusion cannot be achieved at the cost of another's exclusion.

Executive summary – concerns with shared surfaces

Removal of the kerb edge puts the safety and security of blind and partially sighted people seriously at risk, and undermines their confidence, independence and mobility.

Many blind and partially sighted people travel independently using a mobility aid – a long cane or a guide dog, along with any remaining vision. The kerb edge, or other tactile demarcation between footway and carriageway, is a fundamental clue for orientation within the street environment. Without a kerb or tactile demarcation, this becomes very difficult.

In addition, the shared-surface approach proposes that users of the street negotiate priority and movement through eye contact – which puts blind and partially sighted people at an immediate disadvantage.

There is concern that blind and partially sighted people, and other vulnerable groups, will become increasingly excluded from shared-surface areas or be exposed to extremely hazardous situations when using them. Our concerns are shared by other organisations represented on the Joint Committee on Mobility of Blind and Partially Sighted People, including the Royal National Institute of the Blind, St. Dunston's, the National Federation of Blind People, the National League of Blind and Disabled People, the National Association of Local Societies for Visually Impaired People, the Social Care Association, the Circle of Guide Dog Owners and Assistance Dogs UK.

It is therefore suggested that clarification of the meaning of 'shared space' is required, along with guidance on how and when the term should be used and in what contexts shared space works well. More detailed guidance is also required on the development and implementation of shared space, in particular with vulnerable road users in mind. The 'evidence gaps' need to be addressed, particularly in relation to how vulnerable road users navigate shared space and how their views are taken forward in future shared-space designs. Vulnerable, particularly disabled, road users need to be involved at earlier stages in the development of shared space; there needs to be continued engagement with users in the development and implementation of shared space.

8.10. Summary

This chapter has outlined many of the concerns associated with providing inclusive mobility throughout the transport system and within the public realm. Focusing on features that directly or closely affect the geometric design in terms of the location, dimensions and configuration of the facilities, the matters addressed will provide an introduction to the extensive literature and sources of information needed for effective design throughout the transport system. In addition, the material presented will provide some background to recently emerging concerns over the provision of pedestrian facilities, especially for people with cognitive disabilities.

Box 8.2 Key findings of the research on shared-space and shared-surface projects based on JCMBPS (2005)

Key findings of the research

1.0 Safety issues as evidenced by accidents and hazardous situations, such as danger or feeling of lack of safety using shared-surface areas, including reported conflict with buses and cyclists, intimidation by traffic passing close and perceived extreme difficulty in crossing carriageway safely. These experiences were primarily a result of

- the lack of demarcation between safe and unsafe areas with the removal of the distinction between footway and carriageway
- difficulty locating and using crossing points, owing to the removal of signal-controlled crossing
- street design or the use of materials that make it hard for blind and partially sighted people to orientate themselves.

2.0 Reduced confidence and increased anxiety. The problems and associated risks with using shared surfaces in town centres are undermining the confidence and mobility of blind and partially sighted people.

3.0 Avoidance of the shared-surface areas. Many of the blind and partially sighted people in our focus groups said they tended to avoid town centres with shared-surface areas, to the detriment of independence, freedom and quality of life.

4.0 Concerns of other disabled people. People with hearing impairments and learning disability have concerns about shared surfaces. While users and the people with mobility impairments appreciate level surfaces, they also highlighted the need for safe areas away from traffic.

5.0 Inadequate consultation at the planning stage. There has been insufficient consultation between local authorities and blind and partially sighted people and people with other disabilities. Despite some examples of good practice in consultation processes, most people feel their needs and opinions have been ignored and that local authority officers do not have sufficient awareness of visual-impairment issues.

Finding a solution for all – recommendations to improve safety and accessibility of shared-space areas for vulnerable groups.

- Creation or reinstatement of a footway with a kerb, regular dropped kerbs for wheelchair users and properly laid tactile paving.

or

- At least a clear delineation between the carriageway/vehicle area and a footway/pedestrian area through tactile and colour/tone contrast.
- Controlled crossings with audible and tactile signals.
- Installation of guardrails at potential danger points.
- Universal consistency of design features.

REFERENCES

BSI (1995) BS7818:1995. Specification for pedestrian restraint systems in metal (AMD Corrigendum 15047). BSI, London, UK.

BSI (2009) BS 8300. Design of buildings and their approaches to meet needs of disabled people – Code of Practice. BSI, London, UK.

Childs CR, Boampong DK, Rostron H et al. (2009) Effective Kerb Heights for Blind and Partially Sighted People: Research Commissioned by The Guide Dogs for the Blind Association (Guide Dogs). University College London, London, UK.

DfT (Department for Transport) (1991a) TAL 4/91: Audible and Tactile Signals at Pelican Crossings. DfT, London, UK.

DfT (1991b) TAL 5/91: Audible and Tactile Signals at Signal-controlled Crossings. DfT, London, UK.

DfT (1995a) LTN 2/95: The Design of Pedestrian Crossings. The Stationery Office, London, UK.

DfT (1995b) TAL 5/95: Parking for Disabled People. DfT, London, UK.

DfT (2004) LTN 2/04: Adjacent and Shared Use Facilities for Pedestrians and Cyclists. DfT, London, UK.

DfT (2005) Inclusive Mobility: A Guide to Best Practice on Access to Pedestrian and Transport Infrastructure. DfT, London, UK.

DfT (2006a) Carriage of Mobility Scooters on Public Transport – Feasibility Study. DfT, London, UK.

DfT (2006b) Railways for All: The Accessibility Strategy for Great Britain's Railways. DfT, London, UK.

DfT (2011) LTN 1/11: Shared Space. The Stationery Office, London, UK.

DfT (2015) Interim Changes to the Guidance on the Use of Tactile Paving Surfaces: Moving Britain Ahead. DfT, London, UK.

DfT (2017a) Transport Statistics 2016. DfT, London, UK.

DfT (2017b) DB 29/17: Design criteria for footbridges. In Design Manual for Roads and Bridges (DMRB). DfT, London, UK.

Disability Wales (2014) *Design Manual for Roads and Bridges (DMRB) Travel and Transport: Retain Kerbs in Shared Space Says Circulated Guidance from DfT*. Disability Wales, Caerphilly, UK. http://www.disabilitywales.org/travel-and-transport-retain-kerbs-in-shared-space-says-circulated-guidance-from-dft/ (accessed 08/10/2018).

DPTAC (Disabled Persons Transport Advisory Committee) (2007) *Response to the Second Three Year Review of the Department for Transport's Road Safety Strategy*. DPTAC, London, UK.

DVSA (Driver and Vehicle Standards Agency) (2016) Accessibility standards for public service vehicles. *Moving On*. https://movingon.blog.gov.uk/2016/02/16/accessibility-standards-for-public-service-vehicles-2/ (accessed 11/10/2018).

EHRC (Equality and Human Rights Commission) (2011) *Equality Act 2010 Code of Practice: Services, Public Functions and Associations, Statutory Code of Practice*. The Stationery Office, London, UK.

EHRC (2016) Equality Act codes of practice. https://www.equalityhumanrights.com/en/advice-and-guidance/equality-act-codes-practice (accessed 01/10/2018).

HMG (Her Majesty's Government) (1995) Disability Discrimination Act 1995. The Stationery Office, London, UK.

HMG (2010) Equality Act 2010. The Stationery Office, London, UK.

HMG (2016) The Traffic Signs Regulations and General Directions 2016. The Stationery Office, London, UK.

Highways Agency (2010) *Highways Agency DDA Compliance Programme: Design Compliance Assessment Guide, DDA Training Spring 2010*. Highways Agency, London, UK.

Holmes, Lord of Richmond (2015) *Accidents by Design: The Holmes Report on 'Shared Space' in the United Kingdom*. House of Lords, London, UK.

ISO (International Organisation for Standardisation) (1985) ISO 7193. Wheelchairs – overall maximum dimensions. ISO, Geneva, Switzerland.

JCMBPS (Joint Committee on the Mobility of Blind and Partially Sighted People) (2005) *Shared Space in the Public Realm – Policy Statement*. JCMBPS, Reading, UK.

Stait RE, Stone J and Savill TA (2001) *A Survey of Occupied Wheelchairs to Determine Their Overall Dimensions and Weight: 1999 Survey*. Report TRL470. TRL, Wokingham, UK.

TfL (2006) *Accessible Bus Stop Design Guidance, Bus Priority Team*, Technical Advice Note BP1/06. TfL, London, UK.

Schoon, John G
ISBN 978-0-7277-6309-9
https://doi.org/10.1680/pfse.63099.151

Chapter 9
Pedestrian facilities at roundabouts

The objective of both conventional and mini-roundabouts is to ensure the safe interchange of traffic between crossing streams with minimum delay and maximum safety. This is achieved by a combination of geometric layout that, ideally, matches the volume of vehicles in the traffic stream, their speed, applicable siting constraints, and the volume and needs of pedestrian traffic. The descriptions of geometric aspects of roundabouts related to pedestrians in this chapter are mainly based on standards and guidance in TD 16/07 *Geometric Design of Roundabouts* (DfT, 2007a). The design of mini-roundabouts is discussed in Chapter 10.

9.1. General

Elements of roundabout design associated with pedestrian priority in terms of volumes and convenience are, in general, treated in most guidance and advice as secondary considerations. However, accommodation of non-motorised traffic, including pedestrians, at roundabouts is addressed in several instances in the design guidelines and standards.

Roundabouts give priority to vehicles approaching from the right, instead of the usual case of priority junctions where traffic on the main road proceeding straight ahead has priority over turning traffic.

NOTE

No mention is made in the Highway Code *(DfT, 2017) or in current guidance or advice about the priority of pedestrians crossing an approach to a roundabout as would apply at a priority junction. For pedestrians at roundabouts, and assuming that priority rules for vehicles also apply to pedestrians, they no longer have priority over vehicles approaching from behind them and turning left (because such vehicles are approaching from the pedestrian's right) nor over vehicles approaching towards them and turning right (because such vehicles are also approaching from the pedestrian's right) into the road they are about to cross. This puts greater emphasis on the need for pedestrians' abilities and vigilance before crossing and in many cases adds to the distance a pedestrian must walk to continue along a straight desire line.*

As described in TD 16/07 (DfT, 2007a), the main types of roundabout are mini-, compact, normal, grade-separated,

signalled and double roundabouts (the last being a combination of mini, compact or normal roundabouts). Recommendations are given on the selection of roundabout type, geometric layout, visibility requirements and crossfall, with respect to the speed limit on the approach roads, the traffic flow and the level of non-motorised user demand. This standard is applicable to new and improved junctions on trunk roads. Related standards are: TD 54/07 *Design of Mini-roundabouts* (DfT, 2007b) and TD 50/04 *The Geometric Layout of Signal-Controlled Junctions and Signalised Roundabouts* (DfT, 2004).

A more recent development from previous guidance is the new compact roundabout, which has single-lane entries and exits, so that only one vehicle can enter or leave it from a given arm at any one time. Other recent changes relate to

- greater emphasis on non-motorised users
- design hierarchy
- entry paths related to turning movements
- crossfalls
- kerb lines at splitter islands or central reserves
- limitation of visibility
- use of additional signs and markings for larger roundabouts.

As with other aspects of the standards, departures from standard and relaxations require approval of the overseeing organisation.

9.2. Objectives

The principal objective of roundabout design is to minimise delay for vehicles while maintaining the safe passage of all road users through the junction. This is achieved by a combination of geometric layout features that, ideally, are matched to traffic flow *(in most cases, this refers to vehicular traffic flow, not pedestrian traffic)* and speed and to any local topographical or other constraints, such as land availability that apply. Location constraints are often the dominating factor when designing improvements to an existing junction, particularly in urban areas. Designs should be based on forecast demand and capacity determinants, should include associated signs and markings and should ensure that maintenance can be effectively carried out.

9.2.1 Safety of roundabouts

Of the approximately 207 400 personal injury road accidents in Great Britain in 2004 (DfT (2005a) Road Casualties Great Britain, 2004), about 18 000 (8.7%) occurred at roundabouts, of which 0.35% were fatal, whereas 0.88% of all other junction accidents and 2.2% of link accidents were fatal. This indicates the effectiveness of roundabouts in reducing accident severity. It suggests that, on average, roundabouts are safer than other junction types, but not necessarily for all road users or for a particular junction.

Table 9.1 shows the percentage of accidents by type of vehicle and severity for a sample of 1162 roundabouts (TRL Unpublished Report UPR/SE/194/05).

On average, accidents involving a pedestrian accounted for only 3% of the total. However, accident severity is high for pedestrians (23%, compared with 6% for car users). It should be noted that the majority of roundabouts are sited in rural areas with little or no pedestrian demand. Even at urban roundabouts, the number of pedestrians crossing the road within 20 m of the give-way line tends to be low.

■ Roundabouts are often located away from city centres.
■ Pedestrians may prefer to cross away from any flaring, where the road is narrower and traffic movements are more uniform and this may be more than 20 m from the give-way line. When pedestrians do cross the road within 20 m of the give-way line, they are aided by the splitter island, by the lower vehicle speeds and possibly by increased driver alertness in the vicinity of the roundabout.

NOTE
It may also be the case that fewer pedestrians use routes where they encounter roundabouts because of the perceived high speeds of vehicles negotiating them and the difficulty of crossing safely and conveniently. The danger perceived by pedestrians may also be due to drivers looking to their right as they approach the roundabout (as instructed in the Highway Code*) in order to identify vehicles with priority. A pedestrian starting to cross from the left, therefore, may not be identified adequately by the driver. Pedestrian demand may therefore be diminished due to awareness of this danger.*

9.3. Types of roundabout

The three main types of roundabout are normal, mini- and double roundabouts. Aspects of roundabouts particularly relevant to pedestrian and bicycle traffic are also addressed in the *Junctions and Crossings Design Manual* (Sustrans, 2015, Chapter 5). Other forms of roundabout are variants of these basic types, i.e. ring junctions, grade-separated roundabouts and signalled roundabouts. The various types of roundabout are defined as follows.

■ *Normal.* A roundabout having a one-way circulatory carriageway around a kerbed central island 4 m or more in diameter, usually with flared approaches to allow multiple entry, as shown in Figure 9.1.
■ *Compact.* A roundabout having single-lane entries and exits on each arm, as shown in Figure 9.2. The width of the circulatory carriageway is such that it is not possible for two cars to pass one another. Dimensions and configurations are related to the prevailing speed limits;

Table 9.1 Accidents by type of vehicle involved (1999 to 2003) (DfT, 2007a)

	Percentage of accidents	Accident severity (% fatal and serious)
Pedal cycles	8.0	9.5
Powered two wheelers	14.4	19.3
Cars and taxis	76.7	6.0
Public service vehicles	2.6	7.8
Light goods vehicles	6.4	5.6
Large goods vehicles	9.3	8.0
Pedestrians	2.8	22.6

Figure 9.1 Normal roundabout (DfT, 2007a)

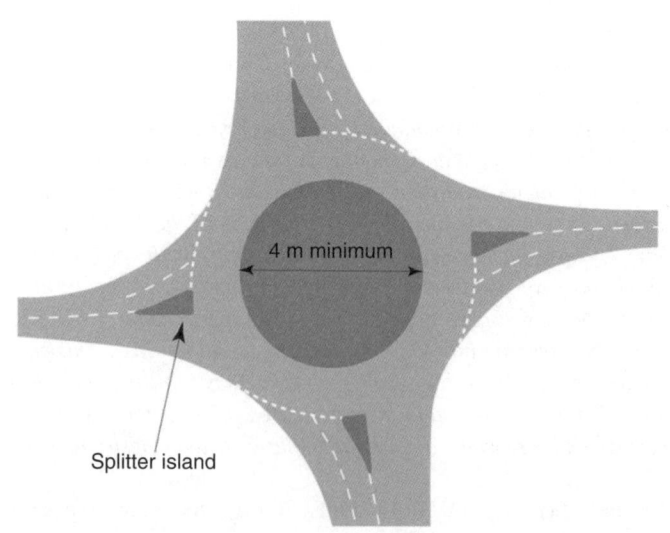

4 m minimum

Splitter island

Figure 9.2 Compact roundabout in an urban area (DfT, 2007a)
© Crown Copyright, 2007

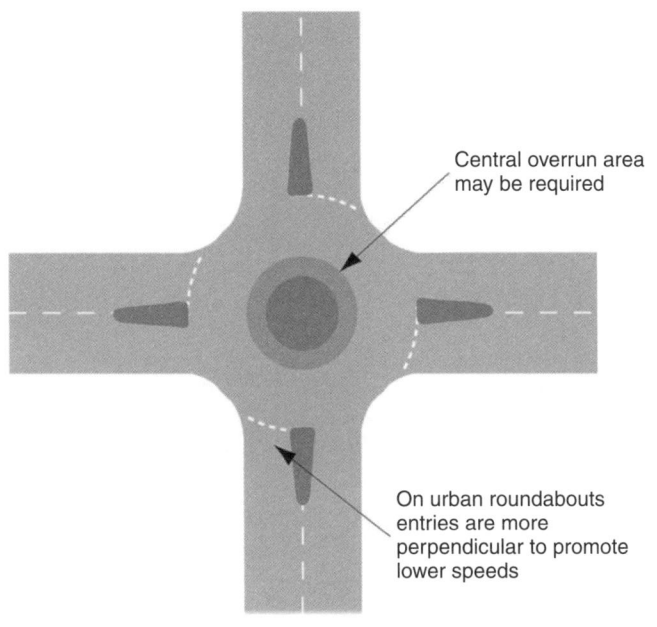

Central overrun area may be required

On urban roundabouts entries are more perpendicular to promote lower speeds

Figure 9.3 Double roundabout with short central link road (DfT, 2007a)
© Crown Copyright, 2007

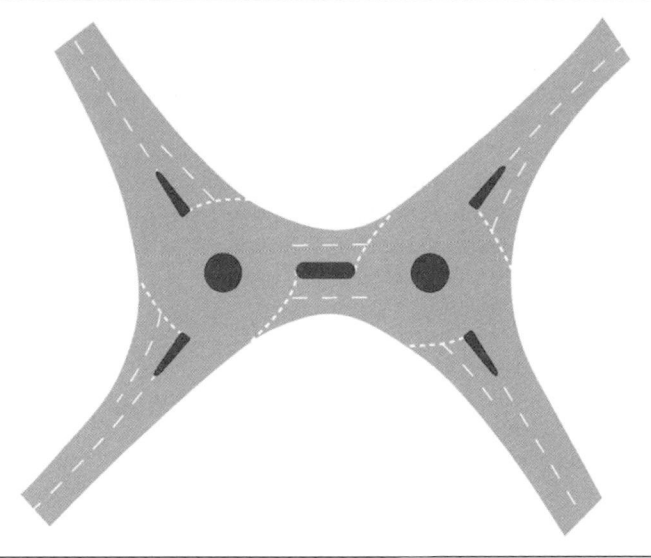

where speed limits exceed 40 miles per hour, the design of compact roundabouts is similar to that for normal roundabouts, but the single-lane entries and exits are retained.

- *Mini-roundabouts.* In place of the kerbed island, a mini-roundabout has a flush or domed circular solid white road marking between 1 and 4 m in diameter, capable of being driven over where unavoidable. Mini-roundabouts are discussed in greater detail in Chapter 10.
- *Grade-separated.* A grade-separated roundabout has at least one approach coming from a road at a different level, and is often used at motorway junctions and other multiple-level intersections.
- *Signalled.* A signalled roundabout has traffic signals on one or more of the approaches and at the corresponding point on the circulatory carriageway itself.
- *Double.* A double roundabout is a junction comprising two roundabouts separated by a short link (see Figure 9.3). The roundabouts may be mini-, compact or normal roundabouts.

NOTE
None of these diagrams of roundabouts gives any indication of provision for pedestrians; this is a possible indication of the level of official concern about the importance of pedestrian travel.

9.4. Siting of roundabouts

The siting of roundabouts may be relevant to the planning of pedestrian routes and to their more detailed design. These factors are outlined below.

In addition to its natural function as a junction, a roundabout may usefully

- facilitate a significant change in road standard, for example, from dual to single carriageways or from grade-separated junction roads to at-grade junction roads, although complete reliance should not be placed on the roundabout alone to act as an indicator to drivers
- emphasise the transition from a rural to an urban or suburban environment (although using one when there are no joining roads is not recommended)
- allow U-turns
- facilitate heavy right-turn flows.

The majority of accidents at major or minor priority junctions are associated with right turns. The inconvenience of banned right turns can be mitigated by providing a roundabout nearby.

Roundabouts are not recommended for at-grade junctions on rural three-lane dual-carriageway roads. Under these conditions, it is difficult to achieve adequate deflection. However, if a grade-separated junction is not achievable, it may be possible to generate suitable deflection by gently curving the approach to the right.

On single-carriageway roads, where overtaking opportunity is limited, roundabouts may be sited so as to optimise the length of straight overtaking sections along the route, according to TD 9/93 *Highway Link Design* (DfT, 2002). They can also provide an overtaking opportunity by having a short length of two lanes on the exit arms. The length of these sections will depend on site conditions.

Roundabouts should preferably be sited on level ground or in sags rather than at or near crests because it is difficult for drivers to appreciate the layout when approaching on an upwards gradient. However, there is no evidence that round-abouts on crests are intrinsically unsafe if correctly signed and where the visibility standards have been provided on the approach to the give-way line. Roundabouts should not be sited at the bottom of or on long descents.

Where several roundabouts are to be installed on the same route, they should be of similar design, in the interests of route consistency and hence safety, to the extent that this is possible with the traffic volumes concerned.

Where a proposed roundabout may affect the operation of an adjacent junction, or vice versa, the interactive effects should be examined. Where appropriate, traffic management measures, such as prohibited turns or one-way traffic orders may be considered. The effects of queueing should be examined to check that additional risk is not generated.

9.5.   Road users' specific requirements
9.5.1   Pedestrians
The types of pedestrian facility available at roundabouts are

- informal crossing
- zebra crossing
- stand-alone signal-controlled crossing (pelican, puffin or toucan)
- grade-separated crossing (underpass for pedestrians, underpass for vehicles or footbridge).

A dropped kerb and tactile paving must be provided at any crossing (DfT, 1995a).

Where possible, the splitter island – extended or widened as necessary – should be used as a pedestrian refuge. An absolute minimum island width of 1.2 m is required, preferably 2.5 m.

NOTE
At least 2.5 m is required to accommodate a person in a wheel-chair with an assistant, or for a person with a child's pushchair, as indicated in Chapter 8.

Figure 9.4 Measurement of distance from roundabout to pedestrian crossing (DfT, 2007a)
© Crown Copyright, 2007
This information is licensed under the Open Government Licence v3.0. To view this licence, visit http://www.nationalarchives.gov.uk/doc/open-government-licence/ **OGL**

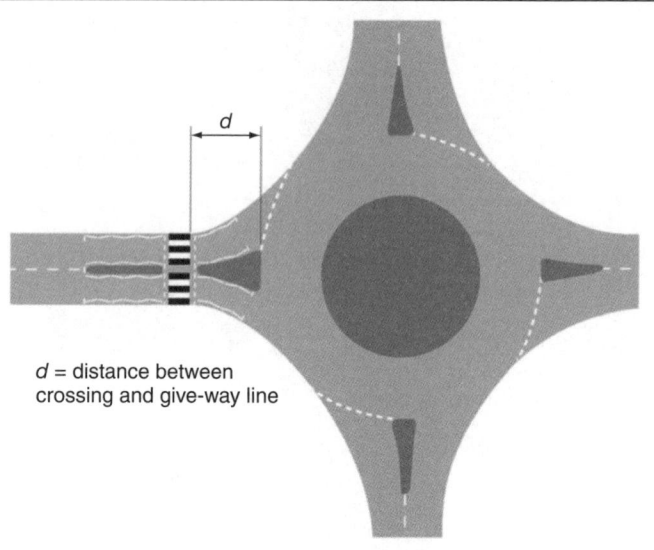

d = distance between crossing and give-way line

For a staggered signal-controlled crossing, 3 m is necessary. These requirements and guidance on facility selection and design are as indicated in Chapters 7, 11 and 12.

If a stand-alone crossing is provided close to the give-way line, there will inevitably be consequences for the operation of the roundabout and possibly for safety. An informal or zebra crossing is normally preferred, as it avoids the possibility that drivers will confuse the green signal with one controlling flow into the roundabout. Where a signal-controlled crossing is located close to the give-way line and drivers could confuse the crossing with the roundabout entry, the line should be supplemented by the use of markings to Diagram 1023 and give-way signs to Diagram 602 of the Traffic Signs Regulations and General Directions 2016.

Where provided, stand-alone pedestrian crossing facilities should be located to suit pedestrian desire lines. If possible, they should be outside of the flared section to keep the crossing short, as shown in Figure 9.4.

NOTE
These requirements make the presence of a roundabout detrimental to pedestrian movement, owing to a combination of a greater distance to walk and the presence of vehicles moving fast as they enter and exit the junction.

Zebra crossings should be located between 5 m and 20 m from the give-way line.

Non-staggered signal-controlled crossings should be sited either at 20 m or more than 60 m from the give-way line. It may be advantageous to use the splitter island (extended as necessary) as a central refuge. The central refuge can also be used to form a staggered crossing. Note that if a puffin crossing is used, a staggered crossing may not be necessary.

On the approach to the roundabout, a distance of 20 m for a signal-controlled crossing will reduce the likelihood of drivers confusing the signal with one controlling flow into the roundabout and will also leave sufficient holding space for vehicles waiting to enter the roundabout. On the exit, a distance of 20 m reduces the likelihood that 'blocking back' will occur where traffic queues extend onto the circulatory carriageway and also helps to ensure that drivers are still travelling slowly as they approach the crossing. If the crossing is staggered, the part on the entry arm can be within the 20–60 m zone.

NOTE
All of these requirements for placing the crossing location will significantly divert pedestrians from desire lines that parallel the directions of the approach and exit carriageways.

Zebra crossings should not be used where the 85th percentile speed exceeds 35 miles per hour (if it does, a signal-controlled crossing will be required). If the 85th percentile speed exceeds 50 miles per hour, serious consideration should be given to speed reduction measures before installing at-grade crossings. Signal-controlled crossings should be equipped with suitable speed measuring and extension equipment, as described in TAL 2/03 (DfT, 2003a) and LTN 1/95 (DfT, 1995b).

The Zebra, Pelican and Puffin Pedestrian Crossing Regulations and General Directions 1997 lay down the requirements for the general layout of these types of crossing. For areas of carriageway that are tapered, especially those including changes in the number of lanes, it is difficult to provide appropriate designs that are not potentially confusing to drivers.

Zig-zag markings are a requirement at zebra, pelican, puffin, toucan and equestrian crossings but must not be used where the crossing is part of a signalised roundabout.

With the exception of zebra crossings, central hatching or chevron markings may be used alongside zig-zag markings in certain conditions, as indicated in Section 15, Chapter 5 of the *Traffic Signs Manual* (DfT, 2003b).

In urban areas, where large numbers of pedestrians are present, the use of guardrails or other means of deterring pedestrians from crossing at inappropriate locations should be considered. Guardrailing should not obstruct drivers' visibility; guardrailing that is designed to provide intervisibility between drivers

and pedestrians is available but should be checked in case blind spots do occur. Further guidance on the use of guardrailing is given in *Inclusive Mobility* (DfT, 2005b).

NOTE
Of considerable importance to the safety of pedestrians is the need for them to see approaching vehicles and to ensure that such vehicles are complying with the signals before moving onto the carriageway from the footway as they start to cross.

Bridges and underpasses may present problems for people with disabilities and should only be used when at-grade crossings are deemed inappropriate, as described in TA 91/05 *Provision For Non-Motorised Users* (DfT, 2005c).

9.6. Design hierarchy
9.6.1 Selection of roundabout type and provision for non-motorised users
The choice of roundabout type is governed by a combination of factors, including

- whether the approach roads are single or dual carriageway (or grade-separated)
- the speed limit on the approach roads
- the level of traffic flow
- the level of non-motorised user (NMU) flow
- other constraints, such as land-take.

Table 9.2 gives the attributes of the different roundabout types, and indicates the normal types of provision for cyclists and pedestrians where there is sufficient demand to justify them. Alternatives are given in TA 91/05 (DfT, 2005c). Grade separation for non-motorised users is the best option at high-speed roundabouts but might not be cost-effective.

9.7. Geometric design – pedestrian aspects
Most of the guidance and advice on geometric design of roundabouts focuses on the needs of motor vehicle traffic. Key dimensions and configurations include those for: inscribed circle diameter; circulatory carriageway; central island; splitter islands; kerb and entry path radii; and flaring and related features.

9.7.1 Pedestrian crossing visibility
Drivers approaching a roundabout with a zebra crossing across the entry must be able to see the full width of the crossing so that they can see whether there are pedestrians wishing to cross. For a signal-controlled crossing, the driver must also be able to see at least one signal head. The visibility required is the desirable minimum stopping sight distance for the design speed of the link, as indicated in TD 9/93 (DfT, 2002) and LTN 2/95 (DfT, 1995a).

Table 9.2 Selection of roundabout type and recommended provision for NMUs (DfT, 2007a)

Roundabout category	Highest class of road on any approach	Highest speed limit within 100 m on any approach: miles per hour	Highest two-way annual average daily traffic flow on any approach	Recommended cyclist provision	Recommended pedestrian provision	Combined cycle and pedestrian provision	Roundabout type
1	Grade-separated entry or exit	Any	Any	Signal-controlled or grade-separated[a]	Signal-controlled or grade-separated[a]	Signal-controlled or grade-separated[a]	Grade-separated
2	Dual carriageway	>40	Any	Signal-controlled or grade-separated[a]	Signal-controlled or grade-separated[a]	Signal-controlled or grade-separated[a]	Normal
3	Single carriageway	>40	>8000	Signal-controlled	Signal-controlled	Signal-controlled	Normal
4	Single carriageway	>40	<8000	Cyclists mix with traffic	Informal		Compact
5	Dual carriageway	≤40	>25 000	Signal-controlled	Signal-controlled	Signal-controlled	Normal
6	Dual carriageway	≤40	16 000–25 000	Signal-controlled	Zebra[b] or signal-controlled	Signal-controlled	Normal
7	Dual carriageway	≤40	<16 000	Informal	Informal or zebra[b]	Informal	Normal
8	Single carriageway	≤40	>12 000	Signal-controlled	Zebra[b]	Signal-controlled	Normal
9	Single carriageway	≤40	8000–12 000	Informal	Informal or zebra[b]	Informal or signal-controlled[b]	Normal or compact
10	Single carriageway	≤40	<8000	Cyclists mix with traffic	Informal	Informal	Compact

[a] Signal-controlled crossing to be provided only if warranted by site-specific conditions; an alternative is grade-separated provision.
[b] Zebra crossings should not be used where the 85th percentile speed exceeds 35 miles per hour.
© Crown Copyright, 2007
This information is licensed under the Open Government Licence v3.0. To view this licence, visit http://www.nationalarchives.gov.uk/doc/open-government-licence/ **OGL**

At the give-way line, drivers must be able to see the full width of a pedestrian crossing (whether signal-controlled, zebra or informal) across the next exit if it is within 20 m of the give-way line on that arm (crossings should not be sited between 20 m and 60 m from the give-way line), as shown in Figure 9.5.

NOTE
Not mentioned in the guidance is the need for pedestrians to be able to see approaching vehicles at a distance that enables them to cross the relevant half of the crossing before a non-compliant vehicle reaches the location of the crossing. See also the comments on intervisibility in Chapter 6.

9.7.2 Other aspects of design
In addition to the main characteristics of the geometric layout of roundabouts, other aspects of design include dimensions related to drivers' forward visibility (stopping sight distance); the position of stop lines, exit visibility and signage. TD 50/04

Figure 9.5 Visibility required at entry to pedestrian crossing at next exit (DfT, 2007a)
© Crown Copyright, 2007

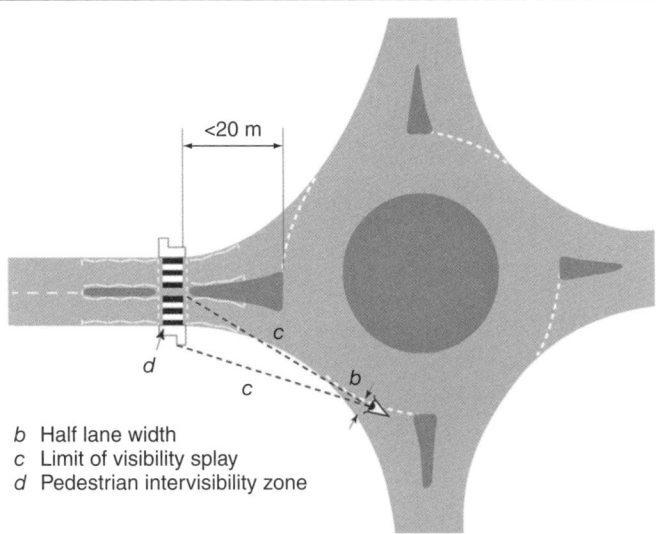

b Half lane width
c Limit of visibility splay
d Pedestrian intervisibility zone

Figure 9.6 Example of Dutch style roundabout (TfL, 2016)
© Crown Copyright, 2016

(DfT, 2004), TD 9/93 (DfT, 2002) and the *Traffic Signs Manual* (DfT, 2003b, Chapter 4) address many of these matters.

NOTE

As described in Chapter 6, this consideration of drivers' intervisibility does not address the pedestrians' need to see an approaching vehicle at an adequate distance to enable a safe crossing, compounded by the Highway Code*'s advice to drivers to look to their right for vehicles as they approach the roundabout.*

In comparison with design in the UK, an example of a roundabout in Holland is shown in Figure 9.6. Provision for pedestrians and cyclists is clearly shown, emphasising desire line continuity and non-motorised user priority.

9.8. Summary

Key features of roundabouts as they apply to pedestrians have been outlined in this chapter. Although designed primarily with motor vehicle traffic capacity and safety in mind, various methods of accommodating non-motorised users, including pedestrians, are also addressed.

Problems for pedestrians, however, remain, in terms of necessary deviation from desire lines and lack of consideration of pedestrians' visibility requirements. These matters should be addressed in the analysis stage to ensure that pedestrians are accommodated to the maximum extent possible.

REFERENCES

Department for Transport (DfT) (1995a) LTN 2/95: *The Design of Pedestrians Crossings*. The Stationery Office, London, UK.

DfT (1995b) LTN 1/95: *The Assessment of Pedestrian Crossings*. The Stationery Office, London, UK.

DfT (2002) TD 9/93: Highway link design, Amendment no. 1. In *Design Manual for Roads and Bridges (DMRB)*. DfT, London, UK.

DfT (2003a) TAL 2/03: *Signal-control at Junctions on High-speed Roads*. DfT, London, UK.

DfT (2003b) *Traffic Signs Manual*. The Stationery Office, London, UK.

DfT (2004) TD 50/04: The geometric layout of signal-controlled junctions and signalised roundabouts. In *Design Manual for Roads and Bridges (DMRB)*. DfT, London, UK.

DfT (2005a) *Road Casualties Great Britain: 2004, Annual Report*. The Stationery Office, London, UK.

DfT (2005b) *Inclusive Mobility: A Guide to Best Practice on Access to Pedestrian and Transport Infrastructure*. DfT, London, UK.

DfT (2005c) TA 91/05: Provision for non-motorised users. In *Design Manual for Roads and Bridges (DMRB)*. DfT, London, UK.

DfT (2007a) TD 16/07: Geometric design of roundabouts. In *Design Manual for Roads and Bridges (DMRB)*. DfT, London, UK.

DfT (2007b) TD 54/07: Design of mini-roundabouts. In *Design Manual for Roads and Bridges (DMRB)*. DfT, London, UK.

DfT (2017) *The Highway Code*. DfT, London, UK.

HMG (Her Majesty's Government) (1997) The Zebra, Pelican and Puffin Pedestrian Crossings Regulations and General Directions 1997. The Stationery Office, London, UK.

HMG (2016) The Traffic Signs Regulations and General Directions 2016. The Stationery Office, London, UK.

Sustrans (2015) *Junctions and Crossings: Design Manual*. Sustrans, Bristol, UK.

TfL (Transport for London) (2016) Junctions and crossings. In *London Cycling Design Standards*, Chapter 5. TfL, London, UK.

TRL Report UPR/SE/194/05. TRL, Crowthorne, UK.

Schoon, John G
ISBN 978-0-7277-6309-9
https://doi.org/10.1680/pfse.63099.159

Chapter 10
Pedestrian facilities at mini-roundabouts

The objective of both conventional and mini-roundabouts is to ensure the safe and efficient negotiation of junctions by motorised and non-motorised traffic. Guidance and advice related to design of mini-roundabouts is based on material in TD 54/07 *Design of Mini-roundabouts* (DfT, 2007a); this forms most of the basis for the descriptions in this chapter. For residential areas, the *Manual for Streets* (*MfS*) (DfT, 2007b) addresses design considerations in this context. Information is also contained in *Mini-roundabouts: Good Practice Guidance* (DfT, 2006).

10.1. General

A mini-roundabout has a central circular solid white road marking (referred to as the 'white circle'), which is flush with the road surface or slightly domed, instead of the kerbed central island of conventional roundabouts. The marking is between 1 and 4 m in diameter. If it is domed, the dome may be up to 125 mm at its highest point, although TD 54/07 (DfT, 2007a) recommends that it does not exceed 100 mm. A mini-roundabout must have an inscribed circle diameter of 28 m or less.

Traffic signs and road markings for mini-roundabouts should be observed in the design process in accordance with the relevant section of the Traffic Signs Regulations and General Directions 2016 and the *Traffic Signs Manual* (DfT, 2003, Chapter 5), including supporting advice on signs and road markings.

10.2. Scope

Advice and guidance in TD 54/07 (DfT, 2007a) sets out requirements for the location and design of mini-roundabouts, including three-arm and four-arm mini-roundabouts and double mini-roundabouts. Also included are advice on: the safety of mini-roundabouts and their suitability for non-motorised road users (NMUs); the siting requirements for mini-roundabouts, including a flow chart for preliminary assessment; and the geometric design features and conspicuity of mini-roundabouts. The main changes in TD 54/07 (DfT, 2007a) from previous advice and guidance relate to

- assessment, safety and accessibility impacts with information on capacity and restrictions on siting

- positioning of the white circle and give-way markings using vehicle swept paths
- forward visibility from approach arms and visibility between arms
- road and lane widths, with restrictions on the number of entry lanes
- height of domed white circles and the use of 'build-outs' and traffic islands
- conspicuity of mini-roundabouts in all conditions.

10.3. Siting and use of mini-roundabouts
10.3.1 Applicability related to speeds and speed limits

Mini-roundabouts must only be used on roads with a speed limit of 30 miles per hour or less and where the 85th percentile dry weather speed of traffic is less than 35 miles per hour within a distance of 70 m from the proposed give-way line on all approaches, unless installed in combination with speed reduction measures.

Where the existing 85th percentile dry weather speed is 35 miles per hour or above and a mini-roundabout is installed in combination with speed reduction measures in anticipation of reducing speeds to the required level, post-installation vehicle speed monitoring must be undertaken. In the event that vehicle speeds remain at 35 miles per hour or above, further speed-reducing measures must be installed.

10.3.2 Road network, land use and traffic characteristics

Mini-roundabouts are suitable as an improvement or remedial measure in a wide range of situations in built-up areas, including residential, business and shopping areas. However, where the forecast two-way annual average daily traffic flow on any arm is less than 500 vehicles, a junction arrangement in accordance with HD 42/05 (DfT, 2005a) should provide the most appropriate design, with less likelihood of unwarranted disruption to the main traffic streams. Mini-roundabouts may be inappropriate for frequent use by long vehicles and some public service vehicles. The dimensions of these vehicles can often lead to relatively difficult manoeuvres within the space usually available. It may be inappropriate to locate mini-roundabouts along routes leading to industrial areas or ports and main bus routes.

A mini-roundabout can be used at an existing junction provided there is adequate visibility and space for kerb realignment.

Mini-roundabouts must not be used at new junctions or direct accesses (see TD 41/95 (DfT, 1995a)), on dual carriageways or at a junction where the forecast two-way annual average daily traffic flow on any arm is less than 500 vehicles per day.

It may be acceptable to introduce a four-arm mini-roundabout where the forecast two-way annual average daily traffic flow on one arm is less than 500 vehicles per day, providing there is a strong expectation of the need to 'give way' on all approaches. This would depend on the circulating movements at each junction. In this instance, the overseeing organisation should be contacted for further advice.

The use of mini-roundabouts is not recommended at or near junctions where turns into or out from side roads are prohibited. This is because drivers do not expect to see vehicles U-turning on mini-roundabouts.

Introducing a mini-roundabout may lead to the reassignment of traffic to and from other routes. There is therefore a need to assess the surrounding network for the traffic and safety implications of introducing a new mini-roundabout.

Roundabouts in urban areas are not always compatible with urban traffic control systems. These systems move vehicles through controlled areas.

10.3.3 Conflicts

Where a three-arm mini-roundabout with single-lane approaches replaces a major or minor priority junction, the junction becomes easier to negotiate, as drivers only have to concentrate on one stream of traffic circulating at low speed from their right. However, as the number of arms or traffic lanes to the mini-roundabout increases, so does the potential for conflict.

NOTE
It should be added here that drivers must also be aware of pedestrians crossing the minor arm of the junction or mini-roundabout, particularly when they are applicable at 'an improvement or remedial measure in a wide range of situations in built-up areas, including residential, business and shopping areas', where many pedestrians engaged on different trip purposes may be expected.

Additional considerations for pedestrians are that the junction might be more difficult to negotiate and thus might be perceived less safe than a priority junction, owing to

- *higher vehicle speeds resulting from larger corner radii*

- *displacement of the crossing location further from the pedestrian desire line*
- *circulatory space for vehicles intruding on, or closer to, pedestrians' former crossing space, thereby causing intimidation and perceived lack of safety to the pedestrian*
- *drivers, when approaching the mini-roundabout, tending to look primarily to their right for approaching vehicles, and so failing to concentrate on pedestrians who may be crossing from their left.*

The presence of two or more approach lanes encourages two-abreast flow through the mini-roundabout, increasing the number of potential conflicts. Division of entries into two lanes may require additional signing and marking to ensure safe and efficient operation. Division of entries into three lanes is not recommended, as this can be associated with high numbers of accidents, particularly involving two-wheeled vehicles.

NOTE
The provision of two lanes also increases the difficulties of pedestrians in crossing conveniently and safely.

When traffic flows are low, drivers might consider conflict with other road users to be unlikely, with the result that they might approach the junction at inappropriate speeds. Drivers can also be caught out by traffic approaching from their right at a relatively high speed. Inadequate or excessive visibility can exacerbate this situation. To make the junction work in a more predictable way, it is best to avoid installing mini-roundabouts where traffic flows or turning proportions differ widely between arms, or where approach speeds are high.

10.3.4 Land requirements

The alignment of the arms and the area of land required are determined by requirements for lateral shift on entry, as well as the requirements relating to visibility to the right. An example of a mini-roundabout and its land requirement is given in Figure 10.1.

NOTE
The alignment of the arms and land required should also be influenced by NMU desire lines and visibility requirements, but this is not mentioned in the guidance.

10.3.5 Four- and five-arm mini-roundabouts

Four-arm mini-roundabouts introduce additional conflicts and can create difficulty for drivers' perception of the layout and turning flows.

NOTE
So, too, difficulties can be created for pedestrians for the same reasons.

Figure 10.1 Example of a mini-roundabout junction (DfT, 2007a)
© Crown Copyright, 2007
This information is licensed under the Open Government Licence v3.0. To view this licence, visit http://www.nationalarchives.gov.uk/doc/open-government-licence/ **OGL**

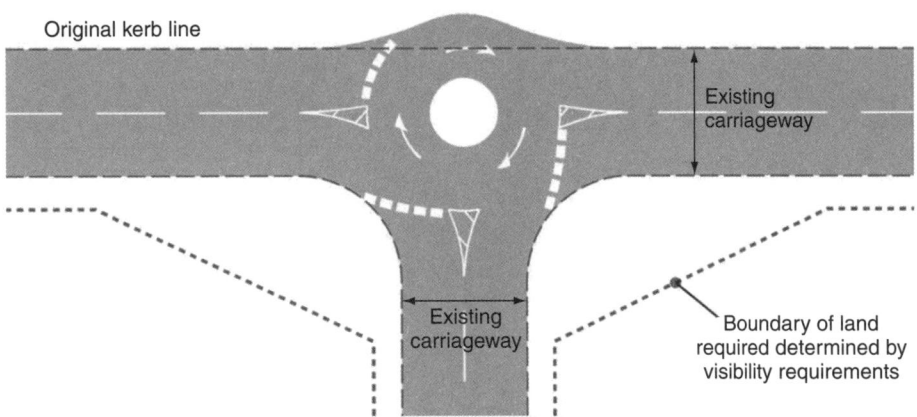

They are not recommended where the sum of the maximum peak hour entry flows for all entry arms exceeds 500 vehicles per hour. Five-arm mini-roundabouts must not be used.

10.3.6 Double mini-roundabouts

Double mini-roundabouts separated by a short link can be effective in improving traffic flows at a pair of closely spaced junctions, a staggered priority junction layout or an existing normal roundabout.

It is important that there is sufficient space for vehicles waiting at the intermediate give-way lines (i.e. those on the short link). Large opposing right-turning movements can produce gridlock at double mini-roundabouts, particularly if the network is congested. An example of a double mini-roundabout layout is shown in Figure 10.2.

The layout should be designed so that it is clear to drivers that they are approaching a double mini-roundabout in order to reduce the possibility of their crossing both roundabouts without noticing the intermediate give-way lines.

10.4. Feasibility of mini-roundabouts

The feasibility of using a mini-roundabout is determined by means of a flow chart, as shown in Figure 10.3. The decision regarding suitability for NMUs follows those for the inscribed circle diameter and the applicable speed limit.

NOTE
Only a brief mention of the needs of NMUs is made in this diagram; no numerical criteria related to them are specified. It would appear that research in this area and development of suitable criteria would be beneficial.

10.5. Road users' specific requirements
10.5.1 Pedestrians
Careful consideration is required where significant numbers of pedestrian crossing movements are likely to take place across any of the arms of a mini-roundabout. In these circumstances, the design organisation must either facilitate uncontrolled crossing movements by providing adequately sized traffic islands, avoiding entries or exits greater than single-lane widths, or provide conveniently situated controlled crossing facilities.

Figure 10.2 Example of a double mini-roundabout junction (DfT, 2007a)
© Crown Copyright, 2007
This information is licensed under the Open Government Licence v3.0. To view this licence, visit http://www.nationalarchives.gov.uk/doc/open-government-licence/ **OGL**

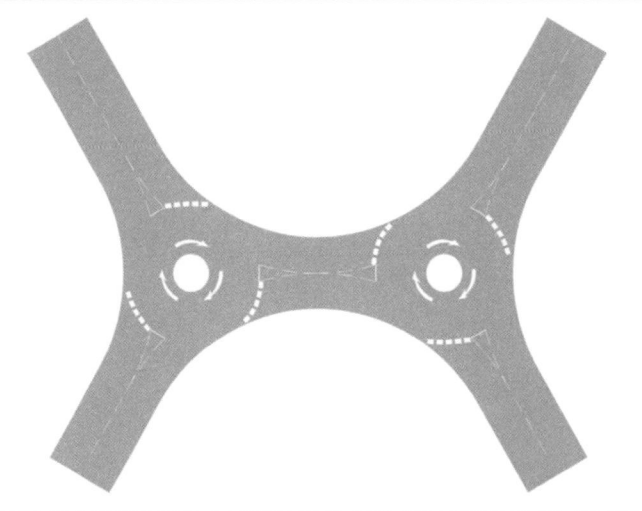

Figure 10.3 Procedures for assessing the feasibility of a mini-roundabout (DfT, 2007a)
© Crown Copyright, 2007
This information is licensed under the Open Government Licence v3.0. To view this licence, visit http://www.nationalarchives.gov.uk/doc/open-government-licence/ **OGL**

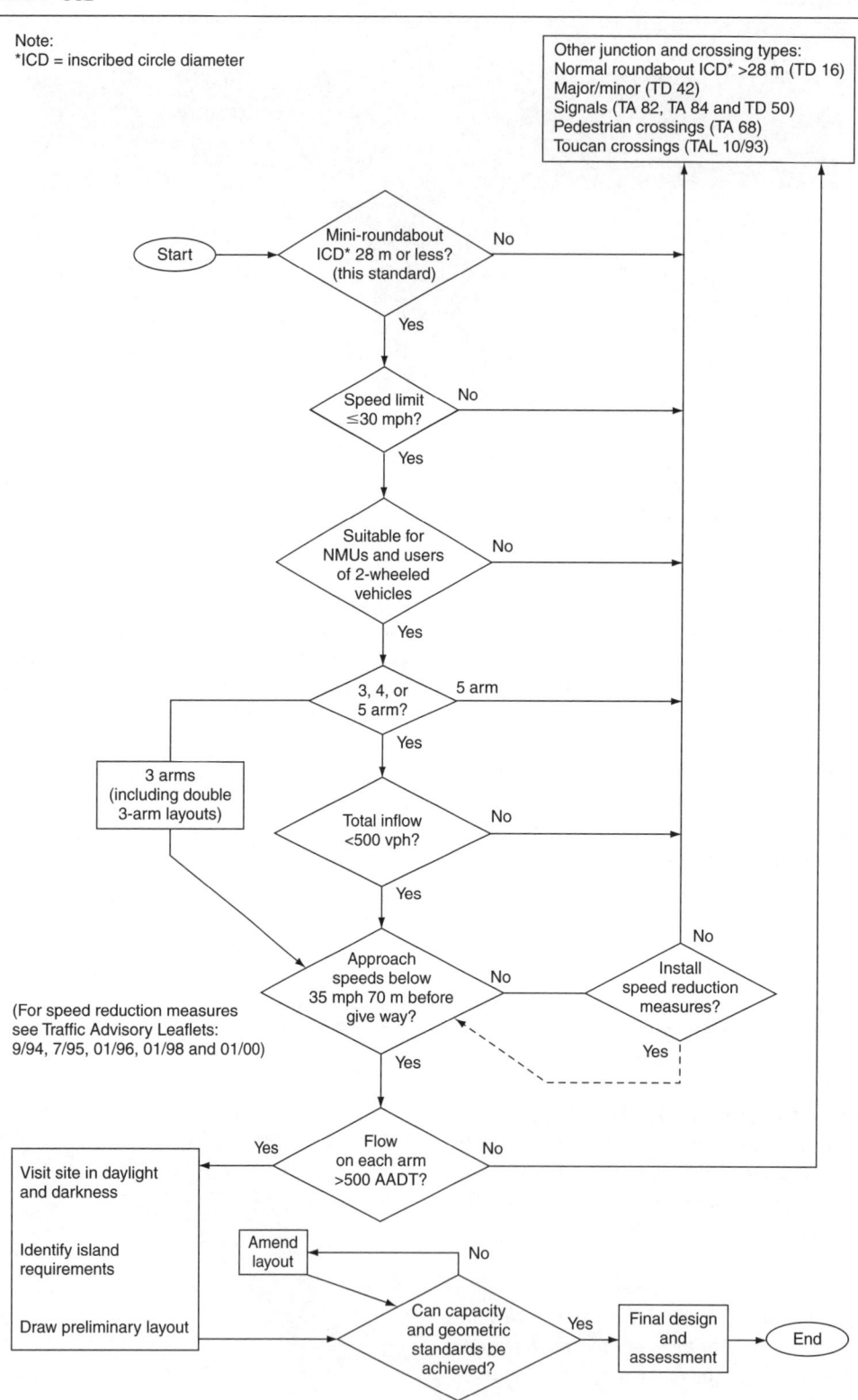

Guidance on providing for pedestrian (and other NMU) requirements is provided by TA 90/05 *Geometric Design of Pedestrian, Cycle and Equestrian Routes* (DfT, 2005b), TA 91/05 *Provision for Non-Motorised Users* (DfT, 2005c) and HD 42/05 *Non-motorised User Audits* (DfT, 2005a).

Introducing a mini-roundabout can significantly affect accessibility for pedestrians. Lack of gaps in entering or exiting traffic streams and difficulty in perceiving turning movements can cause difficulty.

Pedestrian movement is considered significant where flows of pedestrians are high, or likely to include children, older or disabled people (e.g. schools, health facilities, shops, public transport facilities and places of employment). The design organisation should allow for potential pedestrian routes that have been identified as part of a walking strategy or other local policies, or where future development would lead to significant pedestrian movement.

'Continental-style' mini-roundabouts, as outlined in TAL 9/97 (DfT, 1997), may be suitable for flows of between 5000 and 20 000 vehicles per day, and are likely to have a positive impact on pedestrians' and cyclists' safety and comfort because

- their tighter geometries encourage all vehicles to take the junction more slowly
- they provide only one lane on entry and exit on every arm
- the central island is larger relative to the overall size of the junction, as compared with a 'conventional' roundabout, meaning that the entry path curvature of circulating vehicles is increased (they are deviated more, and therefore cannot take the roundabout at higher speeds).

'Continental-style' mini-roundabouts are recommended for use in lower-speed, lower-traffic volume contexts (towards the lower end of the 5000 to 20 000 vehicles per day range).

Advantages for pedestrians are that the tighter geometry allows for pedestrian crossings on desire lines much closer to the entry to the roundabout than would be the case for conventional roundabouts.

10.6. Geometric design features

10.6.1 Inscribed circle diameter

The inscribed circle diameter of a mini-roundabout is the diameter of the largest circle that can be inscribed within the junction kerbs. The maximum inscribed circle diameter of a mini-roundabout is 28 m. At diameters larger than this, a normal roundabout can accommodate the largest design vehicle and must be used. Further guidance is provided in TD 16/07 (DfT, 2007c).

A concentric overrun area may be used if required to increase the deflection and conspicuity; the maximum permitted diameter of the overrun area is 7.5 m. It should be noted that light vehicles are not legally obliged to avoid overrun areas in the same way as they must avoid the white circle of a mini-roundabout; therefore, concentric overrun areas cannot be relied on for the purposes of achieving deflection. Additional road markings must not be placed on or around the edges of a concentric overrun area. The circulatory arrow markings should be placed on the surrounding circulating area and not on the overrun area. Construction materials used for overrun areas should not resemble footways or refuges and measures should be taken, if necessary, to discourage direct pedestrian movements into them. Overrun areas may be sloped up to the white circle at an angle of up to 15°. Further advice on overrun areas is contained in TAL 12/93 (DfT, 1993).

10.6.2 Visibility

Road users approaching the give-way line on any approach to a mini-roundabout need to be sure that it is safe to enter the circulatory area. The conflict point is defined by the construction in Figure 10.4. The time taken for a vehicle to travel from a stationary position at the give-way line to the conflict point is defined as the 'gap acceptance time'. This is a function of the size of the roundabout, and should be taken as 2 s when the distance from the give-way line to the centre of the white circle is 7 m or less; otherwise, it should be taken as 3 s.

NOTE
No mention is made in the guidance regarding pedestrians' gap acceptance time or sight distances; these factors may be important in affecting pedestrians' use of routes that include mini-roundabouts.

The visibility distance *D* shown in Figure 10.4 and Table 10.1 is the minimum sight distance required by a road user approaching the roundabout at a distance *F* from the give-way line. Visibility distances for *D* are measured from the centre of the offside approach lane to the nearside carriageway edge of the arm to the right. This distance enables the driver of an entering vehicle to observe vehicles coming from the right before they reach the conflict point. Distance *D* varies with the 85th percentile 'dry weather' approach speed 70 m before the give-way line on the arm to the right and the 'gap acceptance time'. See TA 22/81 *Vehicle Speed Measurement on All Purpose Roads* (DfT, 1981) for guidance on speed measurement.

NOTE
'Road user' in this context refers to a driver.

The give-way marking used at mini-roundabouts requires road users to give way to circulating traffic at or immediately beyond the line. The distance *F* shown in Figure 10.4 must be a

Figure 10.4 Mini-roundabout visibility distance, *D*, and stopping sight distance, *E* (DfT, 2007a)
© Crown Copyright, 2007
This information is licensed under the Open Government Licence v3.0. To view this licence, visit http://www.nationalarchives.gov.uk/doc/open-government-licence/ **OGL**

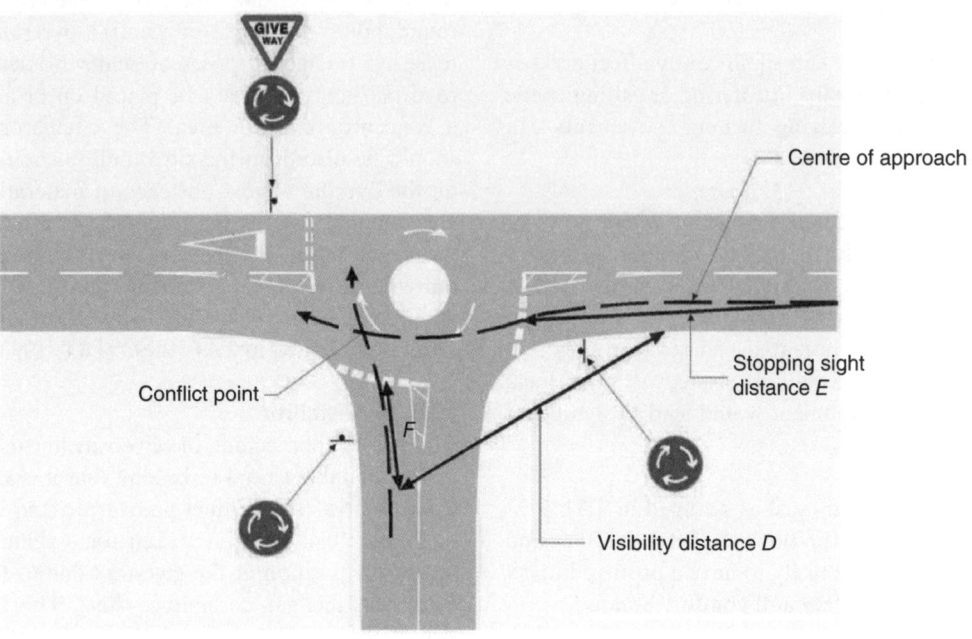

minimum of 9 m, so that the first two vehicles in the approach queue have visibility of traffic coming from the arm on the right, but note should be made of speed reduction and vertical deflection requirements. In difficult circumstances, the distance *F*, shown in Figure 10.4, for an arm may be taken as a relaxation from 9 m to 4.5 m, providing that the maximum peak hour entry flow on the arm is less than 300 vehicles per hour. In exceptionally difficult circumstances, a relaxation to 2.4 m (as the absolute minimum) is permissible, as it enables a road user who has reached the give-way line to see approaching vehicles without encroaching past the give-way line. This will, however, allow only one vehicle at a time to enter safely and requires following drivers to be prepared to stop and look. A distance *F* of 2.4 m must only be used on an arm with a maximum peak hour entry flow of 300 vehicles per hour or less, where there is no entry arm to the left. In such cases the mandatory give-way markings and upright sign must be used to require road users to give way to circulating traffic at the line.

The visibility distance *D* shown in Figure 10.4 must be provided within the whole of the visibility envelope between driver's eye heights of 1.05 and 2.0 m at the centre of the offside approach lane to object heights between 0.26 and 2.0 m at the nearside edge of the arm to the right.

The stopping sight distance *E*, in accordance with Table 10.1 and Figure 10.4, must be provided within the whole of an envelope between eye heights of 1.05 m and 2.0 m at the centre of the path of an approaching vehicle and object heights of 0.26 to 2.0 m at the give-way line.

NOTE

No mention is made in this guidance of the visibility needs and distances for pedestrians to see approaching vehicles adequately. This is particularly important because pedestrians crossing the minor arm do not have priority over vehicles turning into that arm and the Highway Code *makes no mention of drivers having to give such priority to pedestrians at roundabouts and mini-roundabouts, as is the case of a non-signalled priority junction.*

Table 10.1 Minimum visibility distance to the right (DfT, 2007a)

85th percentile speed of arm to the right: miles per hour	*D*: m	
	For a gap acceptance time of 2 s	For a gap acceptance time of 3 s
35	40	55
30	35	50
25	25	40

© Crown Copyright, 2007
This information is licensed under the Open Government Licence v3.0. To view this licence, visit http://www.nationalarchives.gov.uk/doc/open-government-licence/ **OGL**

10.6.3 Pedestrian islands and refuges

Advice on islands and refuges for pedestrian use is given in LTN 1/95 (DfT, 1995b) and in LTN 2/95 (DfT, 1995c), and is also addressed in Chapters 6 and 7.

Traffic islands may be provided to separate opposing streams of traffic and, if appropriate, to serve one or more of the following purposes.

- Assist provision of adequate deflection of the path of vehicles approaching the mini-roundabout.
- Increase conspicuity to drivers approaching the mini-roundabout.
- Pedestrian use.
- Calming.

However, a mini-roundabout must not be used as a speed reduction measure in isolation. Where a mini-roundabout is used within a traffic calming scheme, speed reduction must be achieved by means of suitable speed reduction measures on the approach. If the required speed reduction cannot be achieved, then a mini-roundabout must not be provided.

Islands for separating opposing streams of traffic or deflecting approaching vehicles may be kerbed physical islands or created using prescribed road markings. Solid or raised areas of markings at mini-roundabouts are not permitted, other than for the white circle.

Kerbed islands provided to assist crossing pedestrians must be located within 20 m of the give-way line at the nearest point. Pedestrian facilities located farther than 20 m from the junction might be too far from the desire line; this may cause pedestrians not to use the facility. Pedestrian facilities located over 60 m from a mini-roundabout perform independently of the junction.

The minimum requirement for road markings on the approaches to kerbed physical islands is to provide warning lines according to *Traffic Signs Manual* (DfT, 2003, Chapter 5, Section 8). However, if a kerbed island is at risk of being over-run by approaching, circulating or exiting vehicles, it must be positioned at least 0.5 m clear of any vehicle swept path.

10.7. Summary

Mini-roundabouts have become a firmly established part of traffic infrastructure in the UK. Their use in other countries is very limited; many countries continue to favour the priority or signalled junction in their stead.

From a pedestrian's point of view, they can be of mixed value; useful in that they have a slowing effect on traffic in general but they often result in greater walking distances through diversion from desire lines and they may inhibit pedestrian movement where inscribed circles and road markings emphasise vehicular traffic movement close to space normally used in crossing by pedestrians. The various notes in this chapter summarise many of the unsatisfactory conditions experienced by pedestrians at mini-roundabouts, which could be the subject of new research.

Many authorities have developed their own approaches to justifying and installing mini-roundabouts; each will have a unique set of circumstances, for which the material in this chapter provides an introduction.

REFERENCES

DfT (Department for Transport) (1981) TA 22/81: Vehicle speed measurement on all purpose roads. In *Design Manual for Roads and Bridges (DMRB)*. DfT, London, UK.

DfT (1993) TAL 12/93: *Overrun Areas*. DfT, London, UK.

DfT (1995a) TD 41/95: Vehicular access to all-purpose trunk roads. In *Design Manual for Roads and Bridges (DMRB)*. DfT, London, UK.

DfT (1995b) LTN 1/95: *The Assessment of Pedestrian Crossings*. The Stationery Office, London, UK.

DfT (1995c) LTN 2/95: *The Design of Pedestrian Crossings*. The Stationery Office, London, UK.

DfT (1997) TAL 9/97: *Cyclists at Roundabouts: Continental Design Geometry*. DfT, London, UK.

DfT (2003) *Traffic Signs Manual*. The Stationery Office, London, UK.

DfT (2005a) HD 42/05: Non-motorised user audits. In *Design Manual for Roads and Bridges (DMRB)*. DfT, London, UK.

DfT (2005b) TA 90/05: Geometric design of pedestrian, cycle and equestrian routes. In *Design Manual for Roads and Bridges (DMRB)*. DfT, London, UK.

DfT (2005c) TA 91/05: Provision for non-motorised users. In *Design Manual for Roads and Bridges (DMRB)*. DfT, London, UK.

DfT (2006) *Mini-roundabouts: Good Practice Guidance*. DfT, London, UK.

DfT (2007a) TD 54/07: Design of mini-roundabouts. In *Design Manual for Roads and Bridges (DMRB)*. DfT, London, UK.

DfT (2007b) *Manual for Streets*. Thomas Telford, London, UK.

DfT (2007c) TD 16/07: Geometric design of roundabouts. In *Design Manual for Roads and Bridges (DMRB)*. DfT, London, UK.

HMG (Her Majesty's Government) (2016) The Traffic Signs Regulations and General Directions 2016. The Stationery Office, London, UK.

Schoon, John G
ISBN 978-0-7277-6309-9
https://doi.org/10.1680/pfse.63099.167

Chapter 11

Shared facilities for pedestrians and other non-motorised users*

Pedestrian facilities are often combined with those for other non-motorised users (NMUs), notably cyclists and equestrians. This chapter describes selected geometric design aspects of such combined facilities. Often, the necessity of combining the complementary, and sometimes conflicting, needs of the various users exists in conjunction with the requirement to meet the needs of disabled people. Relevant descriptions in earlier chapters, therefore, also apply here.

The main sources of design information outlined in this chapter are TA 90/05 *The Geometric Design of Pedestrian, Cycle and Equestrian Routes* (DfT, 2005a) and TD 36/93 *Subways for Pedestrians and Pedal Cyclists: Layout and Dimensions* (DfT, 1993a). Background material on wider issues and applications is based on TA 91/05 *Provision for Non-motorised Users* (DfT, 2005b) and LTN 1/04 *Policy, Planning and Design for Walking and Cycling* (DfT, 2004a). Other relevant documents include the section in TA 57/87 *Roadside Features* (DfT, 1989) that covers roadside facilities for ridden horses.

Further matters related to combined pedestrian and cyclist-only facilities in predominantly urban areas are addressed in LTN 1/12 *Shared Use Routes for Pedestrians and Cyclists* (DfT, 2012). Key features of this emphasis, related particularly to geometric design for pedestrians, are described later in this chapter, together with comments on selected updates.

11.1. Geometric design of pedestrian, cycle and equestrian routes (DfT, 2005a)

11.1.1 Scope
Guidance on geometric design for NMU off-carriageway routes associated with trunk road or motorway improvement schemes is provided in TA 90/05 (DfT, 2005a). The advice is also relevant for NMU routes away from trunk roads constructed as part of a trunk road improvement, and for aspects of crossing the trunk road not dealt with in BD 29/04 *Design Criteria for Footbridges* (DfT, 2004b), TD 36/93 *Subways for Pedestrians and Pedal Cyclists: Layout and Dimensions* (DfT, 1993a) and TD 50/04 *The Geometric Layout of Signal-controlled Junctions and Signalised Roundabouts* (DfT, 2004c). Designers should also refer to

guidance on cycle-friendly infrastructure, such as that offered by the Chartered Institute of Highways and Transport (CIHT, 2015a, 2015b) and Sustrans (2015), as well as Chapter 5 of the *Traffic Signs Manual* (DfT, 2003). Issues of route choice are covered in *Provision For Non-Motorised Users Summary* TA 91/05. Designers are also referred to HD 42/05 *Non-motorised User Audits* (DfT, 2005c) and HD 19/15 *Road Safety Audit* (DfT, 2017), both of which set out procedures for ensuring that scheme designs have considered the needs of NMUs.

The geometric design of facilities for pedestrians, cyclists, wheelchair users and equestrians addressed in TA 90/05 (DfT, 2005a) also includes design for power wheelchairs and invalid carriages that conform to current Department for Transport regulations and may be legally used on pedestrians' facilities. Where there is known to be regular use of these latter vehicles, design parameters for cyclists should be used.

As with all highway design, there is a need to balance issues of safety and practicality. Preferred and acceptable minimum values based on the best available evidence are provided but, in exceptional circumstances, flexibility may be warranted in using these figures over short distances and where other measures are used, such as 'SLOW' markings to encourage lower speeds.

11.1.2 Definitions
The following definitions apply to NMU facilities.

- A shared route is an unsegregated facility used by more than one type of NMU, for example, pedestrians and cyclists, or pedestrians, cyclists and equestrians.
- An adjacent-use route is one with clearly defined segregated areas for different types of NMU. Segregation may be achieved using a white line or a physical feature, such as a verge, fence or kerbed level difference.

11.1.3 Design speed
11.1.3.1 Pedestrians
It is stated in TA 90/05 (DfT, 2005a) that, in designing facilities for pedestrian use only, it is not necessary to consider design speed.

NOTE

This is presumably because it is not considered that pedestrians have a meaningful stopping sight distance. However, it should be noted that the walking speed of pedestrians could be of importance, as well as consideration of pedestrians' observation–reaction time when crossing a carriageway or other motorised facility. See Chapter 3 for details.

The advice states that design speed is important if the facility is for use by cyclists and pedestrians, as this will affect other design parameters, such as visibility.

11.1.3.2 Cyclists

Design speeds for cyclists can vary according to different types of user; such speeds can influence the design of shared and adjacent facilities. The design cyclist types identified are: fast commuters; other utility cyclists; inexperienced utility cyclists, who might travel more slowly; regular cyclists; children; and users of specialised equipment.

Different authorities in the UK and overseas have used a range of design speeds, from 10 to 50 km/h. However, cyclists travelling in excess of 30 km/h are less likely to be using off-carriageway facilities.

The design speed of 30 km/h should be adopted for most off-carriageway cycle routes. Where a cyclist would expect to slow down, such as on the approach to a crossing or subway, the design speed may be reduced to 10 km/h over short distances, with the use of 'SLOW' markings.

The design speeds appropriate for different route types are summarised in Table 11.1.

11.1.3.3 Equestrians

The concept of design speed for equestrians is unusual, as there are different speeds at which horses progress, depending on the type of activity being undertaken and the surrounding environment. Tables 11.2 and 11.3 show expected and design speeds, respectively, for different circumstances, namely walking, trotting and cantering. Various ground and handling conditions will significantly affect these speeds.

11.1.3.4 Shared routes

Where routes are shared with other users, the design speed of these routes should be relevant to that of the faster users, as shown in Table 11.4.

11.1.4 Visibility

11.1.4.1 General

The forward visibility for cyclists or equestrians along the road should be considered, such that requirements for an appropriate stopping sight distance (SSD), taking into account eye and

Table 11.1 Design speed of off-carriageway cycle routes (DfT, 2005a)

Item	Design speed: km/h
Acceptable minimum (over short distances)	10
General off-carriageway provision	30

Table 11.2 Expected speeds for equestrian routes (DfT, 2005a)

Situation	Expected speed
Adjacent to carriageway	Walk
On approach to crossing	Walk
Remote from carriageway[a]	Walk
Remote from carriageway[a] for >50 m length	Trot or canter

[a] 'Remote from carriageway' means that road is generally not visible due to screening of planting or visible but more than 6 m from the equestrian route

Table 11.3 Design speed for equestrian routes (DfT, 2005a)

Type of use	Design speed: km/h
Trot or canter	20
Walk	10

Table 11.4 Design speeds where use is shared (DfT, 2005a)

Shared users	User for determining design speed
Pedestrian, cyclist	Cyclist
Pedestrian, equestrian	Equestrian
Cyclist, equestrian	Cyclist
Pedestrian, cyclist, equestrian	Cyclist

object heights, are met. Considerations should also be given to visibility at junctions or crossings, to enable both the NMU to see approaching traffic, and for other users on the main road to see an NMU about to cross.

NOTE

Considerations of observation–reaction times and related values are not included in current practice when estimating the required SSDs. Although SSDs for drivers include perception–reaction times, a similar time (observation–reaction) is not considered for NMUs. Designers may wish to modify this approach in light of the principles and research findings described in Chapters 3, 4, 6 and 7 on pedestrian characteristics and crossings and on the cautionary note on intervisibility in Chapters 6 and 7 on pedestrian crossings.

11.1.4.2 Stopping sight distance on NMU routes

The stopping sight distance (SSD) is the distance for a rider to perceive, react and stop safely in adverse conditions, such as on wet asphalt or where the surface is loose. It is measured in a straight line between any two points on the centre of the road; sighting across the highway boundary line is not permitted. It should, however, be noted that cyclists and equestrians generally have a greater ability to avoid momentary obstructions than does vehicular traffic.

NOTE

No research results seem to be available regarding this observation about the avoidance of obstructions and the reasons for the assertion are not immediately evident, except that NMUs are typically travelling at lower speeds. However, considerations of stability might have some bearing on the safety of NMUs; further investigation appears warranted.

Table 11.5 SSDs for off-carriageway cyclists' routes (DfT, 2005a)

Design speed: km/h	Preferred minimum SSD: m
30	30
10	10

© Crown Copyright, 2005
This information is licensed under the Open Government Licence v3.0. To view this licence, visit http://www.nationalarchives.gov.uk/doc/open-government-licence/ **OGL**

Table 11.6 SSDs for equestrian routes (DfT, 2005a)

Design speed: km/h	Preferred minimum SSD: m
20	30
10	10

© Crown Copyright, 2005
This information is licensed under the Open Government Licence v3.0. To view this licence, visit http://www.nationalarchives.gov.uk/doc/open-government-licence/ **OGL**

The SSDs for cyclists are given in Table 11.5; the corresponding figures for equestrians are shown in Table 11.6.

11.1.4.3 Eye and object heights

Designers should ensure that an object at the minimum SSD is visible from a range of eye heights. For cyclists, an eye height range of 1.0 to 2.2 m should be used, which accommodates a range of cyclists from children and recumbent users to adults, as shown in Figure 11.1. The object height should be taken as

Figure 11.1 Forward visibility for cyclists (DfT, 2005a)
© Crown Copyright, 2005
This information is licensed under the Open Government Licence v3.0. To view this licence, visit http://www.nationalarchives.gov.uk/doc/open-government-licence/ **OGL**

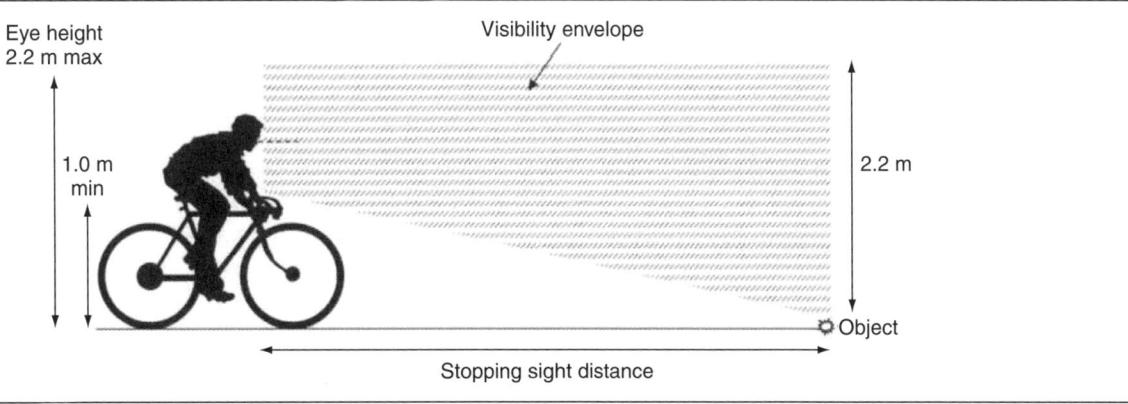

Figure 11.2 Forward visibility for equestrians (DfT, 2005a)
© Crown Copyright, 2005
This information is licensed under the Open Government Licence v3.0. To view this licence, visit http://www.nationalarchives.gov.uk/doc/open-government-licence/ **OGL**

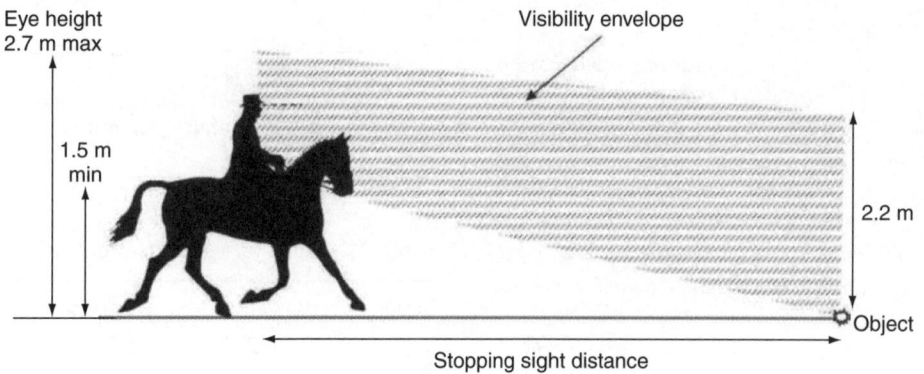

a range from ground level to 2.2 m, as cyclists need to be able to observe deformations, poles and objects that could interfere with safe progress.

For equestrians, the rider's eye height should be taken as 1.5 to 2.7 m. This accommodates a range of horse riders from children on ponies to adults on larger horses, as shown in Figure 11.2. The object height should again be taken as a range from ground level to 2.2 m, so that riders can observe deformations, poles and objects that could interfere with the horse's safe progress.

11.1.4.4 Momentary obstructions
These are important and isolated objects with widths of less than 300 mm that are unlikely to have a significant effect on visibility. For immovable obstructions wider than 300 mm, it may be necessary to provide markings to guide cyclists and equestrians accordingly.

11.1.4.5 Visibility to and from NMU crossing points
Any crossing of a trafficked road should be located such that drivers have full visibility of the NMUs wishing to use the crossing point. A desirable minimum SSD, in accordance with TA 90/05 (DfT, 2005a) should be available for drivers on the highway approaching an NMU crossing point.

NOTE
Again, although SSDs for drivers include perception–reaction times, a similar time (observation–reaction) is not considered for NMUs.

11.1.4.6 Provision of visibility at NMU route junctions
A visibility splay should be provided for NMUs approaching crossings and junctions where they have to stop or give way. Figure 11.3 defines distances x and y. For pedestrians, the preferred distance x is 2.0 m, to allow for the need of some disabled people and users with prams.

Figure 11.3 Visibility splay for NMU route (DfT, 2005a)
© Crown Copyright, 2005
This information is licensed under the Open Government Licence v3.0. To view this licence, visit http://www.nationalarchives.gov.uk/doc/open-government-licence/ **OGL**

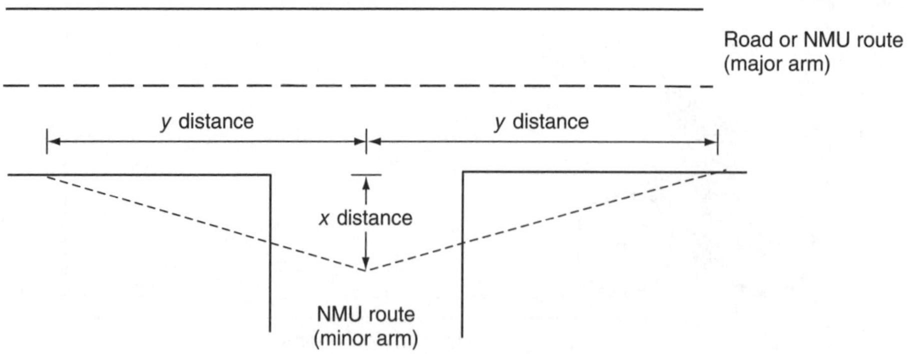

Table 11.7 Minimum distances *x* for NMUs at crossings (DfT, 2005a)

NMU	Preferred: m	Acceptable: m	Minimum for 'jug-handle' crossing: m
Pedestrian	2.0	1.5	N/A
Cyclist	4.0	2.5	1.0
Equestrian	5.0	3.0	N/A

The preferred distance *x* for cyclists is 4.0 m, which equates approximately to the length of two cycles. While every effort should be made to achieve the desirable value of 4.0 m, in practice, the actual distance *x* that can be achieved for existing roads might be limited by the verge widths of the trunk road. In these cases, *x* can be reduced to a minimum of 2.5 m.

Where the crossing is approached by means of a 'jug handle' from a route parallel to the trunk road, the cyclists' approach speed is less than for a route that approaches the crossing at right angles. In these circumstances, *x* can be reduced to 1.0 m. A jug handle is a left-turn diverging lane loop, as defined and illustrated in Paragraph 2.17 and Figure 2/4 of TD 42/95 (DfT, 1995).

A summary of values of *x* is provided in Table 11.7. Note the requirements of TD 42/95 *Geometric Design of Major/Minor Priority Junctions* (DfT, 1995) and conditions affecting equestrians.

Where the main route is a public road, the distance *y* for pedestrians and cycle route crossings should be the same distance *y* identified for vehicles in TD 42/95 (DfT, 1995). However, equestrians require greater visibility, as there is a reaction time between rider perception and the movement of the horse, and it takes additional time for the horse to move fully into the carriageway. Only at this point does it become a visibility hazard to the motorist, and at this stage it will not normally be possible for the rider to turn back or stop. As such, at an equestrian crossing, it is recommended that visibility be provided as shown in Table 11.8. However, as commented later, there are cases where it might be difficult to provide the necessary distance *y*.

NOTE

Although not stated in TA 91/05 (DfT, 2005b), a similar provision with regards to visibility is required by pedestrians (i.e. the reaction time between perception and movement mentioned earlier), particularly disabled people, including wheelchair users, and people crossing with helpers or small children. This is because these groups require greater visibility, as they require a longer observation–reaction time (defined in Chapter 3) for their perception and their movement into the carriageway. Only once they have moved onto the carriageway do they become a visibility hazard to drivers; at this stage it will not normally be possible for these categories of pedestrians to stop or turn back in order to avoid a collision if the driver does not take evasive action. Designers may therefore wish to modify their approach in light of the principles and research findings described in Chapters 3 and 4 and the note on intervisibility in Chapter 6.

A summary of distances *y* is provided in Table 11.8. These should be measured from eye heights of 0.9 to 2.0 m for

Table 11.8 Preferred minimum distances *y* for NMU routes at crossings (DfT, 2005a)

Minor route	85th percentile approach speed on main line: km/h	Main route		
		Mainline carriageway: m	Off-cycleway cycle route	Equestrian route
Pedestrian or cycle	All	All as in TD 42/95 (DfT, 1995)	*[See Table 11.5]*	*[See Table 11.6]*
Equestrian	50	135	*[See Table 11.5]*	*[See Table 11.6]*
	60	165		
	70	211		
	85	270		
	100	345		
	120	At-grade crossing not recommended (See TA 91/05 *[(DfT, 2005b)]* for further details)		

pedestrians, 1.0 to 2.2 m for cyclists and 1.5 to 2.7 m for equestrians. The object height should be taken as 0.26 to 2.0 m, in accordance with TD 9/93 *Highway Link Design* (DfT, 2002).

Where it proves difficult to achieve the visibilities set out in this section, measures that reduce speeds of the major arm, commensurate with the maximum visibility that can be practically provided, should be considered.

11.1.5 Alignment

Practicality is a key consideration in the provision of facilities for NMUs. The presence of obstacles and fragmented design elements should be avoided wherever possible. Principles of overall design are dealt with in TA 91/05 (DfT, 2005b). The key points can be summarised as follows.

- Curves be simple and horizontal.
- At corners and junctions, internal corners should be splayed to assist the passage of wheelchairs and pushchairs.
- The preferred minimum radius for cycle routes is 25 m for a 30 km/h design speed and 4 m for a 10 km/h design speed.
- Severe vertical curves are unlikely to occur; for comfort there should be a preferred minimum crest K value of 5.0 and an acceptable minimum crest K value of 1.6 along off-carriageway cycle routes. K values are defined in TD 9/10.

(DfT, 2010)

11.1.6 Gradient and crossfall

Gradients along new pedestrian routes are considered in HD 39/01 *Footway Design* (DfT, 2001); gradients of NMU routes across footbridges are considered in BD 29/04 *Design Criteria for Footbridges* (DfT, 2004b). Further information is given in *Inclusive Mobility* (DfT, 2005d), and described in Chapter 8.

Where pedestrian and cycle routes are considered together, TA 90/05 (DfT, 2005a) does not make recommendations. However, it states that the preferred maximum gradient of off-carriageway cycle routes is 3%, with an acceptable maximum of 5%. Steeper gradients may be more than 5% in short sections. Also, at the base and top of gradients exceeding 2%, a level plateau at least 5 m long is desirable in advance of give-way or stop lines.

Crossfalls for cycle and equestrian facilities may be similar to the values used for footways, with a preferred maximum of 3%, up to an absolute maximum of 5%, as higher values might create manoeuvring difficulties. Crossfalls greater than 3% can create difficulties for cyclists and wheelchairs when surfaces are icy. Moreover, steep crossfalls can cause movement problems for users of mobility scooters, wheelchairs and bicycles or tricycles.

Crossfall can be either to one side or cambered to both sides. However, on bends, adverse crossfall should be avoided.

11.1.7 Cross-section
11.1.7.1 Factors affecting cross-section

The cross-section of an NMU facility will depend on such factors as: whether it is a shared-use, adjacent-use or unsegregated route; visibility; boundary design; if the route is adjacent to or away from a highway and the need for street furniture within the facility.

Where obstructions are unavoidably present, the widths of routes described in the following sections should be increased by at least the widths of the obstruction. Obstructions at or near the centre line of the route might render the site too hazardous or too narrow to use.

11.1.7.2 Pedestrian-only routes

Surface widths of unbounded pedestrian routes should preferably be 2.6 m, with an acceptable minimum of 2.0 m. A route is considered unbounded when it is not adjacent to a physical barrier, such as a wall or fence, at the edge of the route. Where it is not practicable to provide widths of 2.0 m for the full length of a road, widths of 1.3 m may be provided over short distances.

11.1.7.3 Shared and adjacent-use routes for NMUs

Shared-use facilities should generally be restricted to where flows of either cyclists or pedestrians are low, and hence where the potential for conflict is low. Unsegregated shared facilities have operated satisfactorily down to 2.0 m wide with combined pedestrian and cycle use of up to 200 users per hour. However, the preferred minimum width for an unsegregated facility is 3.0 m. A further point is that the cycle is a vehicle; because riders may be riding at five to ten times (15–30 km/h) faster than the speed of pedestrians (3–4 km/h), separate provision for cycle riders may be advisable. In addition, it is noted (Highways England, 2016, p. 11) that:

Cycle traffic shall be separated from pedestrian and equestrian traffic in order to allow cyclists to travel at the design speed.

(Highways England, 2016, p. 11)

For the strategic road network, the default position is that there should be a separate footway 'where pedestrian demand is high enough to justify it' (Highways England, 2016, p. 6). This is because cycle traffic will normally be travelling at speed. Where there is insufficient pedestrian volume to justify a separate footway, it is quite acceptable and legal for pedestrians to walk in the cycle track. Due care is clearly needed by all categories of road user under such circumstances.

Table 11.9 Surface widths of unbounded line-segregated pedestrian and cycle routes, based on DfT (2005a)

	Cycle route: m	Pedestrian route: m	Total width: m
Preferred minimum	3.0	2.0	5.0
Acceptable minimum	1.5	1.5	3.0

© Crown Copyright, 2005
This information is licensed under the Open Government Licence v3.0. To view this licence, visit http://www.nationalarchives.gov.uk/doc/open-government-licence/ **OGL**

The potential for conflict between users increases where flows of more than one group are high. In this case, it is normally necessary to have some segregation along the route. Route segregation should also be considered where disabled people, people with pushchairs or other vulnerable users are likely to make frequent use of the facility. When determining the method of segregation, consideration should be given to these issues and to site-specific factors. For more detailed information, refer to LTN 2/04 *Adjacent and Shared Use Facilities for Pedestrians and Cyclists* (DfT, 2004d).

The preferred separation between different types of NMU is 1.0 m, with an acceptable separation of 0.5 m. Greater verge widths facilitate maintenance. Verges adjacent to field boundaries and existing hedgerows should be a minimum of 0.5 m wide to allow hedges to overhang the route without interfering with its use.

If the separation described here cannot be provided, segregation may be achieved by use of a post and single rail fence, railings, kerbs or delineator strips. Guardrail should only be used in short lengths because, over any appreciable distance, the risk of cycle handlebars and pedals colliding with them is increased. Fences and guardrail can also trap users on the wrong side. The principles are set out in more detailed in LTN 2/04 (DfT, 2004d) and *Inclusive Mobility* (DfT, 2005d). Table 11.9 provides values for the surface widths of line-segregated pedestrian and cycle routes.

11.1.7.4 Boundary treatments

The above widths for pedestrian and cycle routes should be modified in particular circumstances as follows and as shown in Figure 11.4.

■ For a route bounded on one side, where the boundary height is up to 1.2 m, an extra 0.25 m should be provided to allow for kerb shyness between the route and the barrier.

■ For a route bounded on one side, where the boundary height is greater than 1.2 m, an extra 0.5 m should be

provided to allow for kerb shyness between the route and the barrier.

■ For a route bounded on both sides, an extra 0.25–0.5 m should be provided on each side, as appropriate.

It is preferable to provide physical separation between NMU routes and carriageways. For pedestrians and cyclists, the preferred separation between the NMU route and the carriageway is 1.5 m, with an acceptable separation of 0.5 m. The higher value, 1.5 m, should, where possible, be used on roads with speed limits in excess of 40 miles per hour. If a hard strip is provided, this can be considered part of the separation. When new routes are introduced, street furniture and all vegetation (except grass) within the separation distance should be removed, or the verge widened.

11.1.7.5 Hazards adjacent to NMU routes

Where an NMU route is adjacent to a hazard, such as a ditch – or other water feature – or an embankment with a slope steeper

Figure 11.4 Boundary treatments for NMU routes (DfT, 2005a)
© Crown Copyright, 2005
This information is licensed under the Open Government Licence v3.0. To view this licence, visit http://www.nationalarchives.gov.uk/doc/open-government-licence/ **OGL**

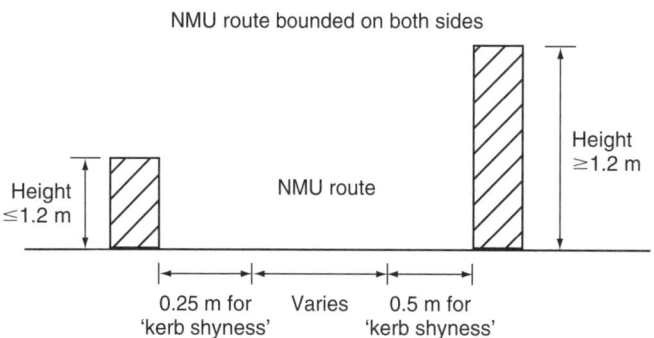

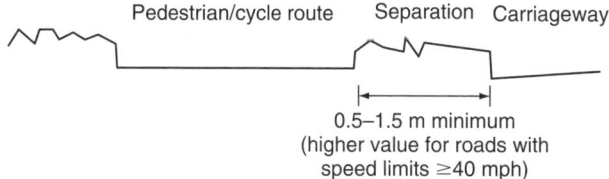

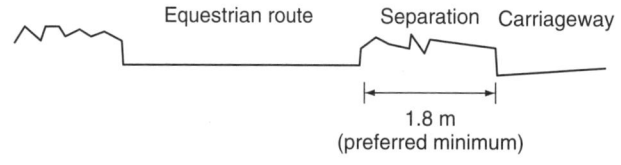

than one in three, a separation greater than that recommended in the previous section should be considered to minimise the risks. Designers should also consider providing physical barriers, such as dense shrubbery, guardrails or fences. Further information is provided in the overseeing organisation's standards for road restraint systems.

These risks are heightened at sharp bends, particularly for cyclists at night if the route is unlit. In such circumstances, consideration should be given to lighting the bend, increasing the separation and the provision of warning signs.

11.1.8   Headroom

For subways and underpasses, guidance on headroom is provided in TD 36/93 *Subways for Pedestrians and Pedal Cyclists: Layout and Dimensions* (DfT, 1993a). The key points for the various NMUs are as follows.

- *Pedestrians.* For obstacles longer than 23 m, a minimum headroom of 2.6 m should be provided. For shorter obstructions, this may be reduced to 2.3 m.
- *Off-carriageway cycle routes.* For obstacles longer than 23 m, a minimum of 2.7 m should be provided.
- *Equestrians.* Desirable headroom for ridden horses is 3.4 m with an absolute minimum of 2.8 m over a short distance. If horses must be led rather than ridden, the headroom may be reduced to 2.8 m over longer distances, such as under bridges. However, this should be avoided where possible, owing to difficulties in controlling horses. Appropriately located mounting blocks and signage should also be provided.

11.1.9   Crossings

11.1.9.1 Visibility and guardrailing

Visibility at crossings is covered in Chapter 3. Advice on the choice of crossing facility within a scheme is given in TA 91/05 (DfT, 2005b).

Where NMUs are in danger of inadvertently entering the carriageway at right angles to the carriageway with limited visibility, or where regular use by unaccompanied children is anticipated, guardrail should be provided to ensure that NMUs slow down before crossing. However, excessive use of guardrailing should be avoided.

11.1.9.2 Pedestrian crossings

Where pedestrian routes cross the carriageway, the desirable minimum crossing provision is a dropped kerb laid flush with the carriageway, with associated tactile paving. Further advice on dropped kerbs is given in TA 57/87 (DfT, 1989). Advice on assessing whether increased crossing provisions is appropriate can be found in TA 68/96 *The Assessment and Design of Pedestrian Crossings* (DfT, 1996) and TA 91/05 (DfT, 2005b).

The ramp gradient across the footway to a dropped kerb should be between 1:12 and 1:20. For narrow footways, a steeper gradient will allow the width of the level strip at the back side of the footway to be maximised. This will make it more comfortable for people with pushchairs or wheelchairs who do not wish to use the crossing.

11.2.   Provision and cross-sections of subways

11.2.1   Factors affecting subway provision and choice of cross-section

The requirements for geometric alignments and cross-sections of subways, access ramps and stairs for use by pedestrians and cyclists are given in TD 36/93 (DfT, 1993a). They are applicable to all schemes for the construction and improvement of trunk roads, including motorways, provided that, in the opinion of the overseeing department, this would not result in significant expense or delay progress.

Many factors affect the choice of whether to provide a subway and, if so, the type of cross-section. Useful factors in considering the provision of a subway include

- volume of pedestrian traffic
- volume of cycle traffic
- type of land use to which access is to be provided
- speed of vehicles along the road and the volume of the traffic, including the proportion of heavy goods vehicles
- location, convenience and safety of alternative routes for pedestrians and cyclists
- use by elderly, visually impaired, disabled people, including wheelchair users, and people with prams and pushchairs
- environmental aspects
- other aspects particularly relevant to the local situation
- cost of subway
- effect of change on local land use over the next 15 years, including any prospective recreational routes for pedestrians and cyclists.

11.2.1.1 Siting of the subway

The line of the subway should be close to the main line of travel for the majority of subway users, in order to maximise its use. The subway should be as short as possible. Where the number of pedestrians is large, an option might be to raise the level of the road, so as to reduce the height and length of pedestrian stairs and ramps.

11.2.1.2 Types of subway

Subways may be designed for use by pedestrians only or by pedestrians and cyclists. Subways for joint use should normally be segregated, preferably by level difference. However, an unsegregated shared surface for both pedestrians and cyclists may be suitable in some situations. Additional headroom may be required where a bridleway passes through a subway.

Personal security aspects are extremely important in encouraging use of the subway. Wide approaches and subway alignments with good through visibility and good lighting within the view of passing pedestrians and passing traffic, will help to minimise pedestrians' fears for their personal safety. Subways and their accesses should be designed to avoid places of concealment, in the interests of personal security. Motorcycle barriers may be necessary in some locations to prevent cars and motorcycles being driven into subways or subway approaches.

11.2.2 Cross-section of subways for pedestrians only

Three types of pedestrian subway may be used.

- *A wide section*, suitable for situations where a subway forms an extension to a footpath system not less than 0.5 m wide carrying large numbers of pedestrians or where, for aesthetic reasons, the normal section is not considered suitable.
- *A normal section*, suitable for the majority of situations.
- *A narrow section*, for situations with small numbers of pedestrians where the normal section could not be justified on cost grounds.

If circular or other shaped sections are proposed, they should circumscribe rectangular sections with dimensions not less than the minimum laid down in TD 36/93 (DfT, 1993a). The minimum height and width for pedestrian-only subways are shown in Table 11.10.

Sight distances of 4.0 m or more should be provided at corners and changes of direction. For calculation purposes, pedestrians can be assumed to be 0.4 m away from an adjacent vertical wall. The visibility envelope should extend from a height of 1.5 m, representative of an adult, to 0.6 m, representative of a child. Inside corners rounded off to a radius of 4.6 m will meet this criterion.

11.2.3 Cross-sections of subways for combined use
11.2.3.1 General

Pedestrians and cyclists can share the same subway and associated ramps. For combined use to be successful, the existing travel lines and those expected in the future should be investigated for both pedestrians and cyclists. Short diversions of one mode might be necessary to encourage the other mode to use the dual facility.

11.2.3.2 Segregated subways

The width for pedestrians should be segregated from the width for cyclists, preferably by level differences. These widths, as well as minimum dimensions for cross-sections, are shown in Figure 11.5 and Table 11.11. Alternatively, segregation can be achieved by means of guardrailing, which would serve as a physical barrier to separate footpath users from cycle track users. Where these measures are not suitable, a raised dividing line and tactile paving should be provided to assist visually impaired people.

The stopping sight distances for cyclists given in Table 11.12 should be provided within the subway and on its approaches. These are illustrated in Figure 11.6. These distances are applicable to design speeds of 10 km/h or less on sharp curves and straight stretches. The geometric requirements for cyclist facilities will, in these cases, also affect the dimensions of the pedestrians' facilities. The design speeds are not significantly affected by gradient. For layout purposes, the line of sight of a cyclist should be taken from a point 1.5 m high, and at least 0.6 m away from the edge of the cycle track. The design of subway walls, wing walls, associated ancillary earthworks and

Figure 11.5 Cross-section of a typical segregated subway for combined use (DfT, 1993a)
© Crown Copyright, 1993
This information is licensed under the Open Government Licence v3.0. To view this licence, visit http://www.nationalarchives.gov.uk/doc/open-government-licence/ **OGL**

Table 11.10 Minimum dimensions for pedestrian-only subways (DfT, 1993a)

Type of subway	Length of subway: m	Height: m	Width: m
Wide	–	2.6	5.0
Normal	<2.3	2.3	3.0
	≥2.3	2.6	3.3
Narrow	–	2.3	2.3

© Crown Copyright, 1993
This information is licensed under the Open Government Licence v3.0. To view this licence, visit http://www.nationalarchives.gov.uk/doc/open-government-licence/ **OGL**

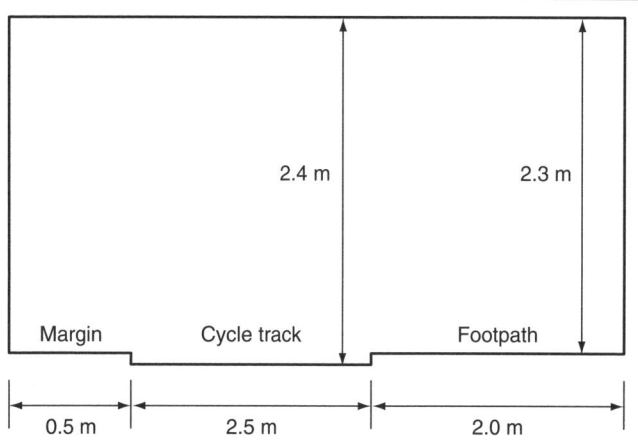

Table 11.11 Minimum dimensions for segregated subways for pedestrians and cyclists (DfT, 1993a)

Length: m	Height: m		Width: m		
	Cycle track	Footpath	Margin between subway wall and cycle track	Cycle track	Footpath
<2.3	2.4	2.3	0.5	2.5	2.0
≥2.3	2.7	2.6	0.5	2.5	2.0

Table 11.12 Stopping sight distance for cyclists (DfT, 1993a)

Design speed: km/h	Minimum stopping sight distance: m	Minimum radius of curvature of walls adjacent to cycle track: m	Minimum radius of curvature of walls adjacent to footpath: m
≤10	4.0	4.6	4.6
≤25	26.0	68.0	28.5

landscape works should take account of these visibility requirements.

11.2.3.3 Unsegregated subways
When the total number of pedestrians and cyclists is small, an unsegregated subway may be acceptable, particularly for short subways with good through visibility. The minimum dimensions for cross-sections are given in Table 11.13. At sites where space is restricted or where the total number of pedestrians and cyclists is very small, the subway width may be reduced to 3.0 m.

An alternative where few cyclists are expected is to provide a narrow pedestrian subway in accordance with Table 11.10. Signs would be required to indicate that the cyclists should dismount before entering the subway and that no cycling is permitted within the subway. It would also be necessary to

Table 11.13 Minimum dimensions for an unsegregated subway for pedestrians and cyclists (DfT, 1993a)

Subway length: m	Height: m	Width: m
<23	2.4	4.0
≥23	2.7	4.0

ensure that the cycle track is legally terminated at either side of the subway.

11.2.3.4 Equestrian use
When bridleways are to be incorporated into subways, the minimum headroom should be 3.7 m, except where suitable facilities for the riders to dismount and remount are provided, when the headroom may be reduced to 2.7 m. Suitable signs should be erected to indicate that equestrians are required to dismount if the latter option is adopted. The minimum width of a subway for equestrians should be 3.0 m.

11.2.4   Access
11.2.4.1 General
Access to subways may be via ramps or stairs, which may be straight or helical. Consideration should be given to providing both ramps and stairs to suit able-bodied people, cyclists, people with prams and pushchairs, those with heavy shopping or luggage, visually impaired people and disabled people, including wheelchair users.

Access ramps or stairs should normally be the same width as the subway, except that when multiple ramps and stairs are connected to a single subway, they may be narrower.

The thresholds of all subway accesses and the tops and bottoms of flights of stairs should be provided with a system of tactile paving to assist visually impaired people. Details are shown in *Inclusive Mobility* (DfT, 2005d).

Figure 11.6 Stopping sight distances for cyclists (DfT, 1993a)
© Crown Copyright, 1993
This information is licensed under the Open Government Licence v3.0. To view this licence, visit http://www.nationalarchives.gov.uk/doc/open-government-licence/ **OGL**

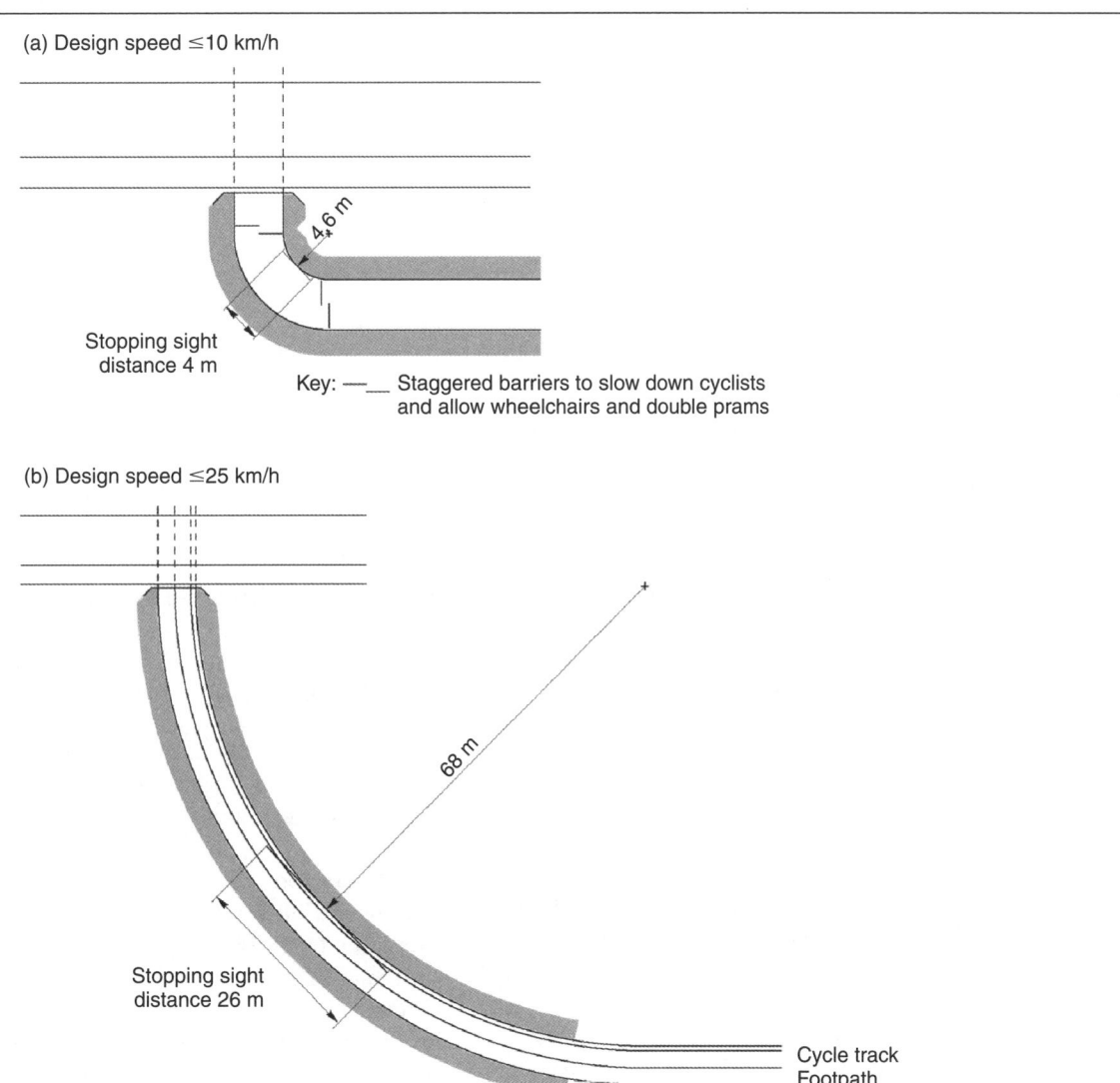

(a) Design speed ≤10 km/h

4.6 m

Stopping sight distance 4 m

Key: ——__ Staggered barriers to slow down cyclists and allow wheelchairs and double prams

(b) Design speed ≤25 km/h

68 m

Stopping sight distance 26 m

Cycle track
Footpath

11.2.4.2 Access ramps

Ramps should not be allowed to run into the subway beyond the threshold, as there could be a risk of cyclists hitting the soffit of the subway.

Landings should be provided at changes of direction and changes of gradient. Landings should be used, even on straight ramps, so that the total rise between landings is less than 3.5 m. Landings should normally be the same width as the ramp and at least 2.0 m long, measured along the centre line of the landing. All landings should be approximately horizontal and appropriately drained.

11.2.4.3 Pedestrian ramps

Gradients of 5% or less are preferred for access ramps where significant numbers of disabled persons or heavily laden shoppers are expected to use the subway. In other situations, gradients shallower than 8% are preferred but gradients up to 10% are permitted for short lengths in exceptionally difficult sites. Stepped ramps may also be considered at exceptionally difficult sites, although wheelchair users find stepped ramps difficult to negotiate.

11.2.4.4 Cycle ramps

From the point of view of limiting speed and reducing cycling effort, access ramp gradients should preferably not exceed

Table 11.14 Dimensions for straight stairs (DfT, 1993a)

Rise, r: mm			Going, g: mm			Pitch: degrees	
Min.	Max.	Optimum	Min.	Max.	Optimum	Max.	Optimum
100	150	130	280	350	300	33	27

Table 11.15 Dimensions for helical stairs (DfT, 1993a)

Rise, r: mm	Going, g: mm			$2r + g$	
Min.	Min. inner going	Min. centre going	Max. outer going	Min.	Max.
150–190	150	250	450	480	800

3%, with a maximum of 5%, or 7% in restricted spaces. For the steeper gradients, barriers should be considered to encourage slower speeds until cyclists have cleared the steep ramps. Staggered barriers may help to reduce speeds.

11.2.4.5 Straight access stairs
Key dimensional features of stairs at subways are summarised in Table 11.14 and are as follows.

- The headroom between any ceiling and stair measured vertically should be not less than the height of the subway.
- The preferred maximum is 20 steps.
- Landings should preferably be the same width as the stair, preferably 1.8 m deep, or 1.2 m in restricted areas.
- There should be not more than three successive flights without a change of direction of 30° or more at a landing.
- All landings should be approximately horizontal and adequately drained.
- Stair flights limited to nine steps are preferred where significant numbers of disabled people are expected to use them.
- Stair treads should be uniform for a subway system, with steps of equal rise.
- Nosings on the stairs should be rounded to a 6 mm radius without overhang, and should be colour-contrasted from the rest of the steps, as shown in Figure 11.7.

11.2.4.6 Helical access stairs
At sites where space is restricted, helical stairs may offer a useful alternative to straight stairs. The dimensions are given in Table 11.15.

11.3. Crossings on trunk roads
An important aspect of walking and other non-motorised modes of travel is the provision of facilities on and adjacent to trunk roads. Vehicular volumes and speeds on these facilities, often in rural areas, tend to be much higher than on many urban routes. Provision for NMUs is addressed in TA 91/05 (DfT, 2005b). This section highlights some of the key design matters described there.

11.3.1 Introduction
The right of all NMUs to use public highways, unless specifically prohibited, is noted, as is the encouragement of modal shift to walking and cycling. Trunk roads often provide important and quick routes for NMUs, despite the discomfort resulting from the proximity of high volumes of fast-moving vehicles. Improved facilities for people walking or cycling and for equestrians, therefore, are of considerable importance.

11.3.2 Design brief and NMU audit
In the scheme development and assessment process, the design brief should take account of HD 42/05 *Non-motorised User*

Figure 11.7 Stair elements (DfT, 1993a)

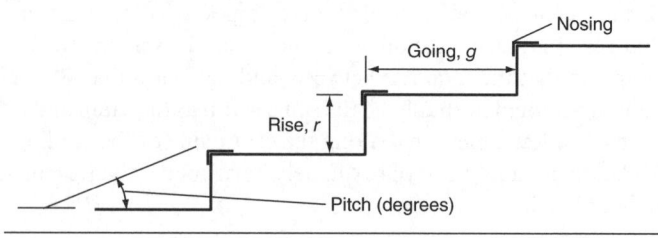

Table 11.16 Hierarchy of provision (DfT, 2005a)

	Hierarchy of provision for:	
	Pedestrians or equestrians	Cyclists
Consider first	Traffic reduction	Traffic reduction
	Speed reduction	Speed reduction
	Reallocation of road space to pedestrians or equestrians	Junction or hazard site treatment, traffic management
	Provision of grade crossings	Redistribution of the carriageway (bus lanes, widened nearside lanes, etc.)
	Improved pedestrian or equestrian routes on existing desire lines	Cycle lanes, segregated cycle tracks constructed by reallocation of carriageway space, cycle tracks away from roads
Consider last	New pedestrian alignment or grade separation	Conversion of footways to unsegregated shared-use cycle tracks alongside the carriageway

Audits (DfT, 2005c), as discussed in Chapter 12. Design organisations are required to apply NMU audit procedures as part of the design process. This audit process will assist in identifying existing conditions and the objectives of new or improved NMU provision. This is important in identifying appropriate solutions. Future usage patterns must also be considered and NMU provision should be designed on the basis of a 15-year design life.

Consultation is essential; this should be carried out with interested parties, including local authorities, The Ramblers (formerly The Ramblers' Association), the Cyclists' Tourist Club, Sustrans and The British Horse Society.

11.3.3 General design principles
A combination of approaches may offer the preferred solution; a hierarchy of provisions is shown in Table 11.16. Five core principles common to NMU routes have been identified in LTN 1/04 *Policy, Planning and Design for Walking and Cycling* (DfT, 2004a): convenience, accessibility, safety (perceived as well as actual), comfort for all types of user, and attractiveness (in ways that will encourage their use).

A basic design principle is that existing right-of-way networks should be preserved as far as possible. Where right-of-way diversions are necessary, such diversions should not normally result in additional journey lengths greater than 10% for NMUs, unless agreed with the overseeing organisation.

NOTE
Not addressed in the official guidance on facilities for NMUs on trunk roads is the fact that vehicle speeds on trunk roads may be particularly high. The need, therefore, to consider NMU sight distances using some numerical consideration, such as observation–reaction times and walking (or cycling and equestrian) crossing times could be especially important. This is discussed in Chapter 3.

11.3.4 Off-carriageway routes
Consideration should also be given to the nature of the route and whether it should be designated an 'off-carriageway route'. These routes may be identified and classified as shown in Table 11.17.

11.3.5 Crossings
Crossing facilities for NMUs should aim for the characteristics described earlier, i.e. safety and comfort; location on desire lines; convenience; capacity; and opportunity in responding quickly and safely to demand from NMUs.

NOTE
In the UK there is no formal advice or guidance for estimating capacity and associated level of service of crossings. Refer to Chapter 13 and the Highway Capacity Manual *(TRB, 2016) for details of the methods used in the USA and Canada.*

The location and geometric design of right-of-way crossings require particular and careful attention because of the interaction of NMUs with vehicles. At informal at-grade crossings, where NMUs are expected to cross without special provision, the difficulty of crossing depends primarily on the width of crossing and the availability of suitable gaps.

Informal at-grade crossings should not be provided on dual carriageways of three or more lanes per carriageway. Table 11.18

Table 11.17 Types of off-carriageway route (DfT, 2005b)

Route type classification	Description
A	Within trunk road verge
B	Land outside, but adjacent to the highway boundary
C	Distant from trunk road
D	Existing right of way
E	Redundant or by-passed road
F	Minor highway
G	Other locations, such as forestry tracks, canal towpaths, abandoned railway lines and farm tracks, which may be in public or private ownership

provides additional criteria to assist in determining whether informal grade crossings are appropriate, based on average annual daily traffic flows. These criteria should be seen as a general guide; local factors should also influence the decision.

Selection of the most appropriate crossing is based on consideration of whether a crossing is required and, if so, its location and type. Crossings should take the shortest safe route and incorporate tactile surfaces, signage, drop kerbs and special markings if appropriate. Adequate visibility is required for both drivers and NMUs, particularly in view of the high speeds likely on trunk roads. Aspects of this are discussed in earlier chapters, particularly referring to TA 90/05 (DfT, 2005a).

Further guidance on provision of NMU facilities is contained in references in earlier chapters to different crossing types and controls. Potential locations for new or improved crossings should be considered in accordance with TA 68/96 *The Assessment and Design of Pedestrian Crossings* (DfT, 1996), which sets out site and option assessments. The shortest safe route, combinations of NMU facilities, signs, markings, dropped kerb provision and adequate visibility are some of the key features. Characteristics of NMU crossing facilities are summarised in Table 11.19 and described in greater detail in TA 91/05 (DfT, 2005b).

Important physical dimensions associated with the crossings are summarised next.

11.3.5.1 Refuge islands

Physical islands on high-speed roads might constitute a hazard; consideration should be given to speed reduction measures in these situations. Any island on a road with a speed limit greater than 40 miles per hour (64 km/h) that is not part of a single-lane dualling design, requires 'departure from standards' approval.

The preferred crossing width for pedestrian refuge islands is 2.0 m (minimum 1.5 m at constrained locations). The preferred crossing width for cyclist refuge islands is 3.0–4.0 m (2.5 m minimum at constrained locations). The length of the refuge should be determined by the frequency and type of use, but should not be less than the width of the connecting cycle facility or less than 2.0 m.

11.3.5.2 Zebra crossings

These crossings offer immediate response to pedestrian demand and provide priority to the pedestrian across the whole crossing. However, they should not be introduced on roads with an 85th percentile speed of 35 miles per hour or above.

Further details on the design of zebra crossings are available in the Zebra, Pelican and Puffin Pedestrian Crossings Regulations and General Directions 1997 and TA 68/96 (DfT, 1996).

Table 11.18 Criteria for suitability of informal at-grade rights of way crossings, based on DfT (2005b)

Road type	Annual average daily traffic flow		
	Normally appropriate	Potentially appropriate[a]	Not normally appropriate
Single carriageway	Below 8000	8000–12 000	Above 12 000
Dual carriageway	Below 16 000	16 000–25 000	Above 25 000
Wide single carriageway	–	Below 10 000	Above 10 000

[a] For 'potentially appropriate' crossings, designers should consider factors, including site-specific factors that assist crossing, such as nearby signals, applicable speed limits, demand characteristics of NMUs, including types of user and journey, overall diversions and delays caused to NMUs, implications of a grade-separated crossing and possible mitigating measures.

Table 11.19 NMU crossing facilities (DfT, 2005b)

Grade	Control	Crossing type
At grade	Informal	Pedestrians and cycle crossing (with or without refuges)
		Cycle-priority crossing
		Equestrian crossing with holding area
	Formal uncontrolled	Zebra crossing
	Formal signalled	Pelican crossing
		Puffin crossing
		Toucan crossing
		Equestrian crossing
		NMU stages at traffic signals
		Advanced stop lines
Grade-separated		Underpasses
		Bridges

© Crown Copyright, 2005
This information is licensed under the Open Government Licence v3.0. To view this licence, visit http://www.nationalarchives.gov.uk/doc/open-government-licence/ **OGL**

11.3.5.3 Signalised crossings in general
The 85th percentile speed must not exceed 50 miles per hour for stand-alone signal-controlled crossings.

11.3.5.4 Pelican crossings
These are used away from junctions and are signal-controlled. The crossing uses farside pedestrian signal heads with a fixed duration green walking figure period and a flashing amber traffic signal and flashing green-figure pedestrian signal, demanded solely by a push-button.

Other features of signal-controlled crossings are discussed in Chapter 7.

11.3.5.5 Drainage, kerbs and related features
Other concerns affecting the geometric design of NMU facilities include the position of drainage features. On dips and on kerbed sections, positive drainage should be provided to prevent ponding. Ditches and gullies hidden in verges are a hazard and should be avoided. Where they are necessary, they should be a minimum of 0.5 m back from the edge of the NMU route to avoid hazards if NMUs accidentally leave the route.

Where an existing kerb segregates a cycle or equestrian route, drainage should take place through 300 mm wide gaps in the extruded kerb. The frequency of the gaps will be influenced by the drainage design for the route.

11.4. Pedestrian–cyclist interaction – an urban area emphasis
This section is based on LTN 1/12 *Shared-Use Routes for Pedestrians and Cyclists* (DfT, 2012), which supersedes LTN 2/86

Shared Use by Cyclists and Pedestrians (DfT, 1986). It should be read in conjunction with LTN 2/08 *Cycle Infrastructure Design* (DfT, 2008) and *Inclusive Mobility – A Guide to Best Practice on Access to Pedestrian and Transport Infrastructure* (DfT, 2005d). For convenient reference, the various paragraph numbers are included in the following sections.

11.4.1   Scope
1.8 [LTN 1/12 (DfT, 2012)] focuses on routes within built-up areas, where the predominant function of the route is for utility transport, and where use by pedestrians and/or cyclists is likely to be frequent. As such, it expresses a general preference for on-carriageway provision for cyclists over shared use. However, it is not meant to discourage shared use where it is appropriate.

1.10 Guidance on introducing cycle routes in rural areas, where urban-style engineering measures can be intrusive, is available from Sustrans (see [Sustrans (2018)]).

1.12 Guidance on cycling in pedestrianised (vehicle restricted) areas is given in LTN 2/08 *Cycle Infrastructure Design* [(DfT, 2008)] and TAL 9/93 *Cycling in Pedestrian Areas* [(DfT, 1993b)].

(DfT, 2012)

11.4.2   The Equality Act
1.13 Shared use schemes are often implemented to improve conditions for cyclists, but it is essential that they are designed to take into account the needs of everyone expected to use the facility.

1.14 Disabled people and older people can be particularly affected by shared use routes. Ultimately, however, it will depend on the quality of the design. Consideration of their various needs is an important part of the design of shared use, and the duties under the Equality Act 2010 are particularly relevant.

(DfT, 2012)

Chapter 8 discusses these matters in more detail.

11.4.3 Pedestrians and shared use

6.6 Designers should aim to ensure that conversion to shared use does not result in the displacement of existing users.

6.7 Conflict between pedestrians and cyclists is not a common occurrence – see [. . .] *Shared Use Operational Review* [(Atkins, 2012)]. Nevertheless, perception of reduced safety is an important issue for consideration, because it has a bearing on user comfort, especially for older and disabled people.

6.10 If a significant proportion of these groups is likely to use the route, it might be necessary to make modifications to meet their particular requirements.

(DfT, 2012)

11.4.4 Segregation

7.1 An unsegregated route is the simplest option – it is relatively inexpensive, the least visually intrusive, easier to maintain and makes good use of the land available where width is limited. [Figure 11.8] shows an example of unsegregated shared use.

(DfT, 2012)

7.2 However, omitting segregation will not always be appropriate, especially on busier routes. Segregation can increase the sense of safety, user confidence and user comfort, and it might be required for a particular scheme to operate satisfactorily.

7.6 Previously, it has been considered good practice to segregate shared-use routes wherever practicable. [. . .] However, designers are increasingly being encouraged to take decisions appropriate to the scheme context rather than adopting certain features as a starting position in the design development process.

(DfT, 2012)

11.4.5 Level surface segregation

7.16 Level surface segregation does not rely on any appreciable difference in level between the pedestrian and cyclist sides. The most common form of level surface segregation feature is the white line.

7.17 Research [(Atkins, 2012) found that,] in general, white line segregation was ineffective in ensuring a high degree of user compliance.

7.18 The cases studied were on relatively narrow routes where the average overall width was 3.5 m. Where width is greater, compliance with white line segregation is less likely to be an issue. Regardless of this, conflict was not found to be a problem. [Table 11.20] shows some of the advantages and disadvantages of level surface segregation.

(DfT, 2012)

One way of displaying the separation of cycle and foot traffic, suggested by Parkin (2018), is by means of colour contrast. Figure 11.9 shows an example in Utrecht in the Netherlands, where this has been done.

11.4.6 Width requirements

7.28 Insufficient width tends to reduce user comfort and increases the potential for conflict between pedestrians and cyclists.

(DfT, 2012)

Therefore the advice given in LTN 2/08 (DfT, 2008) is now superseded in LTN 1/12 (DfT, 2012), which revises the advice on additional clearances related to the cycle track only, as shown in Table 11.21.

7.45 Note that where a shared-use route is unsegregated, any such additions will apply to both sides of the route [see Figure 11.10] because cyclists can use its full width.

(DfT, 2012)

Figure 11.8 Example of unsegregated shared use (DfT, 2012, Figure 7.1)
© Crown Copyright, 2012
This information is licensed under the Open Government Licence v3.0. To view this licence, visit http://www.nationalarchives.gov.uk/doc/open-government-licence/ **OGL**

Table 11.20 Advantages of level surface segregation (DfT, 2012, Table 7.3)

Type	Advantages	Disadvantages
White line	Inexpensive	Not detectable by tactile means
(The Traffic Signs Regulations and General Directions 2002, Diagram 1049)	Minimal width take-up	Often ignored
	Easier to maintain than physically segregated routes	Might be visually intrusive
Raised white line	Detectable by tactile means	Can be difficult to construct properly, which might present a trip or cycle hazard
(The Traffic Signs Regulations and General Directions 2002, Diagram 1049.1)	Inexpensive	
	Minimal width take-up	Often ignored
	Easier to maintain than physically segregated routes	Can impede surface drainage unless gaps are provided
		Might be visually intrusive
Contrasting surfaces, e.g. a block paved footpath alongside an asphalt cycle track	Might be detectable by tactile means	Likely to be ignored
	Minimal width take-up	
	Easier to maintain than physically segregated routes	
Surface texture, e.g. a grass median strip	Detectable by tactile means	Takes up more width than a white line
	Inexpensive	
	Can be easier to maintain than physically segregated routes	

7.34 A width of 3 m should generally be regarded as the preferred minimum on an unsegregated route, although in areas with fewer cyclists or pedestrians a narrower route might suffice. […] The need here for additional width is not clear cut, because the absence of segregation gives cyclists greater freedom to pass other cyclists. It might therefore depend on user flows.

7.35 Note that 3 m is the preferred minimum effective width, and this will be the actual width where the route is not bounded by vertical features [see Figure 11.11].

(DfT, 2012)

Figure 11.9 Separation of footway and cycle way by colour contrast (Parkin, 2018)

7.36 [Figure 11.10] shows an example of unsegregated shared use alongside a typical urban carriageway. In this case, the vertical edge features create the need for additional width [see Table 11.21]. Where a route (segregated or otherwise) passes alongside a high-speed road, it is recommended that the clearance to the kerb is increased as shown to provide a buffer zone.

7.37 Where sign posts or lamp columns are present, they should be located outside the effective width zone where possible.

7.38 On segregated shared-use routes, and where cycle flow is predominantly one-way, the preferred minimum effective width for the cycle track is 2 m. This will allow for the occasional overtaking manoeuvre and will easily accommodate users of cycle trailers, tandems, tricycles, etc. The preferred minimum effective width for a two-way cycle track is 3 m. These effective widths will need additional clearance where track edge constraints such as kerbs or walls are present [see Table 11.21].

Table 11.21 Additional clearances to maintain effective widths for cyclists (DfT, 2012, Table 7.4)

Type of edge constraint	Additional width required to maintain effective width of cycle track: mm
Flush or near-flush surface	No additional width needed
Kerb up to 150 mm high	Add 200
Vertical feature from 150 to 600 mm high	Add 250
Vertical feature above 600 mm high	Add 500

© Crown Copyright, 2012
This information is licensed under the Open Government Licence v3.0. To view this licence, visit http://www.nationalarchives.gov.uk/doc/open-government-licence/ **OGL**

7.39 As a general rule, for any shared-use route (segregated or otherwise) away from the road, it can be assumed that cyclists will want to travel along the route in both directions.

7.46 [Table 11.21] introduces a further amendment to Table 8.2 of LTN 2/08. The additional clearance of 250 mm now applies to vertical features from 150 mm to 600 mm high (previously it was for features from 150 mm to 1200 mm high). This change recognises the fact that features between 600 mm and 1200 mm high can have a similar potential to come into contact with handlebars as do features higher than 1200 mm.

7.47 [Table 11.22] summarises the minimum width recommendations for pedestrians and cyclists.

(DfT, 2012)

7.48 For shared-use routes, capacity is unlikely to be a constraint. Before capacity is reached, a shared use route will tend to become uncomfortable [because of such factors

Figure 11.10 Unsegregated shared use bounded by vertical features (DfT, 2012, Figure 7.7)
© Crown Copyright, 2012
This information is licensed under the Open Government Licence v3.0. To view this licence, visit http://www.nationalarchives.gov.uk/doc/open-government-licence/ **OGL**

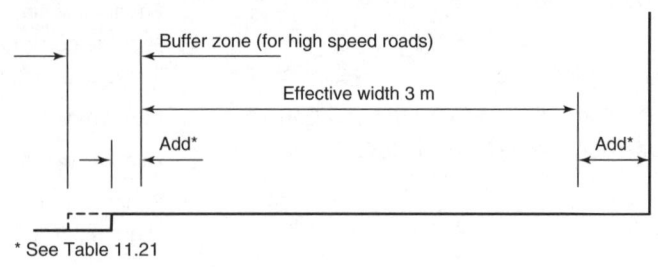

* See Table 11.21

Figure 11.11 Unsegregated shared use (DfT, 2012, Figure 7.6)
© Crown Copyright, 2012
This information is licensed under the Open Government Licence v3.0. To view this licence, visit http://www.nationalarchives.gov.uk/doc/open-government-licence/ **OGL**

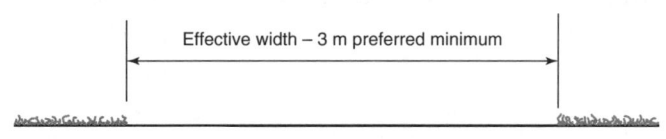

as] the ratio of pedestrians to cyclists, the type of journeys being made and the extent to which people walk in groups. These factors will often differ substantially by time of day and by day of the week.

7.49 [Table 11.23] lists sources of advice on user flows. It can be seen that there is little consistency in the values – when expressed in terms of users per hour per metre width, they range from 25 to 180. [LTN 1/12 (DfT, 2012)], therefore, does not include any recommended flow values.

(DfT, 2012)

Table 11.22 Summary of minimum widths (DfT, 2012, Table 7.5)

Type		Minimum widths
Unsegregated shared use		3 m preferred (effective)[a]
Segregated shared use	Pedestrian path unbounded on at least one side, e.g. segregated by white line	1.5 m (actual)
	Pedestrian path bounded on both sides	2 m (actual)
	One-way cycle track	2 m preferred (effective)[a]
	Two-way cycle track	3 m preferred (effective)[a]

[a] Additional width is needed where there are edge constraints, see Table 11.21.
© Crown Copyright, 2012
This information is licensed under the Open Government Licence v3.0. To view this licence, visit http://www.nationalarchives.gov.uk/doc/open-government-licence/ **OGL**

Table 11.23 Sources of advice on user flows (DfT, 2012, Table 7.6)

Source document	Suggested flows	Comments
LTN 2/86 *Shared Use by Cyclists and Pedestrians* (DfT, 1986)	Combined peak flows of 180 cyclists and pedestrians per hour per metre width for routes with a 500 mm clearance margin to the carriageway	This figure was derived from surveys of routes with level surface segregation
Greenways Handbook (Countryside Agency, 2001a)	200 users per hour	No indication of route width given
Sign Up for the Bike – Design Manual for a Cycle-friendly Infrastructure (CROW, 1993)	25 pedestrians per hour per metre width on traffic-free paths away from town centres	
Countryside Agency – *How People Interact on Off Road Routes* (Countryside Agency, 2001b), *How People Interact on Off Road Routes: Phase 2* (Countryside Agency, 2003)	At least 100 users per hour on 3 m path	Actual and perceived conflict was found to be low at these flow levels
Federal Highway Administration (2006)	150 users per hour on 3 m path	Level of service C, taken from look-up table, assuming average modal split

ª Additional width is needed where there are edge constraints, see Table 11.21.

7.51 The figures of 180 and 25 users per hour average out as three users per minute, and one user about every 2.5 minutes, respectively. [...] In practice, flow is not uniform. Group size itself might then become the dominant factor, especially where width is limited.

7.52 It can be seen from [Table 11.23] that, at a value of, say, 120 users per hour per metre width (2 per minute), the flow appears to be quite conservative, even on an unsegregated route at its minimum recommended width. It seems likely, therefore, that, on wider routes, flows considerably in excess of 180 users per hour per metre width might be comfortably accommodated.

7.53 The last row in [Table 11.23] refers to level of service (LOS). This is a concept aimed at quantifying how well a route performs for its users. Level of service has been used outside the UK and is being developed for use by Transport for London. The Sustrans report *The Merits of Segregated and Non-Segregated Traffic-Free Paths* (Phil Jones Associates, 2011) includes a review of LOS measures in other countries. It identifies a number of models and provides guidance on their possible application to designing shared use in the UK (but with caveats).

(DfT, 2012)

In the USA, a flow of 150 users per hour is defined as LOS C, as indicated in a look-up table (Federal Highway Administration, 2006). See also Chapter 13 for the more detailed approach given in the *Highway Capacity Manual* (TRB, 2016).

11.4.7 Provision alongside carriageways

7.54 Where a footway is converted to shared use, care is required to ensure the route is not unduly obstructed by lighting columns, signs and other street furniture. The cycle track should normally be located on the carriageway side of a segregated shared-use route. This avoids placing pedestrians between cyclists and motor vehicles, it makes it easier for cyclists to leave or join the carriageway and it reduces the potential for cyclists coming into conflict with drivers or pedestrians exiting from private premises along the route.

7.55 While placing cyclists nearest the carriageway is generally preferred when segregation is present, it can increase the potential for conflict between cyclists and people waiting at bus stops. Section 8.10 of LTN 2/08 *Cycle Infrastructure Design* [(DfT, 2008)] suggests that conflict might be reduced by swapping the footway and cycle track positions over so that cyclists pass behind the bus shelter (where present) and any people waiting. However, these crossover points can become areas of conflict, and the resulting markings add to visual intrusion. In view of this, it might be better in such situations to simply dispense with segregation altogether.

7.6 Previously, it has been considered good practice to segregate shared-use routes wherever practicable. This approach appears to have been based on a presumption that there is considerable potential for conflict between pedestrians and cyclists on unsegregated routes. However, designers are increasingly being encouraged to take decisions appropriate to the scheme context rather than adopting certain features as a starting position in the design development process.

7.56 Where the route is frequently interrupted by side roads or [...] crossovers to private driveways, shared use might be a less attractive option because of the increased potential for conflict at these locations and the need for cyclists to keep slowing down. In such cases, designers should consider whether cyclists would be better served by being kept in the carriageway.

7.57 Cyclists should have priority at vehicle crossovers to private accesses. However, where the cycle track crosses the mouth of a side road or an access road to commercial properties, giving cyclists priority needs careful consideration because of the potential consequences of a driver failing to recognise the need to give way. [...] A cycle track crossing can only be signed to give cyclists priority if it is located on a road hump (see Direction 34(2) of The Traffic Signs Regulations and General Directions 2002).

7.58 A cycle track without priority can also cross on a road hump [...]. This is a useful option – it avoids the potential problems of giving cyclists priority while, in many cases, cyclists will be able to cross without stopping, especially where visibility is good. Placing the crossing on a hump can also improve conditions for pedestrians. Other measures to consider include narrowing the carriageway, highlighting the crossing with contrasting surfacing and tightening nearby kerb radii. Cycle track crossings are discussed in more detail in LTN 2/08 *Cycle Infrastructure Design* [(DfT, 2008)].

(DfT, 2012)

11.5. Summary

With the encouragement of NMUs as an important part of environmental and health goals and stated governmental policies, the provision of joint facilities is increasing. As a result, the efficient use of often limited space will be required. The outline of major aspects of the needs and response in terms of geometric design presented in this chapter is intended both to provide information about current design guidance and to alert designers to the need for greater attention and potential research needs for safely and conveniently accommodating the needs of users. Only by these means and related encouragement will users be adequately provided for.

REFERENCES

Atkins (2012) *Shared-Use Operational Review*. DfT, London, UK.

CIHT (Chartered Institute for Highways and Transport) (2015a) *Planning for Walking*. CIHT, London, UK.

CIHT (2015b) *Designing for Walking*. CIHT, London, UK.

Countryside Agency (2001a) *Greenways Handbook*. Natural England, Worcester, UK.

Countryside Agency (2001b) *How People Interact on Off Road Routes*, CRN32. Natural England, Worcester, UK.

Countryside Agency (2003) *How People Interact on Off Road Routes*, CRN69. Natural England, Worcester, UK.

CROW (Centre for Research and Contract Standardisation in Civil Engineering) (1993) *Sign Up for the Bike – Design Manual for a Cycle-friendly Infrastructure*. CROW, Ede, the Netherlands.

DfT (Department for Transport) (1986) LTN 2/86: *Shared Use by Cyclists and Pedestrians*. The Stationery Office, London, UK.

DfT (1989) TA 57/87: Roadside features. In *Design Manual for Roads and Bridges (DMRB)*. DfT, London, UK.

DfT (1993a) TD 36/93: Subways for pedestrians and pedal cyclists: layout and dimensions. In *Design Manual for Roads and Bridges (DMRB)*. DfT, London, UK.

DfT (1993b) TAL 9/93: *Cycling in Pedestrian Areas*. DfT, London, UK.

DfT (1995) TD 42/95: Geometric design of major/minor priority junctions. In *Design Manual for Roads and Bridges*. DfT, London, UK.

DfT (1996) TA 68/96: The assessment and design of pedestrian crossings. In *Design Manual for Roads and Bridges*. DfT, London, UK.

DfT (2001) HD 39/01: Footway design. In *Design Manual for Roads and Bridges*. DfT, London, UK.

DfT (2002) TD 9/93: Highway link design. In *Design Manual for Roads and Bridges*. DfT, London, UK.

DfT (2003) *Traffic Signs Manual*. The Stationery Office, London, UK.

DfT (2004a) LTN 1/04: *Policy, Planning and Design for Walking and Cycling*. DfT, London, UK.

DfT (2004b) BD 29/04: Design criteria for footbridges. In *Design Manual for Roads and Bridges (DMRB)*. DfT, London, UK.

DfT (2004c) TD 50/04: The geometric layout of signal-controlled junctions and signalised roundabouts. In *Design Manual for Roads and Bridges (DMRB)*. DfT, London, UK.

DfT (2004d) LTN 2/04: *Adjacent and Shared Use Facilities for Pedestrians and Cyclists*. DfT, London, UK.

DfT (2005a) TA 90/05: The geometric design of pedestrian, cycle and equestrian routes. In *Design Manual for Roads and Bridges (DMRB)*. DfT, London, UK.

DfT (2005b) TA 91/05: Provision for non-motorised users. In *Design Manual for Roads and Bridges (DMRB)*. DfT, London, UK.

DfT (2005c) HD 42/05: Non-motorised user audits. In *Design Manual for Roads and Bridges (DMRB)*. DfT, London, UK.

DfT (2005d) *Inclusive Mobility: A Guide to Best Practice on Access to Pedestrian and Transport Infrastructure*. DfT, London, UK.

DfT (2008) LTN 2/08: *Cycle Infrastructure Design*. The Stationery Office, London, UK.

DfT (2010) TD 9/10: Road link design. In *Design Manual for Roads and Bridges (DMRB)*. DfT, London, UK.

DfT (2012) LTN 1/12: *Shared Use Routes for Pedestrians and Cyclists*. The Stationery Office, London, UK.

DfT (2017) HD 19/15: Road safety audit. In *Design Manual for Roads and Bridges (DMRB)*. DfT, London, UK.

Federal Highway Administration (2006) *Shared-use Path Level of Service Calculator*. Report no. FHWA-HRT-05-138. Federal Highway Administration, Research, Development, and Technology. McLean, VA, USA.

Highways England (2016) *Interim Advice Note 195/16: Cycle Traffic and the Strategic Road Network*. Highways England, London, UK.

HMG (Her Majesty's Government) (1997) The Zebra, Pelican and Puffin Pedestrian Crossings Regulations and General Directions 1997. The Stationery Office, London, UK.

HMG (2002) The Traffic Signs Regulations and General Directions 2002. The Stationery Office, London, UK.

HMG (2010) Equality Act 2010. The Stationery Office, London, UK.

Parkin J (2018) *Designing for Cycle Traffic: International Principles and Practice*. Thomas Telford, London, UK.

Phil Jones Associates (2011) *The Merits of Segregated and Non-segregated Traffic-free Paths: A Literature-based Review*. Phil Jones Associates, Birmingham, UK. https://www.sustrans.org.uk/sites/default/files/file_content_type/phil_jones_associates_report_-_september_2011.pdf (accessed 15/10/2018).

Sustrans (2015) *Junctions and Crossings: Design Manual*. Sustrans, Bristol, UK.

Sustrans (2018) Designing Traffic Free Routes. https://www.sustrans.org.uk/our-services/what-we-do/route-design-and-construction/route-design-resources/designing-traffic-free (accessed 15/10/2018).

TRB (Transportation Research Board) (2016) *Highway Capacity Manual*, 6th edn. National Academy of Sciences, Washington, DC, USA.

* This chapter contains extracts from LTN 1/12: *Shared Use Routes for Pedestrians and Cyclists*. The Stationery Office, London, UK.

Schoon, John G
ISBN 978-0-7277-6309-9
https://doi.org/10.1680/pfse.63099.189

Chapter 12
Audits for non-motorised user (NMU) facilities*

The aim in auditing of highway facilities is to ensure that proposals for new schemes, as well as reviews of selected existing ones, are critically examined and reported on by those experienced in road safety. Importantly, issues identified 'on site' comprise the essence of the auditing process. In conjunction with the design process, such identification and remedial action is essential to the safety and convenience of road users.

The auditing of designs involving NMUs can have many advantages during the geometric and general design process. Elements of the auditing process particularly relevant to the design of pedestrian facilities are outlined in this chapter. Chapter 15 includes an example of an audit where observed issues are identified and action is taken to improve the design – a useful exercise during the design process and one that can use the expertise of a number of professionals to improve the design.

12.1.   Background

Road safety audits are required for trunk roads, including motorways; the relevant procedures are described in HD 19/15 *Road Safety Audit* (DfT, 2017), which superseded HD 19/03 (DfT, 2003) and includes amendments made in 2017. Additionally, an emphasis on NMUs (consisting of pedestrians, cyclists and equestrians) is contained in HD 42/05 *Non-motorised User Audits* (DfT, 2005a). The latter provides most of the material discussed in this chapter. Also included is an outline of auditing related to residential streets, as described in the *Manual for Streets* (DfT, 2007). Essentially similar in many respects, and adopted by many local authorities for application to local roads, is the CIHT (2008) publication entitled *Road Safety Auditing*.

An important consideration related to the auditing process is that it can prove useful in the earlier stages of undertaking the analysis and design of any NMU facilities. For example, HD 19/15 (DfT, 2017, Annex A, List A4) asks the following.

- Have pedestrian and cycle routes been provided where required?
- Do shared facilities take account of the needs of all user groups?
- Can verge strips dividing footways/cycleways and carriageways be provided?
- Where footpaths have been diverted, will the new alignment permit the same users free access?
- Are footbridges/subways sited to attract maximum use?
- Is specific provision required for special and vulnerable groups? (i.e. the young, older users, mobility and visually impaired?)
- Are tactile paving, flush kerbs and guardrailing proposed? Is it specified correctly and in the best location?
- Have all NMU needs been considered, especially at junctions?
- Are these routes clear of obstructions such as signposts, lamp columns etc.?
- Have equestrian needs been considered?
- Does the scheme involve the diversion of bridleways?

(DfT, 2017)

12.2.   Basis for audits

12.2.1   The Road Traffic Act 1988

Highway safety auditing has assisted in ensuring that new and modified highway schemes have addressed safety issues to the greatest extent possible. Since the early 1980s, and as later required by the Road Traffic Act 1988, which states 'in constructing new roads, [each relevant authority] must take such measures as appear to the authority to be appropriate to reduce the possibilities of such accidents when the roads come into use', safety audits have been conducted.

12.2.2   Extent of problems

Major areas of risk for NMUs associated particularly with geometric design, as based on a study of problems identified within road safety audit reports (Belcher *et al.*, 2015) include, for pedestrians

- intervisibility between pedestrians and drivers
- conflicts between pedestrians and cyclists at shared-use facilities
- locations where vehicles can overrun footways
- crossing side roads,

Table 12.1 Facilities problems (Proctor *et al.*, 2001)

Description	Percentage accepted by clients for attention
Obsolete road signing	100
Pedestrian visibility restricted	100
Future maintenance problem	92
Poor design of build-outs or bus stops	83
Obstruction of view by street furniture	80
Pedestrian view obstructed	76
Confusing location of crossing	75
Absence of pedestrian crossing points	75
Missing or wrongly used road surface colour	71
Inadequate tactile paving	67
Features inconspicuous	60
Poor design of splitter island or refuge	57
Inadequate signing	54

and for people with disabilities

- difficulties in distinguishing between footways and carriageways
- conflict between impaired pedestrians and cyclists
- difficulties where wheelchair users are unable to cross roads or can become 'trapped' in the road.

The extent to which auditors' reports are accepted as important to the design of a scheme varies considerably. However, a list of issues most frequently accepted by clients as needing attention is shown in Table 12.1. Signing and visibility issues feature extensively in this list. A further study, involving 1571 separate safety audit comments, indicated that about a third of comments referred to vulnerable road users, as shown in Figure 12.1,

Figure 12.1 Number of comments for each road user (Proctor *et al.*, 2001)

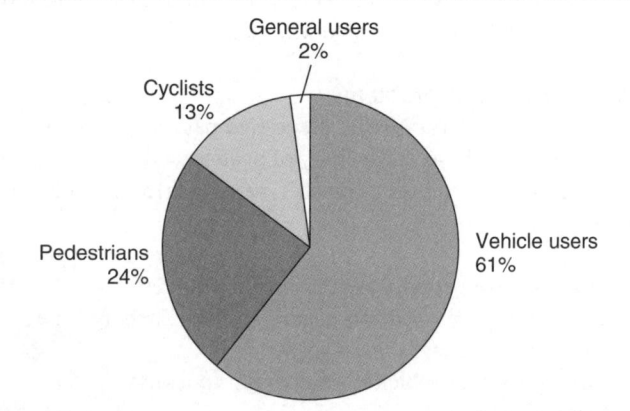

of which 12% related specifically to pedestrians with sight impairment and a further 11% to those with mobility impairment (Proctor *et al.*, 2001).

12.3. Objectives and procedures
12.3.1 The road safety audit (DfT, 2017)
Pertinent aspects of the intent of HD 19/15 (DfT, 2017) include Sections 1.4, 1.5 and 2.18, as follows:

1.4 Although road safety has always been considered during scheme preparation, there have been instances where details of the design have contributed to collisions and/or incidents on newly opened schemes. Design teams do not necessarily contain staff with collision investigation or road safety engineering experience and consequently they may not foresee potential factors pertaining to collision causation.

1.5 The road safety audit procedure has been developed to ensure that operational road safety experience is applied during the design and construction process in order that the number and severity of collisions is kept to a minimum. Road safety auditors identify and address problem areas using the experience gained from highway design, road safety engineering, collision analysis and road safety related research. The overseeing organisations' aim is that the monitoring of road safety audited schemes will result in more informed designs, leading to schemes that rarely require road safety related changes after opening.

2.18 Road safety audit is not a technical check that the design conforms to standards and/or best practice guidance. Design organisations are responsible for ensuring that their designs have been subjected to the appropriate design reviews (including, where applicable, non-motorised user (NMU) audits [(DfT, 2005a)]) prior to road safety audit.

(DfT, 2017)

As noted, not all technical checks ensure that designs conform to standards. However, auditors should examine the overall geometry of the scheme. All users of the highway should be considered, including pedestrians, cyclists, equestrians, those working on the highway and motor vehicle users. Particular attention should be given to vulnerable road users, such as the very young, older people and the mobility- and visually impaired.

NOTE
It would seem that many of the features of an audit should be those that are also considered and conducted in a responsible and realistic design process. It must be realised that many of the road safety matters that a designer addresses are directly related to technical matters, including geometry and the needs of a variety of users.

Four stages of the audit process are required for trunk road schemes (DfT, 2017), the salient features of which are described briefly as follows.

- *Stage 1*. Road safety audits will be undertaken at the completion of preliminary design (for example at the order publication report stage), before publication of draft orders and, for developer-led highway improvement schemes, before planning consent is applied for.
- *Stage 2: Completion of detailed design*. At this stage, the road safety audit team is concerned with the more detailed aspects of the highway improvement scheme. The road safety audit team will be able to consider geometry (such as the layout of junctions and highway cross-sections), street furniture (such as the position of traffic signs and road restraint systems), carriageway markings, street lighting provision and other issues.
- *Stage 3: Completion of construction*. This audit should be undertaken when the highway improvement scheme is substantially complete and preferably before the works are opened to road users. Road safety auditors are required to examine the highway improvement scheme from all users' viewpoints and may decide to drive, walk or cycle through the scheme, as well as considering motorcycle and equestrian use, to assist their evaluation and ensure that they have a comprehensive understanding.
- *Stage 4: Monitoring*. The overseeing organisation will arrange for evidence-led collision monitoring of road safety audited highway improvement schemes. Stage 4 road safety audits should be undertaken by individuals with the appropriate training, skills and experience, as identified in paragraphs 2.76 to 2.84 of HD 19/15 (DfT, 2017). When a highway improvement scheme is opened to road users, monitoring, in the form of Stage 4 road safety audits, of the number of personal injury collisions that occur must be carried out, so that any road safety problems can be identified and remedial action taken as soon as possible.

An important aspect of the audit process is that the audit team must be independent of the design team to maximise the objective nature of the audit and obtain an overall perspective of the scheme.

12.3.2 NMU audit (DfT, 2005a)

The objective of the auditing process is to encourage consideration of the needs of NMUs in all highway schemes, by means of a policy that supports safety and accessibility. The standards of HD 42/05 (DfT, 2005a) set out the requirements for conducting NMU audits on schemes affecting trunk roads and motorways.

An NMU audit is defined as a systematic process applied to highway schemes, by which the design team identifies scheme objectives for NMUs, documents the design decisions affecting them, and reviews designs and construction to assess how well objectives have been achieved. Thus, NMU audits should

- encourage the design team to take all reasonable opportunities to improve the service offered to NMUs
- prevent conditions for NMUs being worsened by the introduction of highway schemes
- document design decisions that affect NMUs.

An NMU audit is not intended to add significant work to the design process. Rather, it is a means of documenting design decisions in a formal and consistent manner. The extent of work required to carry out the NMU audit process will vary depending on the scale and type of scheme under consideration.

12.3.3 Mandatory sections, exemptions and scope

Parts of HD 42/05 (DfT, 2005a) are mandatory, although exemptions may be allowed:

1.11 Unless exemption has been agreed [...], this standard must apply to all highway schemes on trunk roads, including motorways, for which the Highways Agency, Welsh Assembly Government or the Department for Regional Development Northern Ireland is the highway authority. This standard does not apply in Scotland.

1.12 The scope must include work carried out under agreement with the overseeing organisation resulting from developments alongside or affecting the trunk road.

(DfT, 2005a)

Also, as in other cases of standards implementation, approval for departure from standards should be obtained from the overseeing organisation.

12.3.4 Related publications

As well as HD 42/05, reference should also be made to TA 90/05 *Geometric Design of Pedestrian, Cycle and Equestrian Routes* (DfT, 2005b) and TA 91/05 *Provision for Non-motorised Users* (DfT, 2005c) for further information.

Concerning combined pedestrian and cyclist facilities in predominantly urban areas, LTN 1/12 *Shared Use Routes for Pedestrians and Cyclists* (DfT, 2012) advises that a well-designed shared-use scheme should not need separate user audits for cyclists or pedestrians and recommends that user audits as a part of the design process would help to ensure the provision of scheme objectives. Furthermore, community street audits involving a range of pedestrians with different needs can be an effective way of informing the design and getting community 'buy-in' for the scheme.

In addition, supplementary (non-*DMRB*) guidance that may be of assistance includes *Inclusive Mobility* (DfT, 2005d), *Cycling by Design* (Transport Scotland, 2010), *Guidelines for Cycle Audit and Cycle Review* (IHT, 1998), *Cycle-friendly Infrastructure: Guidelines for Planning and Design* (IHT, 1996) and *Guidelines for Providing for Journeys on Foot* (IHT, 2000).

12.3.5 Definitions

Highway schemes for which audits are applicable include road layout changes and their features and appurtenances, but refurbishment, replacement and most maintenance are not included. Definitions of terms unique to the auditing process include the following.

- *NMU audit leader*. A member of the design team, with the appropriate training, skills and experience, who has responsibility for overseeing the NMU audit process and for liaison with the project sponsor and design team leader.
- *NMU context report*. The first stage of an NMU audit. The NMU context report is a simple statement of background information on current or potential NMU issues relevant to the scheme. The NMU context report should ensure that the design team have the necessary information to take appropriate decisions on design elements that might affect NMUs.
- *NMU audit report*. An NMU audit report is produced for each relevant design stage of a scheme, as agreed by the project sponsor. The NMU audit report sets out the objectives of the scheme for NMUs, as well as the objectives of the design stage. It also documents decisions taken in relation to providing for NMU needs during the design stage, and notes any failures to meet objectives and considerations for subsequent design stages.

12.4. NMU characteristics

12.4.1 Principles

Scheme designs should reflect the principle that people using a non-motorised mode have the same basic concerns as any transport user. For routes to be viable for NMUs, they should

- not give rise to road safety or personal safety concerns
- directly facilitate the desired journey without undue deviation or difficulty
- link origins and destinations
- be attractive and comfortable to use
- be accessible to disabled users and people with children and pushchairs
- be continuous and not subject to severance or fragmentation.

Potential for conflict exists between some of these requirements. Different individuals will also have different requirements; for example, an adult pedestrian's desire for a direct route might suggest an at-grade crossing, whereas the need for a safe crossing for child pedestrians in the same location might suggest the need for a grade-separated crossing.

An individual's transport needs may also vary depending on other factors, such as journey purpose; for example, a commuter's need for directness compared with the needs of someone walking for leisure. Moreover, in contrast with designing for motorised users, the designer cannot assume any given level of competence, recognition of signs or familiarity with traffic law and conventions on the part of the NMU. Also, in addition to variation between individuals, the designer must also consider variation between the types of user.

A list of common problems experienced by NMUs is included in Annex A of HD 42/05 (DfT, 2005a). Reference should also be made to TA 91/05 (DfT, 2005c).

12.5. The NMU audit process

12.5.1 Scope of NMU audit

An NMU audit should be used as a design tool during scheme development to assist the project sponsor and design team in ensuring that the needs of all road users are met in the scheme design. It is not a process applied by independent scrutineers.

An NMU audit should consider the implications of schemes for NMU accessibility, safety, comfort and convenience. It does not duplicate a road safety audit. While issues of both road safety and personal safety for NMUs should be included within an NMU audit, these should be balanced against consideration of all elements likely to affect NMU travel. An NMU audit is a continuous process, unlike a road safety audit, which is staged, and should minimise the number of NMU issues identified at the road safety audit.

An NMU audit is not a technical design check. Rather, it should be carried out from the user's perspective and offer an opportunity to assess the value of a proposed design to the end user.

An NMU audit must actively involve all members of the design team. The NMU audit leader must act as a focal point and be responsible for managing the process and quality of outputs.

12.5.2 Stages of NMU audit

An NMU audit consists of two elements

- the collation of background information of relevance to NMUs, and the presentation of that information in an NMU context report, leading to agreement on the design stages for which an NMU audit report is required

consideration of NMUs within the design process and following construction. This consideration is to be documented with an NMU audit report for each design stage that has been specified by the project sponsor.

The NMU context report must be produced at the earliest possible stage in a scheme, ideally where scheme objectives are defined and prior to preliminary design.

The most likely stages for completion of NMU audit reports (subject to the agreement of the project sponsor) are as follows.

- *Preliminary design*. During development of the preliminary design and prior to public consultation and the publication of draft orders (if required).

- *Detailed design*. During development of the detailed design.
- *Completion of construction*. Prior to, or shortly after, scheme opening.

For smaller schemes, where design stages are combined, an NMU audit should be applied to the combined stage.

The NMU audit process for schemes for which no exemption has been granted is shown in Figure 12.2.

12.5.3 NMU context report
The NMU context report must provide a summary of all available information relevant to existing and potential patterns of use by NMUs within the design life of the scheme. The NMU

Figure 12.2 Summary of the NMU audit process (DfT, 2005a)
© Crown Copyright, 2005
This information is licensed under the Open Government Licence v3.0. To view this licence, visit http://www.nationalarchives.gov.uk/doc/open-government-licence/ **OGL**

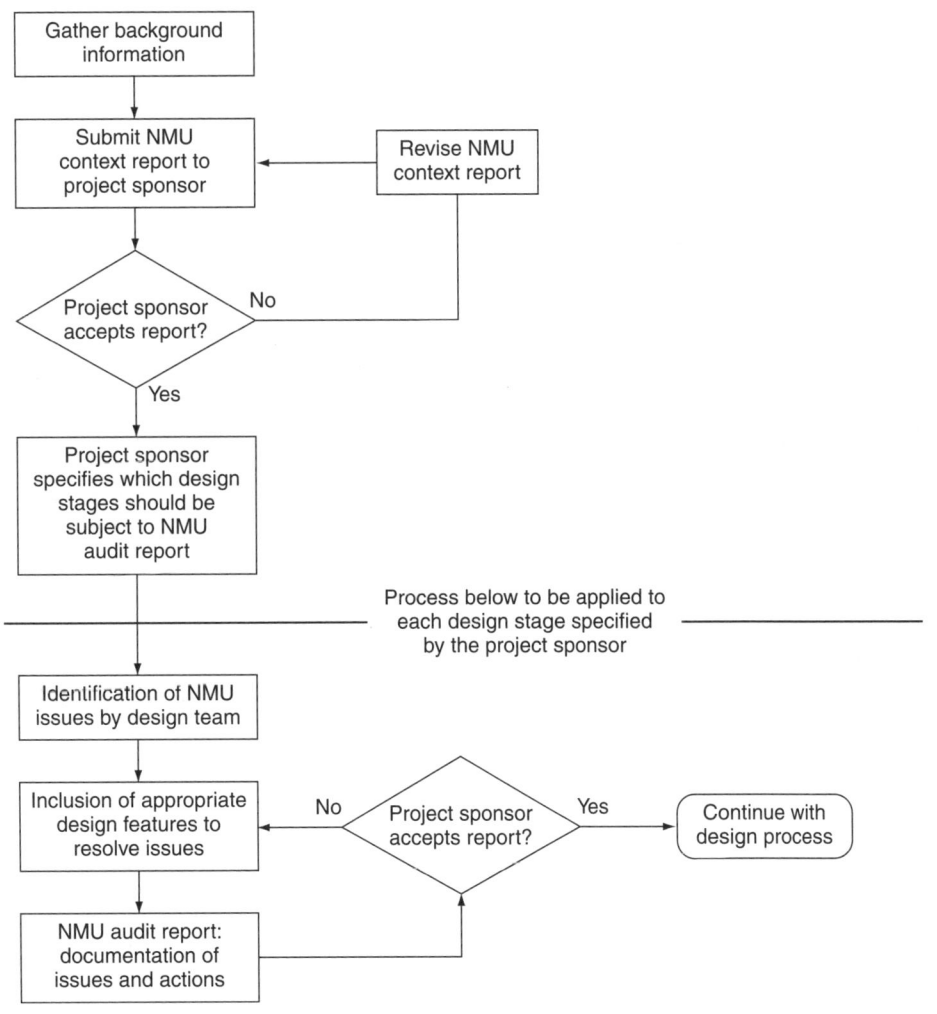

context report must also set out opportunities and objectives to improve conditions for NMUs.

Compilation of the NMU context report need not be an excessively time-consuming task, particularly for small schemes. The objective of the context report is to ensure that the design team and project sponsor have sufficient information to allow them to fully consider the interests of NMUs within the scheme design.

Information presented in the NMU context report may include, but need not be confined to

- flows of NMUs
- flows and speeds of motorised traffic
- existing and future land use
- desire lines
- overseeing organisation or other highway authority policies and strategic objectives for NMUs in the area affected by the scheme
- trip generators
- public rights of way and permissive routes
- information from user groups on the routes that they use within the area
- National Cycle Network routes and other routes provided for NMUs
- transportation modelling (if appropriate)
- accident data
- the views of relevant user groups, highway authorities, the police and public transport operators
- regular events, e.g. time trials, that might increase flows of NMUs.

Much of the necessary information for an NMU context report may already be available, for example if an environmental assessment has been carried out, as set out in *Pedestrians, Cyclists, Equestrians and Community Effects* (DfT, 1993). However, in some cases, it may be necessary to gather further information. It is recommended that the process described in the following sections should be carried out in compiling the NMU context report. For smaller schemes, a less detailed approach may be appropriate. The steps may include the following.

- *Step 1.* On a map or graphical representation of the study area, plot information pertaining to the existing situation. This may include such data as
  - peak and off-peak motorised vehicle flows along the trunk route
  - peak and off-peak motorised vehicle flows across the trunk route
  - speeds of motor vehicles
  - peak and off-peak NMU flows along the trunk route

  - other peak and off-peak NMU flows in the area
  - NMU accident records, including all casualties and reports of non-injury accidents if available
  - potential routes and desire lines not currently used, e.g. due to personal safety or road safety fears.
- *Step 2.* Consider the existing NMU and motorised traffic flows and speeds, and consider how these are expected to change over the design life of the highway scheme.
- *Step 3.* Consider the strategic context of the area and any planned developments. On a map or graphical representation of the study area, plot such features as
  - locations of relevant strategic NMU routes within the study area, defined by the highway authority or user groups
  - locations of routes used by NMUs for which no reasonable alternative exists
  - locations of any known planned developments, changes to land use or other factors that might affect the flows of NMUs or motorised traffic within the study area during the design lifetime of the scheme (making reference to strategic and local planning documents)
  - routes of public transport services and the location of interchanges.
- *Step 4.* From the plotted data and other information, identify such features as
  - key NMU desire lines
  - locations of potential conflict
  - locations with high motorised traffic flows
  - locations with high NMU flows
  - local rights of way
  - National Cycle Network or local cycle routes.
- *Step 5.* Make a note of any information or data that might be significant to the project but is not available at this time.
- *Step 6.* Propose the overall objectives for NMUs within the scheme as a whole.

Examples of NMU context reports are provided in Annex B of HD 42/05 (DfT, 2005a).

12.6. Guidance and prompts in the auditing process

Examples of guidance and prompts are presented in HD 42/05 (DfT, 2005a) for the different stages of

- *frequent problems* – issues common to more than one group of NMU; additional issues for pedestrians, cyclists, mobility impaired people and equestrians
- *preliminary design* – convenience, attractiveness and environmental quality, public transport, accessibility, safety and consistency

- *detailed design* – convenience, attractiveness and environmental quality, public transport, accessibility, safety and consistency
- *completion of construction* – consideration of the needs of pedestrians, cyclists and equestrians, as well as the more vulnerable of these groups, such as those with various impairments, children and older people.

A selection of the kinds of NMU issues to be considered in the detailed design stage are listed next. It should be noted that this list is not comprehensive and should be added to for specific design projects. The original paragraph numbers have been retained for ease of identification in the original document.

12.6.1 Convenience

120 How have NMU routes been designed to optimise the balance between safety and convenience?

121 How have NMU routes been designed to closely align with desire lines without deviation?

122 How have connections to other NMU routes been considered throughout the design of the NMU route?

123 How have connections to origins/destinations and NMU facilities been considered throughout the design of the NMU route?

124 What priority has been given to NMUs throughout the design of the NMU route?

125 Are widths along the whole route, including crossings, adequate for all classes of NMU, including wheelchair users, to be served?

126 Is adequate headroom available on all NMU routes?

127 Is tactile information provided at all appropriate points on pedestrian routes?

128 Are NMU routes given priority over private accesses?

129 Are cyclists and horse riders able to use the routes without dismounting?

(DfT, 2005a, Annex A)

12.6.2 Attractiveness and environmental quality

130 How have the aesthetic qualities of the NMU route been considered throughout its design?

(DfT, 2005a, Annex A)

12.6.3 Public transport

131 How have direct and obvious connections to public transport services been considered throughout the design of the NMU route?

(DfT, 2005a, Annex A)

12.6.4 Accessibility

132 What maximum gradients have been allowed to ensure that NMUs can manoeuvre themselves throughout the route with ease, safety and control?

133 What surfacing textures and tones have been considered throughout the route, including non-slip surfaces and colour contrast at appropriate points?

134 What (lateral and vertical) clearances have been provided to ensure that NMUs can manoeuvre themselves through corners, dips and crests throughout the route?

135 Are dropped kerbs specified at all appropriate points on NMU routes?

136 Are appropriate rest points provided for NMUs?

137 Are ramps provided as alternatives to steps?

138 How have the special needs of vulnerable NMU groups been considered throughout the design of the NMU route? Special provisions may have been included for vulnerable NMU groups such as

- people with mobility impairments
- people with visual impairments
- people with hearing impairments
- children and younger people
- older people
- adults with young children.

139 How have the NMU facilities been designed and located to ensure that all different types of NMU may use them? Special provisions may have been made for different types of NMU such as

- pedestrians
- cyclists
- equestrians.

(DfT, 2005a, Annex A)

12.6.5 Safety

140 What (lateral and vertical) clearances have been provided to deliver conflict-free shared use along the non-segregated sections of the NMU route?

141 How has adequate intervisibility between various types of NMU been provided?

142 How have NMUs been separated and protected from motorised traffic throughout the design of the NMU route?

143 How have NMUs on the route been made visible to the motorised traffic on the trunk road?

144 How have NMUs been protected from headlight glare?

145 How has personal security of NMUs been provided for throughout the design of the NMU route?

146 Is direction signing for NMUs adequate?

(DfT, 2005a, Annex A)

12.6.6   Consistency

147 What measures have been included to ensure the NMU route provides a consistent level of service regardless of changes in natural lighting levels? (Dawn, day, twilight, night...)

148 What measures have been included to ensure the NMU route provides a consistent level of service regardless of weather? (Wind, dry, wet, hot, cold...)

(DfT, 2005a, Annex A)

12.6.7   Consultation and site visits

It is important that the NMU audit process is based on a combination of desk assessment, site visits and consultation.

Consultation with interested parties, particularly local authorities, user groups, residents' groups and the police, is valuable in assisting in identifying issues and opportunities for NMUs. These stakeholders can contribute to the quality of the scheme design and should be consulted as early as is practical in the development of designs.

User groups can contribute significant information, particularly in cases where the use of a mode, or the needs of people with certain disabilities, is not within the direct experience of those undertaking the design. It is recommended that such groups are consulted at every appropriate stage of the design process in order that the design team is aware of their views as designs are progressed. Local representatives should be contacted where possible. These may be affiliated to national groups, selected details of which are included in Annex D of HD 42/05 (DfT, 2005a) and may initially be contacted via those organisations. The local authority may also be able to assist with contacts.

Such consultation may also be useful in the validation of any collected data. It is desirable that site visits take place in a range of weather and lighting conditions, particularly at those times of day when it is anticipated that NMU flows are likely to be highest. The location of the scheme and of nearby trip generators, such as schools, will assist in anticipating both the type and time of use, e.g. commuter, leisure or education, and the nature of the user and, hence, the most appropriate times to conduct site visits.

An NMU audit at completion of construction must, as a minimum, include site visits during daylight and after dark.

When conducting a site visit, the member of the design team should walk any NMU routes to be affected by the scheme. It is recommended that only when this opportunity to observe and exercise judgement has been taken should detailed notes be made using the audit prompts described in Annex A of HD 42/05 (DfT, 2005a).

12.6.8   Example

Annex B of HD 42/05 (DfT, 2005a) provides an example of an NMU context report; major features of this example are described in Chapter 15.

12.7.   Audits for residential streets

12.7.1   Interaction between design and auditing processes

Recent reviews and new guidelines for residential roads and streets in the UK are discussed in the *Manual for Streets* (DfT, 2007), and address the matter of road safety audits as one element of new development. An important aspect in this regard is the stress on a need for innovation based on sound evidence, supported by numerical values where available.

12.7.2   Application of audit

The road safety audit procedure, while mandatory for trunk road schemes, does not have to be applied by local highway authorities (DfT, 2007). It is therefore important to note that the design team for residential roads and streets retains responsibility for the scheme, and is not governed by the findings of the road safety audit. A proposed scheme, therefore, cannot be thought of as passing or failing the road safety audit process.

Furthermore, the need for road safety audits has been linked to the highway authority's duty and Section 39.3(c) of the Road Traffic Act 1988. While a policy of carrying out a road safety audit is in conformity with this general duty, it is important to note that there is no statutory requirement to carry out road safety audits, nor must they be undertaken in any particular way (DfT, 2007).

12.7.3   Potential areas of improvement

In reviewing the current road safety auditing process, the following points are made (DfT, 2007).

12.7.3.1 Innovative design against prior experience

There have been a number of reported problems with the process in the context of streets, particularly where designers are aiming to move beyond normal road standards in order to create more distinctive and better-quality environments. One concern is that the normal audit process is based on the experience of the audit team. By definition, therefore, it is difficult to assess innovative designs from first principles.

12.7.3.2 Evidence against opinion

As far as possible, the assessments of each factor should be based on evidence, rather than simply the auditor's opinion.

Factual assessments of risk factors can be based on such research reports as *TRL Report 184 – Accidents at Three-arm Priority Junctions on Urban Single-carriageway Roads* (Summersgill *et al.*, 1996) and the SafeNET software package (TRL, 2006), which, as described in Chapter 14, can estimate the number of accidents on a network based on vehicular and pedestrian traffic flow and highway geometry.

12.8. Summary

This chapter has outlined several of the key areas of highway auditing as it applies to non-motorised users (NMUs) and as addressed in both trunk road and residential street situations. Some overlap is evident between road safety audits and NMU audits. Consequently, awareness of guidance and advice on both types of audit would be required in the comprehensive analysis and design of pedestrian facilities. Since all designs are required to undergo an audit, the coordination of design and auditing, and the documentation of design decisions and justification for them, are essential in the success of the geometric design and engineering process.

REFERENCES

Belcher M, Proctor S and Cook C (2015) *Practical Road Safety Auditing*, 3rd edn. Thomas Telford, London, UK.

CIHT (Chartered Institution of Highways and Transportation) (2008) *Road Safety Auditing*. CIHT, London, UK.

DfT (Department for Transport) (1993) Pedestrians, cyclists, equestrians and community effects. In *Design Manual for Roads and Bridges (DMRB)*, vol. 11, section 3, part 8. DfT, London, UK.

DfT (2003) HD 19/03: Road safety audit. In *Design Manual for Roads and Bridges (DMRB)*. DfT, London, UK.

DfT (2005a) HD 42/05: Non-motorised user audits. In *Design Manual for Roads and Bridges (DMRB)*. DfT, London, UK.

DfT (2005b) TA 90/05: Geometric design of pedestrian, cycle and equestrian routes. In *Design Manual for Roads and Bridges (DMRB)*. DfT, London, UK.

DfT (2005c) TA 91/05: Provision for non-motorised users. In *Design Manual for Roads and Bridges (DMRB)*. DfT, London, UK.

DfT (2005d) *Inclusive Mobility: A Guide to Best Practice on Access to Pedestrian and Transport Infrastructure*. DfT, London, UK.

DfT (2007) *Manual for Streets*. Thomas Telford, London, UK.

DfT (2012) LTN 1/12: *Shared Use Routes for Pedestrians and Cyclists*. The Stationery Office, London, UK.

DfT (2017) HD 19/15: Road safety audit. London. In *Design Manual for Roads and Bridges (DMRB)*. DfT, London, UK.

HMG (1988) Road Traffic Act 1988. The Stationery Office, London, UK.

IHT (1996) *Cycle-friendly Infrastructure: Guidelines for Planning and Design*. IHT, London, UK.

IHT (1998) *Guidelines for Cycle Audit and Cycle Review*. IHT, London, UK.

IHT (2000) *Guidelines for Providing for Journeys on Foot*. IHT, London, UK.

Proctor S, Belcher M and Cook P (2001) *Practical Road Safety Auditing*. Thomas Telford, London, UK.

Summersgill I, Kennedy JV and Baynes D (1996) *TRL Report 184 – Accidents at Three-arm Priority Junctions on Urban Single-carriageway Roads*. TRL, Crowthorne, UK.

Transport Scotland (2010) *Cycling by Design*. Transport for Scotland, Glasgow, UK.

TRL (Transport Research Laboratory) (2006) *SafeNET (Software for Accident Frequency Estimation for Networks)*. TRL, Crowthorne, UK.

Pedestrian Facilities, Second edition

Schoon, John G
ISBN 978-0-7277-6309-9
https://doi.org/10.1680/pfse.63099.199

Chapter 13
Pedestrian flow and capacity: the *Highway Capacity Manual**

Extensive work has been carried out since the 1950s in the USA in quantifying space needs and flow for a wide variety of transport modes in urban settings. The pedestrian mode was included in significant detail in the *Highway Capacity Manual* (HCM) in 2000 (TRB, 2000) and has progressed with *HCM 2010* (TRB, 2010) and, in 2016, the *HCM* 6th edition (TRB, 2016).

The geometric design matters addressed in these editions of *HCM* include footway widths, road crossings, queueing and dimensions of the related infrastructure to be evaluated and designed and, of increasing importance, the interaction of pedestrians and cyclists. In Europe, work in this area has been carried out, as in, for example, *Vorrang für Fussgänger* (Thaler, 1993). The publication *Pedestrian Comfort Level Guidance* (TfL, 2010) also addresses movement and space requirements, based on pedestrians' perceptions, in terms of pedestrian comfort level on footways and crossings.

The 6th edition of the *HCM* adds a subtitle: '*A Guide for multimodal mobility analysis*'. This emphasises a focus on evaluating the operational performance of several modes simultaneously, and, importantly, their interactions. Pertinent tools added include enhanced methods for analysing pedestrian transit facilities as well as the interaction of pedestrians with motor vehicles. Although not directly applicable to the UK's pedestrian and other traffic conditions, the key elements are described in this chapter as an indicator of the potential methodology and the variables in the analysis. The emphasis here is on pedestrian traffic, as well as aspects of interaction with bicycle traffic. Both of these conditions are addressed in terms of uninterrupted flow (i.e. not including signalisation), in order to illustrate key principles underlying provision of combined facilities, as discussed in Chapter 11.

NOTE
In terms of use outside the USA, the HCM *states:*

Although there is considerable value in the general methods presented, their use outside of North America requires

additional emphasis on calibrating the equations and the procedures to local conditions as well as recognizing major differences in the composition of traffic; in driver, pedestrian, and bicycle characteristics; and in typical geometrics and control measures.

(TRB, 2000)

The 2010 and the 2016 editions of the Highway Capacity Manual *are published using imperial units, as is customary in the USA. Because values, equations and procedures must suit local conditions, imperial units have been retained in this chapter and may be converted to metric dimensions during recalibration. Conversion factors are*

- *1 in = 25.4 mm, 1 mm = 0.034 in*
- *1 ft = 0.305 m, 1 m = 3.281 ft*
- *1 ft² = 0.093 m², 1 m² = 10.76 ft².*

13.1. Pedestrian characteristics
13.1.1 Pedestrian space requirements
A simplified body ellipse of 20 in × 24 in is used as the basic space for a single pedestrian, as shown in Figure 13.1 (TRB, 2016). This represents the practical minimum for standing pedestrians. In evaluating a pedestrian facility, an area of 2.35 ft² is used as a buffer zone for each standing pedestrian.

A walking pedestrian requires a forward space. This critical dimension determines the speed of walking and the number of pedestrians who are able to pass a point in a given time. The forward space is divided into a pacing zone and a sensory zone, as shown in Figure 13.2.

13.1.2 Pedestrian walking speeds
Pedestrian walking speed is highly dependent on the characteristics of the walking population. The proportion of older pedestrians (65 years old or more) and children, as well as the trip purpose, affects walking speed. A national study (AASHTO, 2011) found the average walking speed of younger (13–60 years old) pedestrians crossing streets to be significantly different from that of older pedestrians (4.74 ft/s compared with

Figure 13.1 Pedestrian body ellipse (TRB, 2013)

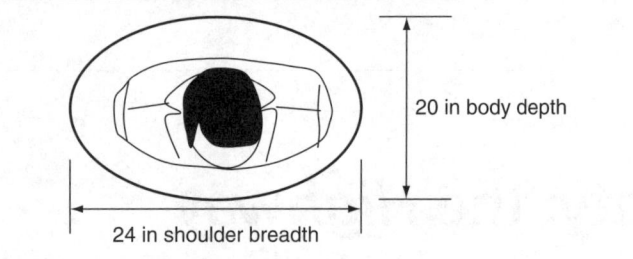

20 in body depth

24 in shoulder breadth

Figure 13.2 Pedestrian walking space requirements (Fruin, 1990)

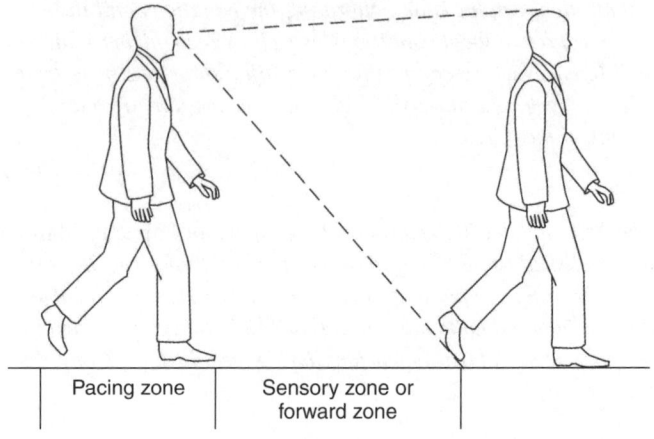

Pacing zone | Sensory zone or forward zone

Figure 13.3 Relationships between pedestrians speed and density (TRB, 2016, Exhibit 4-14)
Reprinted with permission from *Highway Capacity Manual, 6th Edition: A Guide for Multimodal Mobility Analysis*, 2016, the National Academy of Sciences, courtesy of the National Academies Press, Washington, DC.

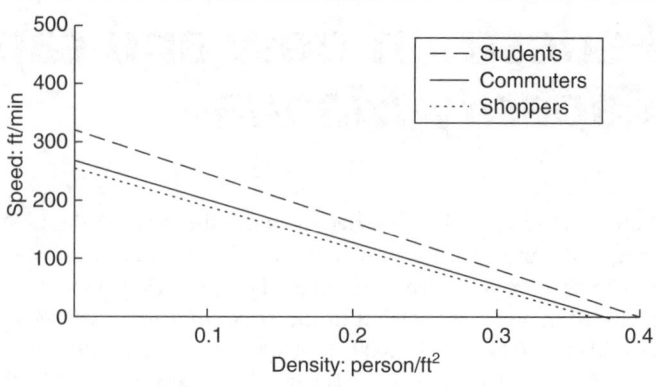

A pedestrian start-up time of 3.0 s is a reasonable midrange value for evaluating crosswalks at traffic signals. A capacity of 23 person/(min ft) or 1380 p/(h ft) is a reasonable value for a pedestrian facility if local data are not available.

13.1.4 Capacity
At capacity, a walking speed of 2.4 ft/s is considered reasonable. Figure 13.3 shows a typical distribution of free-flow walking speeds.

13.2. Pedestrian flow parameters
13.2.1 Speed, flow and density relationships
13.2.1.1 Speed–density relationships
The fundamental relationship between speed, density and volume for directional flow in facilities with no cross flows, where pedestrians are constrained to a fixed walkway width (because of walls or other barriers) is analogous to that for vehicular flow. As volume and densities increase, pedestrian speed declines. As density increases, the degree of mobility afforded to the individual pedestrian declines, as does the average speed of the pedestrian stream (Pushkarev and Zupan, 1975), as indicated in Figure 13.3.

Flow-density relationships, illustrated in Figure 13.3, can be expressed as

$$v_{ped} = S_{ped} \times D_{ped}$$

where

v_{ped} = unit flow rate (person/(min ft))
S_{ped} = pedestrians speed (ft/min)
D_{ped} = pedestrian density (person/ft^2).

4.25 ft/s). The 15th percentile speed, used in the *Manual on Uniform Traffic Control Devices for Streets and Highways* (Federal Highway Administration, 2018) for timing the pedestrian clearance times interval at traffic signals, was 3.03 ft/s for older pedestrians and 3.77 ft/s for younger pedestrians.

Walking speed appears to depend extensively on the proportion of older people in the walking population. Typically, where older pedestrians make up less than 20% of the total volume, walking speed is approximately 4.0 ft/s. If older pedestrians constitute greater than 20% of the volume, the speed decreases to 3.3 ft/s. Uphill gradients decrease walking speeds by about 0.33 ft/s for each 10% of gradient. On footways, the free-flow speed of pedestrians is approximately 5 ft/s. Other factors, such as a higher than normal proportion of young children, will also decrease the average walking speeds.

13.1.3 Pedestrian start-up time
At crosswalks, pedestrians might not step off the kerb immediately when the 'Walk' indication appears, in part because of the perception–reaction time and in part to make sure that no vehicles have moved or are about to move into the crosswalk area. This hesitation is termed 'start-up time' and is used in evaluating pedestrian crosswalks at traffic signals.

Figure 13.4 Relationships between pedestrian flow and space (TRB, 2016, Exhibit 4-15)
Reprinted with permission from *Highway Capacity Manual, 6th Edition: A Guide for Multimodal Mobility Analysis*, 2016, the National Academy of Sciences, courtesy of the National Academies Press, Washington, DC.

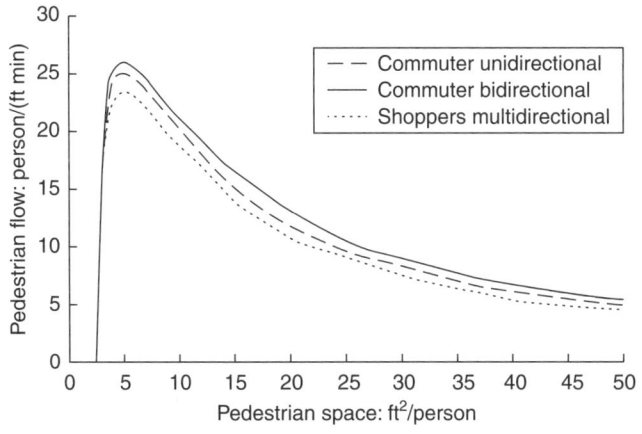

Figure 13.5 Relationships between pedestrian speed and flow (TRB, 2016, Exhibit 4-16)
Reprinted with permission from *Highway Capacity Manual, 6th Edition: A Guide for Multimodal Mobility Analysis*, 2016, the National Academy of Sciences, courtesy of the National Academies Press, Washington, DC.

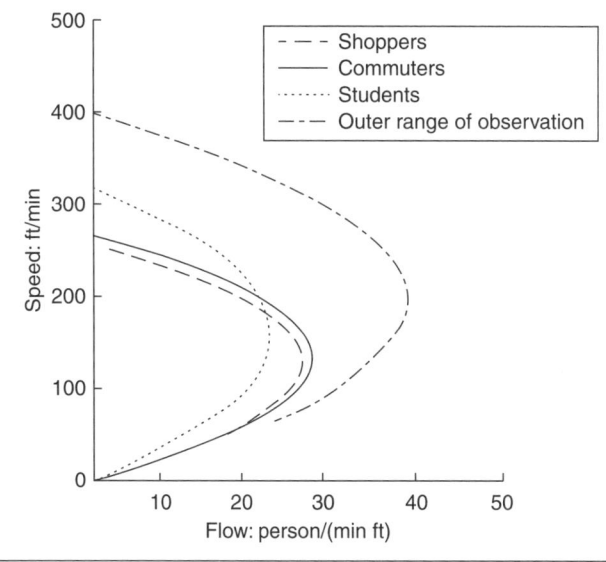

The flow variable in this expression is the unit width flow, defined earlier. An alternative, more useful expression, using the reciprocal of density, or space, shown in Figure 13.4 (TRB, 2016), is

$$v_{ped} = S_{ped}/M$$

where M is the pedestrian space (in square feet per person).

Conditions at maximum flow represent the capacity of the walkway. All observations of maximum unit flow fall within a narrow range of density, with the average space per pedestrian varying between 5 and 9 ft^2. As space reduces to less than 5 ft^2/person, the flow rate declines quickly. All movement stops at the maximum space allocation of 2–4 ft^2/person.

These relationships show that pedestrian traffic can be evaluated quantitatively by using basic concepts similar to those of vehicular traffic analysis. At flow rates closer to capacity, an average of 5–9 ft^2/person is needed for each moving pedestrian. However, at this level of flow, the limited area available restricts pedestrian speed and freedom to manoeuvre.

13.2.1.2 Speed–flow relationships

Figure 13.5 shows the relationship between pedestrian speed and flow. These curves, similar to vehicle flow curves, show that when there are few pedestrians on a walkway (i.e. low flow levels) there is space available to choose higher walking speeds. As flow increases, speed declines because of closer interactions between pedestrians; when a critical level of crowding occurs,

movement becomes more difficult, and both flow and speed decline (Pushkarev and Zupan, 1975).

13.2.1.3 Speed–space relationships

Figure 13.6 confirms the relationships of walking speed and available space, and suggests some points of demarcation for developing level of service (LOS) criteria. The outer range of observations shown indicates that, at an average space of less than 16 ft^2/person, even the slowest pedestrians cannot achieve their desired walking speeds. Faster pedestrians, who walk at speeds of up to 6 ft/s are not able to achieve that speed unless the average space is 43 ft^2/person or more (Pushkarev and Zupan, 1975).

13.2.1.4 Relationships of space and flow to level of service

Although not included in the sixth edition of *HCM*, a graphical indication of level of service (LOS) included in the earlier editions related to pedestrian space and flow rates can be informative. Figure 13.7 (TRB, 2010) indicates key features of these relationships.

13.2.2 Flow on urban sidewalks and streets

Most sidewalks (footways in UK terms) will experience cross flows, stationary pedestrians, spillover outside the walkway and other issues. High cross flows will disrupt speed–flow relationships, resulting in reduced pedestrian's speeds and flows.

Figure 13.6 Relationship between pedestrian speed and space (TRB, 2016, Exhibit 4-17)

Reprinted with permission from *Highway Capacity Manual, 6th Edition: A Guide for Multimodal Mobility Analysis*, 2016, the National Academy of Sciences, courtesy of the National Academies Press, Washington, DC.

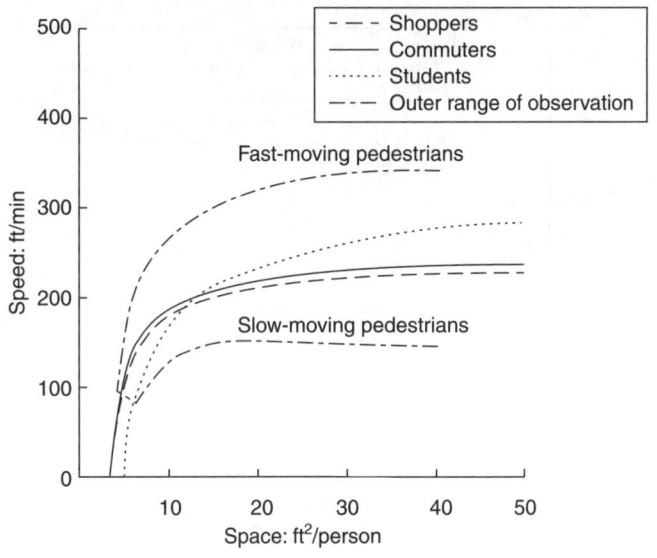

Moreover, stationary or turning pedestrians reduce flow by manoeuvring and decrease the available walkway width. Finally, pedestrians may choose to walk outside the prescribed walkway when high densities are reached. Thus, in practice, facilities will often break down, with pedestrians spilling into the street (carriageway in UK terms), before a maximum flow rate is reached.

Because of these conditions, many pedestrian facilities will reach effective failure at densities far less than the facility's capacity. Analyses of facilities must, therefore, take these factors into consideration.

13.2.3 Pedestrian type and trip purpose

Within any group, or between groups, considerable flow differences can result from different trip purposes, along with adjacent land use, age, mobility and other factors. Some typical characteristics are that

- commuters walk faster than shoppers
- shoppers walk slower and tend to stop and window-shop, and are often carrying shopping bags
- older or very young people tend to walk more slowly than others.

These characteristics can materially affect the basic speed, volume and density curves.

13.2.4 Influences of pedestrians on each other

Photographic studies (Hall, 1966) show that pedestrian movement on sidewalks is affected by the presence of other pedestrians. Salient features include the following.

- At 60 ft^2/person, pedestrians may walk in a checkerboard pattern.
- Up to 100 ft^2/person may be the *minimum* for completely free movement.
- At 130 ft^2/person, individual pedestrians are no longer influenced by others (Hall, 1966).
- If directional split is 90% against 10%, capacity reductions of about 15% can occur; this reduction results from the minor flow using more than its proportionate share of the walkway. For stairways, more severe effects may occur.
- A pedestrian's ability to cross a pedestrian stream is impaired at space values less than 35 ft^2/person, as shown in Figure 13.8.
- Below 25 ft^2/person, most crossing movements encounter conflicts.
- The ability to pass slower pedestrians is unimpaired above 35 ft^2/person, as shown in Figure 13.8, but becomes progressively more difficult as space allocations reduce to 18 ft^2/person, which is the point at which passing becomes virtually impossible (Fruin, 1990; Khisty, 1985).

13.2.5 Pedestrian platoons

Often, pedestrian movement is typified by the formation of platoons; this phenomenon must be related to average flow and to fluctuations in minute-by-minute flow rates in order to reflect design conditions adequately.

The average flow rate is of limited usefulness, unless reasonable time periods are specified. Even during peak 15 min periods, the peak 1 min flow may exceed the average flow by at least 20% and sometimes up to 75%, depending on the traffic patterns for which a facility is designed. However, it is not prudent to design for extreme 1 min flows that occur only 1% or 2% of the time. A relevant time period should be determined through closer evaluation of the short-term fluctuations of pedestrian flow.

On sidewalks, random fluctuations are exaggerated by the interruption of flow and queue formation caused by traffic signals. Transit facilities can cause surges in demand. Until they disperse, pedestrians in such groups move together as a platoon.

Platoons can also form when passing is impeded, owing to insufficient space; faster pedestrians must slow down behind slower pedestrians.

The scatter diagram of Figure 13.9 compares the platoon flow rate (i.e. the rate of flow of walking platoons of pedestrians)

Figure 13.7 Pedestrian level of service (LOS) related to graphical representation of pedestrian space and flow rates (TRB, 2010) Reprinted with permission from *Highway Capacity Manual* 2010, the National Academy of Sciences, courtesy of the National Academies Press, Washington, DC.

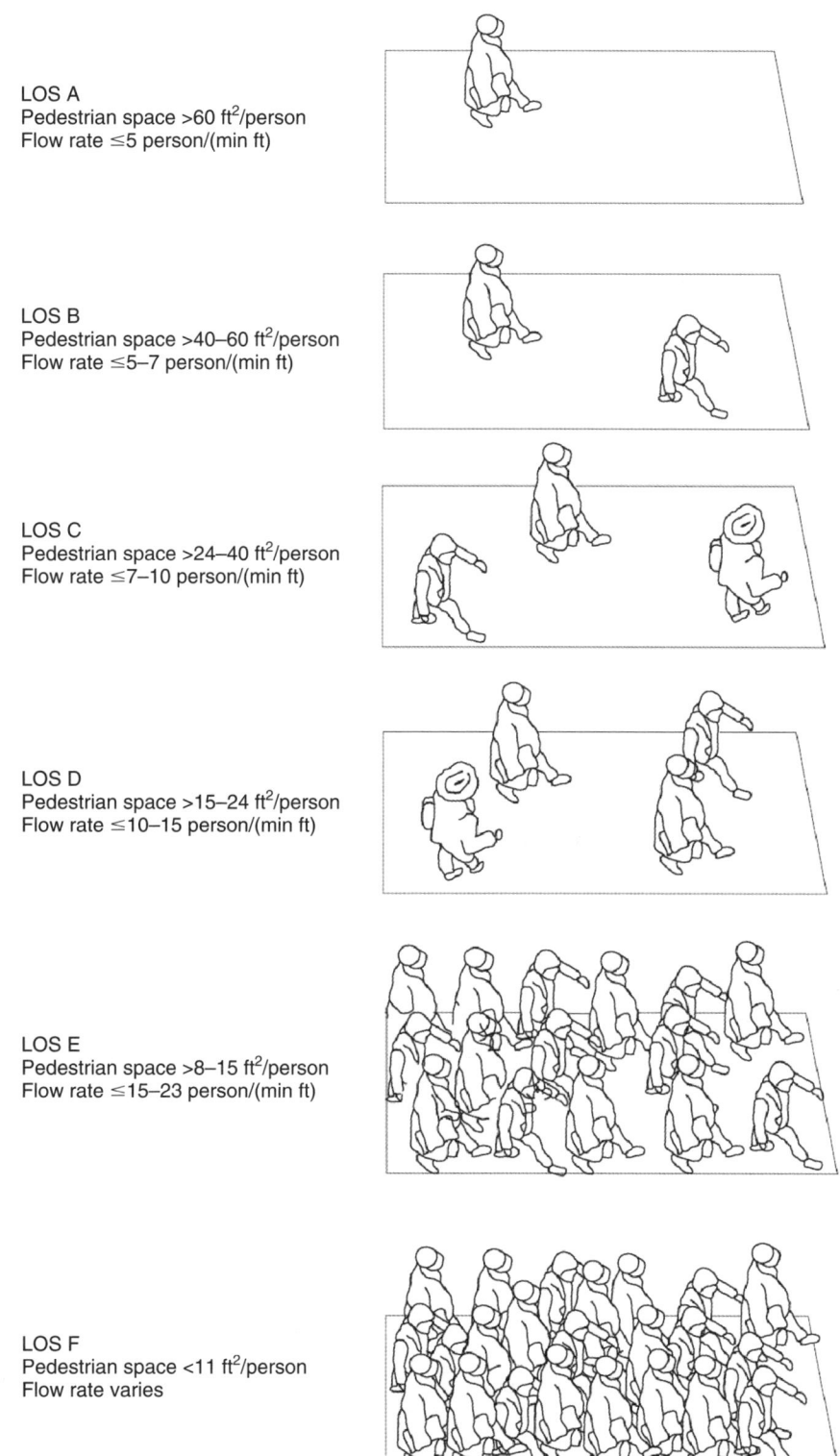

LOS A
Pedestrian space >60 ft^2/person
Flow rate ≤5 person/(min ft)

LOS B
Pedestrian space >40–60 ft^2/person
Flow rate ≤5–7 person/(min ft)

LOS C
Pedestrian space >24–40 ft^2/person
Flow rate ≤7–10 person/(min ft)

LOS D
Pedestrian space >15–24 ft^2/person
Flow rate ≤10–15 person/(min ft)

LOS E
Pedestrian space >8–15 ft^2/person
Flow rate ≤15–23 person/(min ft)

LOS F
Pedestrian space <11 ft^2/person
Flow rate varies

Figure 13.8 Cross-flow traffic probability of conflict (TRB, 2016, Exhibit 4-18)
Reprinted with permission from *Highway Capacity Manual, 6th Edition: A Guide for Multimodal Mobility Analysis*, 2016, the National Academy of Sciences, courtesy of the National Academies Press, Washington, DC.

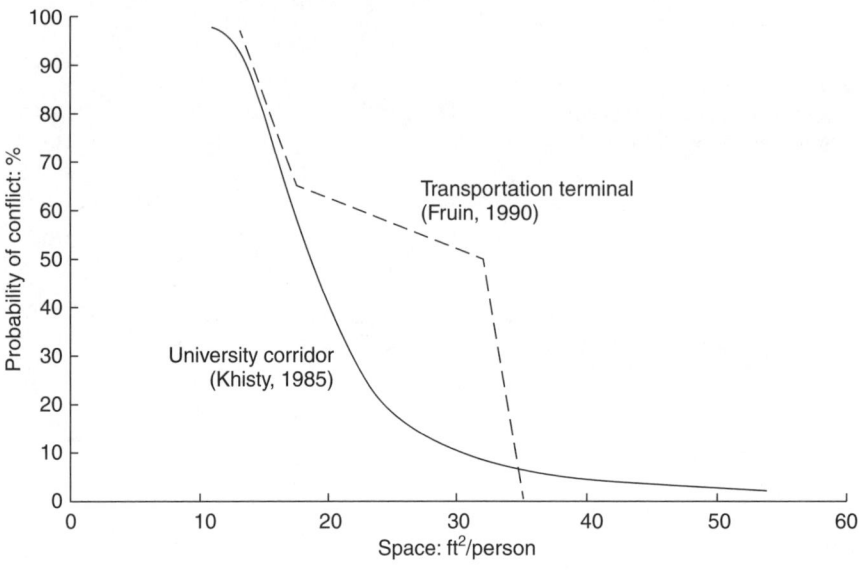

Figure 13.9 Relationship between platoon flow and average flow (TRB, 2016, Exhibit 4-21)
Reprinted with permission from *Highway Capacity Manual, 6th Edition: A Guide for Multimodal Mobility Analysis*, 2016, the National Academy of Sciences, courtesy of the National Academies Press, Washington, DC.

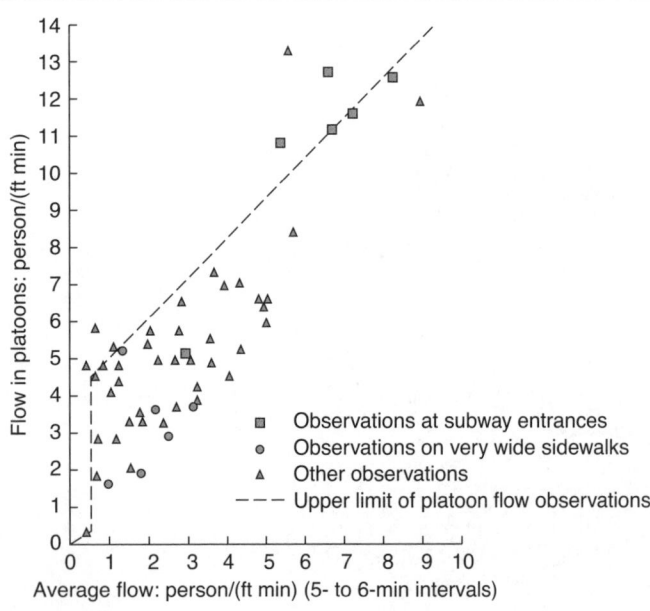

with the average flow rate for durations of 5 to 6 min. The dashed line approximates the upper limit of platoon flow observations (Pushkarev and Zupan, 1975).

13.3. Pedestrian methodology – urban street segments

This section describes key features of the sixth edition of the *HCM* (TRB, 2016) for determining level of service (LOS) for urban streets, as one of its frequent applications.

13.3.1 Scope of the methodology

Key areas of interest include the following.

- *Target travel mode.* This addresses travel by walking in an urban street and is not designed to evaluate travel by means such as motor scooters or other wheeled devices.
- *Typical pedestrian focus.* Pedestrians with disabilities are not included in the method, but should be subject to separate procedures designed to comply with the US Access Board guidelines related to the requirements of the Americans with Disabilities Act of 1990.

13.3.2 Spatial limits

Unless otherwise noted, all variables identified in this section are specific to the subject side of the street. A segment includes, for example, a length of sidewalk and adjoining intersection or intersections. A link includes the length of sidewalk between the intersections.

Table 13.1 Qualitative description of pedestrian space (TRB, 2016, Exhibit 18-15)

Pedestrian space: ft²/person	Description
>60	Ability to move in desired path; no need to alter movements
>40–60	Occasional need to adjust path to avoid conflicts
>24–40	Frequent need to adjust path to avoid conflicts
>15–24	Speed and ability to pass slower pedestrians restricted
>8–15	Speed restricted; very limited ability to pass slower pedestrians
≤8	Speed severely restricted; frequent contact with other users

Reprinted with permission from *Highway Capacity Manual, 6th Edition: A Guide for Multimodal Mobility Analysis*, 2016, the National Academy of Sciences, courtesy of the National Academies Press, Washington, DC.

This typical performance will focus on the performance of the segment (i.e. the link and boundary intersection concerned). However, in some situations, an evaluation of just the link, excluding the intersections, is appropriate.

13.3.3 Segment-based evaluation

This process consists of ten steps, culminating in the determination of the segment LOS. Key characteristics of the process are that

- pedestrian space and a LOS score determine the segment LOS
- the worst LOS resulting from the pedestrian space and the LOS score determine the overall pedestrian LOS.

13.3.3.1 Performance measures

Performance measures are pedestrian travel speed, space and LOS score. The LOS score is an indication of the typical pedestrian's perception of the overall segment travel experience and is also considered a performance measure useful to elected officials, policy makers, administrators or the public. The LOS is based on pedestrian space and pedestrian LOS score. Table 13.1 provides a qualitative description of the pedestrian space to evaluate sidewalk performance from a circulation-area perspective.

13.3.4 Required data and sources

Key items of data, which affect pedestrian performance resulting from the geometric design, in addition to pedestrian flow rate on the segment and facility being considered, include

- downstream sidewalk width
- width of outside through lane, bicycle lane, outside shoulder, parking lane or presence of kerb
- presence of sidewalk (pedestrians are assumed to walk in the street if there is no sidewalk)
- total walkway width
- buffer width and spacing of objects in buffer

- other data, including distance to nearest signalised crossing, legality of midsegment pedestrian crossing and proportion of sidewalk adjacent to window, building or fence.

Depending on the purpose of the evaluation, sources of information may include site measurements or, in the case of design for future facilities, estimated or proposed values.

13.3.4.1 Performance measures

The main measure here is pedestrian delay, which comprises

- delay to pedestrians travelling through the boundary intersection along a path parallel to the segment centreline
- delay to pedestrians by pedestrians who cross the subject intersection at the nearest signalised crossing
- pedestrian waiting delay at an uncontrolled crossing.

13.3.5 Pedestrian LOS for the segment and link
13.3.5.1 Overview of the methodology

The pedestrian LOS score for the signalised intersection is obtained from the pedestrian methodology in Chapter 19 of the *HCM* (TRB, 2016). Incorporating this score, the ten steps mentioned earlier for determining the segment LOS are shown in Figure 13.10.

However, as mentioned earlier, in many practical cases, only the link characteristics are needed, and serve adequately to inform officials and technical personnel of the merits of a particular link – i.e. portions of a sidewalk between intersections, without the need to evaluate signalised crossings. *Therefore, in such cases only Steps 6 and 7 of the process are needed to conduct the evaluation,* the key features of which are shown in Box 13.1.

The pedestrian level of service for the link is then determined using the LOS score $I_{p,link}$ from Step 6. This score is compared with the link-based LOS thresholds given in the rightmost two

Figure 13.10 Pedestrian methodology steps for urban street segments (TRB, 2016, Exhibit 18-17)
Reprinted with permission from *Highway Capacity Manual, 6th Edition: A Guide for Multimodal Mobility Analysis*, 2016, the National Academy of Sciences, courtesy of the National Academies Press, Washington, DC.

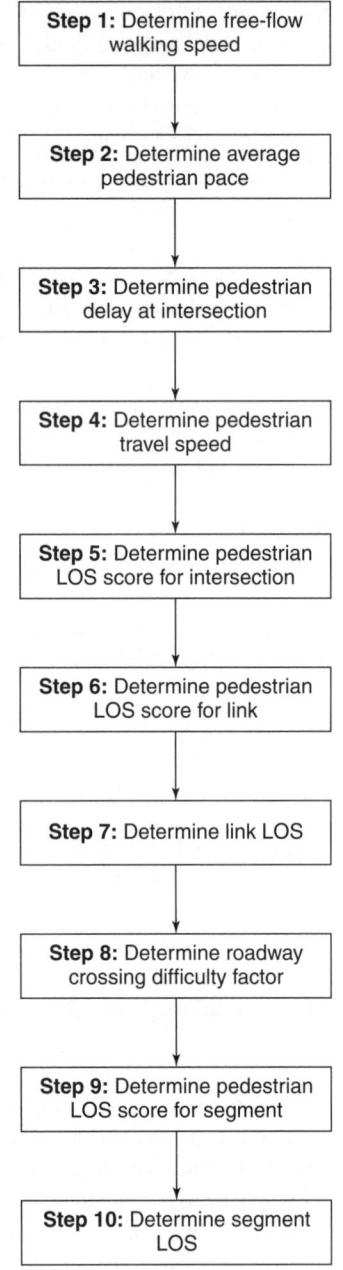

Box 13.1 Calculating level of service for a link

Step 6. Determine pedestrian score for the link

$$I_{p,link} = 6.0468 + F_w + F_v + F_s$$

where

$I_{p,link}$ = pedestrian LOS score for the link
F_w = cross-section adjustment factor
F_v = motorised vehicle volume adjustment factor
F_s = motorised vehicle speed adjustment factor

Step 7. Determine link LOS

Further conditions may have to be investigated in order to validate the use of the foregoing equation for $I_{p,link}$, primarily related to available walkway width. Figure 13.11 shows examples of the variables involved, in this case with specific values, and indicates both a total walkway width, W_T, and an effective walkway width W_E.

13.4. Off-street pedestrian and bicycle facilities

13.4.1 Overview

Off-street pedestrian and bicycle facilities are used only by non-motorised modes and are not considered part of an urban street or transit facility. Those facilities within about 35 ft of an urban street are generally considered off-street, although the precise definition of 'off-street' varies. The 35 ft threshold is based on studies of pedestrian and bicycle facilities (Dowling *et al.*, 2008; Rouphail *et al.*, 1998; Zegeer *et al.*, 2008), in which it was found that motor vehicle traffic influenced pedestrian and bicycle quality of service on facilities at least this distance from the roadway.

Capacity and LOS estimation procedures are described for the following types of service

- walkways, such as paved paths, ramps and plazas, and streets reserved for pedestrian traffic, either full or part-time
- stairways that are a part of a longer pedestrian facility
- shared-use paths separated physically from highway traffic for use by pedestrians, bicyclists, runners, in-line skaters, etc.
- exclusive off-street paths, physically separated from highway traffic for the exclusive use of bicycles.

13.4.2 LOS concepts and criteria

By definition, motorised traffic effects are absent from off-street facilities. Instead, quality of service for off-street facilities reflects the interaction of facility users with each other.

columns of Table 13.2 to determine the LOS for the specified direction of travel along the subject link. For example, if the value of $I_{p,link}$ were to be calculated as 3.25 (first column), the link would be operating at LOS C (see Figure 13.7).

Table 13.2 LOS scores related to segment and link-based values (TRB, 2016, Exhibit 18-2)

Segment-based pedestrian LOS score	Segment-based LOS by average pedestrian space: ft²/person						Link-based pedestrian LOS	
	>60	>40–60	>24–40	>15–24	>8.0–15	≤8.0	Link-based LOS score	LOS
≤2.00	A	B	C	D	E	F	≤1.50	A
>2.00–2.75	B	B	C	D	E	F	>1.50–2.50	B
>2.75–3.50	C	C	C	D	E	F	>2.50–3.50	C
>3.50–4.25	D	D	D	D	E	F	>3.50–4.50	D
>4.25–5.00	E	E	E	E	E	F	>4.50–5.50	E
>5.00	F	F	F	F	F	F	>5.50	F

In cross-flow situations, the LOS E/F threshold is 13 ft²/person.
Reprinted with permission from *Highway Capacity Manual, 6th Edition: A Guide for Multimodal Mobility Analysis*, 2016, the National Academy of Sciences, courtesy of the National Academies Press, Washington, DC.

Figure 13.11 Width adjustment for fixed objects (TRB, 2016, Exhibit 18-18)
Reprinted with permission from *Highway Capacity Manual, 6th Edition: A Guide for Multimodal Mobility Analysis*, 2016, the National Academy of Sciences, courtesy of the National Academies Press, Washington, DC.

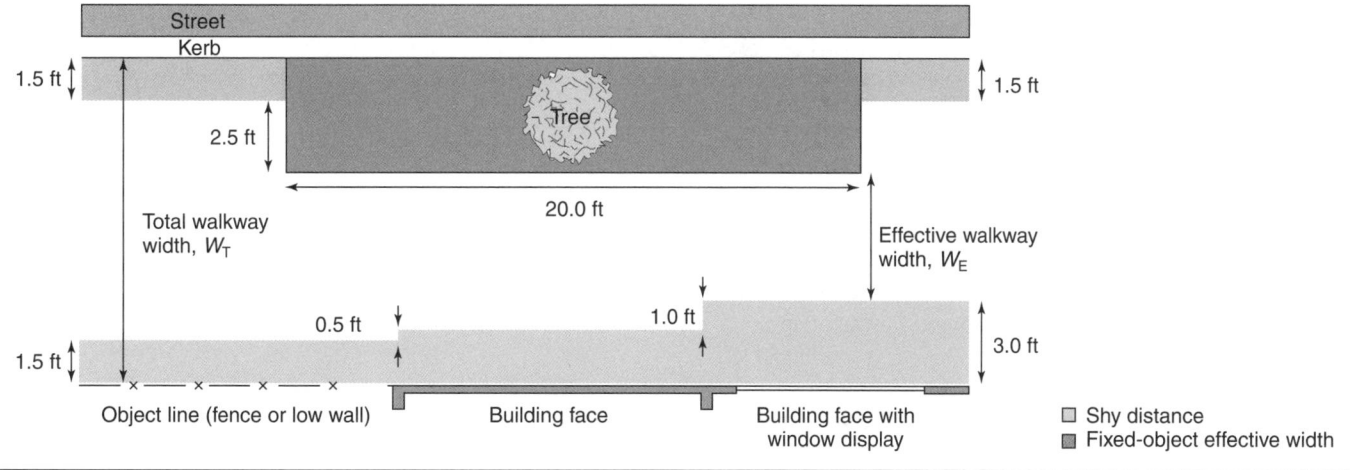

Three criteria apply to the analysis, depending on the travel mode and off-street facility type

- pedestrian space measure in square feet per pedestrian for pedestrians on exclusive facilities
- bicycle meeting and passing events per hour on share use facilities
- bicycle LOS score, incorporating meetings per minute, active passings per minute, presence of a centreline, path width and delayed passings.

The LOS thresholds are based on available user perception research and, in other cases, on expert judgement; it does not reflect whether a facility complies with the Americans with Disabilities Act of 1990 or other standards. A 'pedestrian' is considered to be someone walking: therefore, the pedestrian LOS does not necessarily reflect the quality of service for joggers, persons in wheelchairs, or others who could be considered typical pedestrians.

13.4.3 Walkways

The walkway LOS tables apply to paved pedestrian paths, zones (e.g. exclusive of pedestrian streets), walkways and ramps with up to 5% grade and walking zones through plazas. Table 13.3 applies to random flow; Table 13.4 applies when platoons of pedestrians occur, for example, when a signalised crosswalk is at one end of the segment being analysed, or to pedestrian movement at special events.

Cross flows can occur at the intersections of two approximately perpendicular pedestrian streams. Because of the increased number of conflicts that occur between pedestrians, walkway

Table 13.3 Random-flow level of service criteria for walkways (TRB, 2016, Exhibit 24-1)

Level of service	Average space: ft²/person	Related measures			Comments
		Flow rate: person/(min ft)[a]	Average: ft/s	v/c ratio[b]	
A	>60	≤5	>4.25	>0.21	Ability to move in desired path, no need to alter movements
B	>40–60	>5–7	>4.17–4.25	>0.21–0.31	Occasional need to adjust path to avoid conflicts
C	>24–40	>7–10	>4.00–4.17	>0.31–0.44	Frequent need to adjust path to avoid conflicts
D	>15–24	>10–15	>3.75–4.00	>0.44–0.65	Speed and ability to pass slower pedestrians restricted
E	>8–15[c]	>15–23	>2.50–3.75	>0.65–1.00	Speed restricted, very limited ability to pass slower pedestrians
F	≤8[c]	Variable	≤2.50	Variable	Speed severely restricted, frequent contact with other users

This table does not apply to walkways with steep grades (>5%).
[a] Pedestrians per minute per foot of walkway width.
[b] v/c ratio = flow rate/23, based on random flow: level of service is based on average space per pedestrian.
[c] In cross-flow situations, the level of service E–F threshold is 13 ft²/person.
Reprinted with permission from *Highway Capacity Manual, 6th Edition: A Guide for Multimodal Mobility Analysis*, 2016, the National Academy of Sciences, courtesy of the National Academies Press, Washington, DC.

Table 13.4 Platoon-adjusted level of service criteria for walkways (TRB, 2016, Exhibit 24-2)

Level of service	Average space: ft²/person	Related measure Flow rate[a]: person/(min ft)[b]	Comments
A	>530	≤0.5	Ability to move in desired path, no need to alter movements
B	>90–530	>0.5–3	Occasional need to adjust path to avoid conflicts
C	>40–90	>3–6	Frequent need to adjust path to avoid conflicts
D	>23–40	>6–11	Speed and ability to pass slower pedestrians restricted
E	>11–23	>11–18	Speed restricted, very limited ability to pass slower pedestrians
F	≤11	>18	Speed severely restricted, frequent contact with other users

[a] Rates in the table represent average flow rates over a 5 min period. Flow rate is directly related to space; however, level of service is based on average space per pedestrian.
[b] Pedestrians per minute per foot of walkway width.
[c] In cross-flow situations, the level of service E–F threshold is 13 ft²/person.
Reprinted with permission from *Highway Capacity Manual, 6th Edition: A Guide for Multimodal Mobility Analysis*, 2016, the National Academy of Sciences, courtesy of the National Academies Press, Washington, DC.

capacity is lower at cross flows than at other parts of the walkway. The LOS E–F threshold is 13 ft²/person, as indicated in the notes for Tables 13.4 and 13.5. Levels of service for average space and flow rates for stairways and shared-use paths are shown in Tables 13.5 and 13.6, respectively.

13.4.4 Exclusive off-street facilities

The steps required for estimating the LOSs for off-street pedestrian facilities are shown in Figure 13.12; the required variables and calculation formulae for Steps 1 to 5 are given in Box 13.2.

Table 13.5 Level of service criteria for stairways (TRB, 2016, Exhibit 24-3)

| Level of service | Average space: ft²/person | Related measures | | Comments |
		Flow rate: person/(min ft)[a]	v/c ratio[b]	
A	>20	≤5	≤0.33	No need to alter movements
B	>17–20	>5–6	>0.33–0.41	Occasional need to adjust path to avoid conflicts
C	>12–17	>6–8	>0.41–0.53	Frequent need to adjust path to avoid conflicts
D	>8–12	>8–11	>0.53–0.73	Limited ability to pass slower pedestrians restricted
E	>5–8	>11–15	>0.73–1.00	Very limited ability to pass slower pedestrians
F	≤5	Variable	Variable	Speeds severely restricted, frequent contact with other users

[a] Pedestrians per minute per foot of walkway width.
[b] v/c ratio = flow rate/15. Level of service is based on average speed per pedestrian.
Reprinted with permission from *Highway Capacity Manual, 6th Edition: A Guide for Multimodal Mobility Analysis*, 2016, the National Academy of Sciences, courtesy of the National Academies Press, Washington, DC.

Table 13.6 Level of service criteria for shared-use paths (TRB, 2016, Exhibit 24-4)

Level of service	Event rate/h	Related measure Bicycle service flow rate per direction: bicycles/h	Comments
A	≤38	≤28	Optimum conditions, conflicts with bicycles are rare
B	>38–60	>28–44	Occasional good conditions, few conflicts with bicycles
C	>60–103	>44–75	Difficult to walk two abreast
D	>103–144	>75–105	Frequent conflicts with cyclists
E	>144–180	>105–131	Conflicts with cyclists frequent and disruptive
F	>180	>131	Significant user conflicts, diminished experience

An 'event' is a bicycle meeting or passing a pedestrian.
Bicycle service flow rates (i.e. flow during the peak 15 min) are shown for reference and are based on a 50/50 directional split of bicycles; level of service is based on number of events per hour and applies to any directional split.
Reprinted with permission from *Highway Capacity Manual, 6th Edition: A Guide for Multimodal Mobility Analysis*, 2016, the National Academy of Sciences, courtesy of the National Academies Press, Washington, DC.

Figure 13.11 illustrates a portion of a sidewalk or a walkway. Typically, a walkway operational analysis evaluates the portion of the walkway with the narrowest effective width because this section forms the constraint on pedestrian flow. A design analysis identifies the minimum effective width that must be maintained along the length of the walkway to avoid pedestrian queueing or spill over.

13.4.5 Stairways

A minor pedestrian flow in the opposing direction can result in reduced capacity disproportionate to the magnitude of the reverse flow. Consequently, a small reverse flow could be assumed to occupy a pedestrian lane or 300 in of the stair's width. For a stairway with an effective width of 60 in, a small reverse flow could consume half its capacity. The allowance for

small reverse flow, when used, is included as part of the W_0 term in Step 1 of Box 13.2.

13.4.6 Pedestrians on shared-use paths
13.4.6.1 Required steps
The LOS for pedestrians on shared-use paths is based on the number of events during which a pedestrian either meets an oncoming bicyclist or is passed by a bicyclist. As the number of events increases, the pedestrian LOS decreases because of reduced comfort. Figure 13.13 shows the required steps; the required variables and calculation formulae for Steps 1 to 3 are given in Box 13.3.

Shared-use paths, used by non-motorised modes, often serve recreational areas where vehicular traffic is restricted. Bicycles,

Figure 13.12 Flowchart for analysis of exclusive off-street pedestrian facilities (TRB, 2016, Exhibit 24-7)
Reprinted with permission from *Highway Capacity Manual, 6th Edition: A Guide for Multimodal Mobility Analysis*, 2016, the National Academy of Sciences, courtesy of the National Academies Press, Washington, DC.

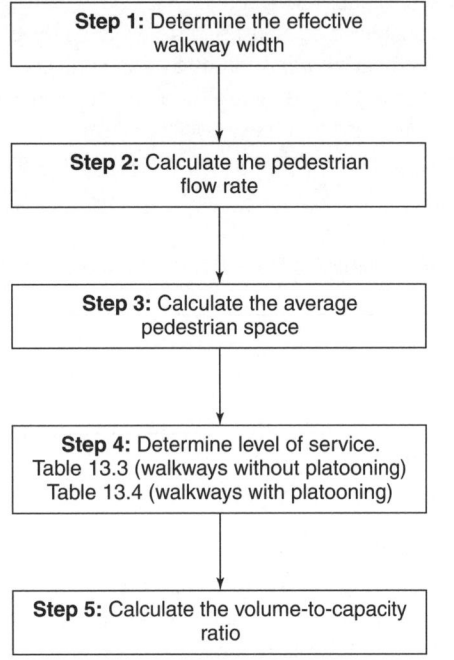

Figure 13.13 Flowchart for analysis of off-street pedestrian LOS on shared-use paths (TRB, 2016, Exhibit 24-10)
Reprinted with permission from *Highway Capacity Manual, 6th Edition: A Guide for Multimodal Mobility Analysis*, 2016, the National Academy of Sciences, courtesy of the National Academies Press, Washington, DC.

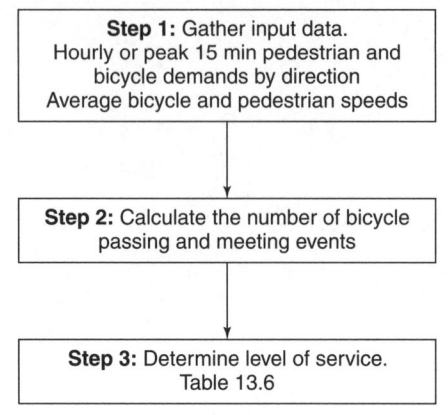

because of their higher speeds have a negative effect on LOS on shared-use paths.

13.4.6.2 Example
To illustrate the application of the LOS concept in a practical example, Box 13.4 contains the relevant problem, calculations and resulting level of service for pedestrian LOS on a shared-use path and an exclusive path.

Box 13.2 Calculating level of service for off-street pedestrian facilities

Step 1. Determine effective walkway width
In walkways and cross-flow areas, the *effective walkway width* is the portion of a walkway that can be used effectively by pedestrians. Various types of obstructions and linear features can reduce the walkway area that pedestrians can effectively use. The effective walkway width at a given point along the walkway is computed as

$$W_E = W_T - W_0$$

where

W_E = effective walkway width (ft)
W_T = total walkway width at a given point along the walkway (ft)
W_0 = sum of fixed-object effective widths and linear-feature shy distances at a given point along the walkway (ft).

Step 2. Calculate pedestrian flow rate
In walkways and cross-flow areas, an hourly pedestrian demand is used as an input to the analysis. Consistent with the general analysis procedures used throughout the *HCM*, hourly demand is usually converted into peak 15 min flows, so that LOS is based on the busiest 15 consecutive minutes during an hour

$$v_{15} = \frac{v_h}{4 \times PHF}$$

where

v_{15} = pedestrian flow rate during peak 15 min (person/h)
v_h = pedestrian demand during analysis hour (person/h)
PHF = peak hour factor.

However, if peak 15 min pedestrian volumes are known, the highest 15 min volume can be used directly without the application of a PHF. Next, the peak 15 min flow is converted into a unit flow rate (pedestrians per minute per foot of effective path width)

$$v_p = \frac{v_{15}}{15 \times W_E}$$

where v_p is pedestrian flow per unit width (person/(ft min)), and all other variables are as previously defined.

Step 3. Calculate average pedestrian space

The service measure for walkways is *pedestrian space*, the inverse of density

$$\text{Space} = \frac{1}{\text{Density}}$$

Pedestrian space can be directly observed in the field by measuring a sample area of the facility and determining the maximum number of pedestrians at a given time in that area. Pedestrian space is related to pedestrian speed and unit flow rate

$$A_p = \frac{S_p}{v_p}$$

where

A_p = pedestrian space (ft^2/p)
S_p = pedestrian speed (ft/min)
v_p = pedestrian flow per unit width (p/(ft min)).

Step 4. Determine level of service

- *Walkways with random pedestrian flow.* Select LOS corresponding to the average space; related measures of flow rate, average speed and *v/c* ratio are shown in Table 13.3.
- *Walkways with platoon flow.* If platooning occurs, Table 13.4 should be used, selecting the LOS corresponding to the average space and the flow rate.

Step 5. Calculate volume to capacity (*v/c*) ratio

Compute using the following values of capacity for various exclusive pedestrian facilities

- walkways with random flow: 23 person/(min ft)
- walkways with platoon flows (average over 5 min): 19 person/(min ft)
- cross-flow areas: 17 person/(min ft) (sum of both flows)
- stairways: 15 person/(min ft) in the ascending direction.

Box 13.3 Calculating level of service for shared-use paths

Step 1. Gather input data
- Hourly or peak 15 min pedestrian and bicycle demand by direction.
- Average pedestrian and bicycle speeds.

Step 2. Calculate number of bicycle passing and meeting events
The level of service for shared-use paths is based on hindrance. The pedestrian LOS is based on the frequency with which the average pedestrian is met and overtaken by bicyclists. The methodology does not account for events with other users, such as skate-boarders.

The average numbers of passing and meeting events per hour are calculated as

$$F_p = \frac{Q_{sb}}{PHF}\left(1 - \frac{S_p}{S_b}\right)$$

$$F_m = \frac{Q_{ob}}{PHF}\left(1 + \frac{S_p}{S_b}\right)$$

where

F_p = number of passing events (event/h)
F_m = number of meeting events (event/h)
Q_{sb} = bicycle demand in the same direction (bicycle/h)
Q_{ob} = bicycle demand in the opposing direction (bicycle/h)
PHF = peak hour factor
S_p = mean pedestrian speed on path (mph)
S_b = mean bicycle speed on path (mph).

These equations do not account for the range of bicycle speeds encountered in practice but because of their limited overlap between the speed distribution of bicyclists and pedestrians the difference is negligible.

For one-way paths, there are no meeting events, so only F_p, the number of passing events, needs to be calculated. Paths 15 ft or wider may effectively operate as two adjacent one-way facilities, in which case F_m may be set to zero.

If peak 15 min volumes by direction are known, they should be substituted for the Q_{sb}/PHF and Q_{ob}/PHF terms in these equations. If only two-directional volumes are known, a directional distribution factor can be applied to the two-directional volumes to estimate the directional volumes. The LOS results are highly sensitive to the choice of directional factor, and field measurements of the directional distribution are recommended if possible.

Meeting events allow direct visual contact, so opposing-direction bicycles tend to cause less hindrance to pedestrians. To account for this, a factor of 0.5 is applied to the meeting events on the basis of theory (Pushkarev and Zupan, 1975). Where sufficient data are available on the relative effects of meetings and passings on hindrance, this factor can be calibrated to local conditions.

Because the number of events calculated in the previous step was based on hourly demand, a PHF must be applied to convert them to the equivalent demand for the 15 min condition. The total number of events is

$$F = (F_p + 0.5 F_m)$$

where F is the total number of events per hour and the other events are as defined earlier.

Step 3. Determine level of service
Table 13.6 is used to determine the LOS based on the total events per hour calculated in Step 2. Unlike the case for the exclusive pedestrian facilities, the LOS F threshold does not reflect the capacity of a shared-use path but rather a point at which the number of bicycle meeting and passing events results in a severely diminished experience for the pedestrians.

Box 13.4 Example of pedestrian LOS on a shared-use path and an exclusive path (TRB, 2016)

The facts

The parks and recreation department responsible for an off-street shared-use path has received several complaints from pedestrians that the volume of bicyclists using the path makes walking on the path an uncomfortable experience. The department wishes to quantify path operations and, if necessary, evaluate potential solutions. The following information was collected in the field for this path.

- Q_{sb} = bicycle volume in same direction = 100 bicycle/h.
- Q_{ob} = bicycle volume in opposing direction = 100 bicycle/h.
- v_{15} = peak 15 min pedestrian volume = 100 pedestrians.
- PHF = peak hour factor = 0.83.
- S_p = average pedestrian speed on path = 4.0 ft/s (2.7 mph).
- S_b = average bicycle speed on path = 16.0 ft/s (10.9 mph).
- No pedestrian platooning was observed.

Step 1. Gather input data

The shared-use path pedestrian LOS methodology requires pedestrian and bicycle speeds and bicycle demand, all of which are available from the field measurements just given.

Step 2. Calculate number of bicycle passing and meeting events

The number of passing events F_p is determined as

$$F_p = \frac{Q_{sb}}{PHF} \left(1 - \frac{S_p}{S_b}\right)$$

$$F_p = \frac{100 \, \text{bicycle/h}}{0.83} \left(1 - \frac{4.0 \, \text{ft/s}}{16.0 \, \text{ft/s}}\right)$$

$$F_p = 90 \, \text{event/h}$$

The number of meeting events F_m is determined as

$$F_m = \frac{Q_{ob}}{PHF} \left(1 + \frac{S_p}{S_b}\right)$$

$$F_m = \frac{100 \, \text{bicycle/h}}{0.83} \left(1 + \frac{4.0 \, \text{ft/s}}{16.0 \, \text{ft/s}}\right)$$

$$F_m = 151 \, \text{event/h}$$

The total number of events is calculated as

$$F = (F_p + 0.5 F_m)$$

$$F = (90 + 0.5(151))$$

$$F = 166 \, \text{event/h}$$

Step 3. Determine shared-use path pedestrian level of service

The shared-use path LOS is determined from Table 13.6. The value of F, 166 event/h, falls into the LOS E range. Because this LOS is rather low, what would happen if a parallel, 5 ft wide, pedestrian-only path were provided?

Step 4. Compare exclusive-path pedestrian level of service

Step 4.1. *Determine effective walkway width*

Assuming no obstacles exist on or immediately adjacent to the path, the effective width would be the same as the actual width, or 5 ft. If common amenities, like trash cans and benches, are to be located along the path, they should be placed at least 3 ft and 5 ft, respectively, off the path to avoid affecting the effective width. These distances are based on data given in the *HCM*.

Step 4.2. *Calculate pedestrian flow rate*

Because a peak 15 min pedestrian volume was measured in the field, it is not necessary to use the equation in Step 2 of Box 13.2 to determine v_{15}. The unit flow rate for the walkway v_p is determined as

$$v_p = \frac{v_{15}}{15 \times W_E}$$

$$v_p = \frac{100}{15 \times 5}$$

$$v_p = 1.33 \text{ person/min}$$

Step 4.3. *Calculate average pedestrian space*

Average pedestrian space is determined (including applying a conversion from seconds to minutes) as

$$A_p = \frac{S_p}{v_p}$$

$$A_p = \frac{(4.0 \text{ ft/s})(60 \text{ s/min})}{(1.33 \text{ person/min})}$$

$$A_p = 180 \text{ ft}^2/\text{p}$$

Step 4.4. *Determine level of service*

Because no pedestrian platooning was observed, Table 13.3 should be used to determine LOS. A value of 180 ft^2/min corresponds to LOS A.

13.5. Summary

This introduction to pedestrian and bicycle aspects of the sixth edition of the *Highway Capacity Manual* has provided an indication of some of the principles and numerical methods of evaluating and possibly modifying the geometric design of pedestrian facilities. Although the basic parameters and dimensions are based on US experience, the concepts and methods can have application to the design of facilities in the UK and elsewhere. In particular, as mentioned in the next chapter, the use of LOS criteria and the associated descriptive variables enables more consistent comparisons between alternative designs, and can help to highlight specific areas of improvement.

REFERENCES

AASHTO (American Association of State Highway and Transportation Officials) (2011) *A Policy on the Geometric Design of Highways and Streets.* AASHTO, Washington, DC, USA.

Dowling RG, Reinke DB, Flannery A *et al.* (2008) *Multimodal Level of Service Analysis for Urban Streets.* NCHRP Report 616. TRB, Washington, DC, USA.

Federal Highway Administration (2018) *Manual on Uniform Traffic Control Devices for Streets and Highways.* Federal Highway Administration, Washington, DC, USA.

Fruin JJ (1990) *Pedestrian Planning and Design, Revised Edition.* Elevator World, Mobile, AL, USA.

Hall ET (1966) *The Hidden Dimension.* Doubleday, Garden City, NY, USA.

Khisty CJ (1985) Pedestrian cross flow characteristics and performance. *Environment and Behavior* **17(6)**: 679–695.

Pushkarev B and Zupan JM (1975) *Urban Space for Pedestrians: A Report of the Regional Plan Association.* MIT, Cambridge, MA, USA.

Rouphail N, Hummer J, Milazzo J and Allen D (1998) *Capacity Analysis of Pedestrian and Bicycle Facilities: Recommended Procedures for the 'Pedestrians' Chapter of the* Highway Capacity Manual. Report FHWA-RD-98-107. Federal Highway Administration, Washington, DC, USA.

Thaler R (1993) *Vorrang für Fussgänger.* Verkehrsclub Österreich, Wien, Austria. (In German.)

TfL (Transport for London) (2010) *Pedestrian Comfort Level Guidance for London.* TfL, London, UK.

TRB (Transportation Research Board) (2000) *Highway Capacity Manual.* National Academy of Sciences, Washington, DC, USA.

TRB (2010) *Highway Capacity Manual*, 5th edn. National Academy of Sciences, Washington, DC, USA.

TRB (2013) *Transit Capacity and Quality of Service Manual.* TCHRP Report 165. TRB, Washington, DC, USA.

TRB (2016) *Highway Capacity Manual: A Guide for Multimo-dal Mobility Analysis*, 6th edn. National Academy of Sciences, Washington, DC, USA.

US Congress (1990) Americans with Disabilities Act of 1990. US Congress, Washington, DC, USA.

Zegeer JD, Vandehey MA, Blogg M, Nguyen K and Ereti M (2008) *Default values for highway capacity and level of service analyses*. NCHRP Report 599. TRB, Washington, DC, USA.

*Reprinted with permission from *Highway Capacity Manual, 6th Edition: A Guide for Multimodal Mobility Analysis*, 2016, the National Academy of Sciences, courtesy of the National Academies Press, Washington, DC.

Schoon, John G
ISBN 978-0-7277-6309-9
https://doi.org/10.1680/pfse.63099.217

Chapter 14

Simulation and computerised models for pedestrian facilities

The need for computerised methods stems from often complex and extensive data, design variables and inputs, computational methods and required outputs that characterise the analysis and design of pedestrian facilities.

The ability to structure and investigate the implications of various scenarios – featuring a potential range of geometric alternatives – is a key requirement, and computerised methods can assist this process. For complex transport terminals and other movement and assembly areas with large passenger flows, computerised models may be the only way to quickly and accurately explore alternative designs.

14.1. Overview of selected simulation models

Computerised methods for conducting transportation planning and aspects of urban vehicular traffic flow are used widely. Among the latter, aimed at operational analysis planning, and design (including, in certain cases, accident prediction) for vehicular traffic, and developed by TRL with continuing updates are

- ARCADY (Assessment of Roundabout Capacity and Delay) (TRL, 2018a)
- PICADY (Priority Intersection Capacity and Delay)
- ARCADY & PICADY combined as junction components
- Transyt (Traffic Network and Isolated Intersection Study Tool) (TRL, 2018b). This may be combined with other simulation methods, including Vissim and Aimsun (see later in this chapter for these latter two methods)
- OSCADY PRO (Optimised Signal Capacity and Delay, Phase-based Rapid Optimisation of Traffic Signals) (TRL, 2018c) is a phase-based traffic signal optimisation program for isolated junctions.

For pedestrian activity, software and simulation methods are less common, particularly when relating pedestrian movement to measures of effectiveness and levels of service. Nevertheless, several products and methods have been developed and are in use. The descriptions here outline, for illustrative purposes, several methods, and provide examples of key features of the outputs.

- First, Vissim (PTV AG, 2018) and Legion (Bentley, 2018), both proprietary software models, simulate the interaction of pedestrian and vehicular traffic under various scenarios. The examples shown in this chapter illustrate the evaluation of pedestrian and vehicular interaction and the movement of large volumes of people.
- Second, models that evaluate the safety of pedestrians at junctions, typified by Pedestrian and Bicyclist Intersection Safety Indices (Federal Highway Administration, 2007) and PEDSAFE 2013 (Zegeer et al., 2013).

Other approaches are also in use. For example, Pedestrian Environment Review Software (TRL, 2018d) is a walking audit tool that audits links, crossing waiting areas and spaces and other aspects of a pedestrian's environment. Complementary software products are: Cycling Environment Review Software (TRL, 2018e) and Rate My Street (TRL, 2010). The method and output of these approaches is largely qualitative, thereby enabling the 'feel' of pedestrian and cycling environments to be gauged and assessed.

The methods described in this chapter, however, focus on quantitative approaches. As with many simulation and modelling methods, their functions and outputs may overlap. The models' adaptation by the analyst and designer can assist in obtaining results responsive to individual demand characteristics and design needs, as long as shortcomings in the available data, models specification and application are recognised. A summary of key features of the models and the extent to which they address geometric design issues that affect the interaction of pedestrians and vehicles is shown in Table 14.1.

14.2. Vissim (PTV AG, 2018)

Vissim is a commercially available proprietary product, described as a microscopic, time-step, behaviour-based model. It is available in several different versions with varying capabilities, depending on specific objectives. Multimodal, it models

217

Table 14.1 Comparison of key features of selected pedestrian safety and movement models

Model name	Major purpose	Basis	Application to geometric design of pedestrian facilities
Vissim	Model multimodal traffic interaction	Simulation of behavioural characteristics of traffic elements, including pedestrians	Includes pedestrian speed, gap acceptance, delay and related performance measures at non-signalled and signalled crossings
SpaceWorks (Legion) and Legion for Aimsun	Pedestrian movement and assembly simulation	Simulation of behavioural characteristics of traffic elements, including pedestrians	Analysis of key pedestrian metrics, e.g. journey times, density (level of service), speed, occupancy and evacuation time to guide geometric design decisions
Ped ISI	Prioritise intersections in terms of pedestrian accident incidence		Indicates relative pedestrian accidents at specific intersections as input to PEDSAFE (below)
PEDSAFE	Assist in selecting countermeasures against accident occurrence	Regression equations for estimating priority indices	Indicates countermeasures, including geometric changes, to reduce accidents
SafeNET	Predict accidents in a street or highway network	Microscopic simulation of pedestrians' preferences and objectives, calibrated by empirical measurements	Modification of geometric design or other determinants to compare between networks

such entities as drivers, pedestrians, bicyclists and motorised vehicles operating on a surface street network. It has been used in a number of cases to model pedestrian and vehicular traffic interaction.

14.2.1 Overview

Interactions among modes can be modelled using Vissim through a set of user-coded rules (e.g. speed control, gap acceptance and yielding behaviour). The resulting network and individual locational performance is sensitive to user behaviour, system status and time. Vissim enables the user to track each individual entity (e.g. a vehicle or a pedestrian) at designated data-collection points. Movement of pedestrians in groups is simulated by motion of Newtonian particles, modified by algorithms to accommodate overlapping. The fundamental forces that simulate person movement are termed 'physico-social' attributes. These consist essentially of physical manifestations of social customs and conventions that apply as a person moves within a group. Such manifestations consist of a driving force in the desired direction of movement and repulsion forces from other pedestrians and physical borders.

Vissim has three major components – an input module, a simulator and an output module. The input module is a Windows-based user interface. The simulator (processor) is used to generate and move traffic, update system status, and collect statistics. The output module can produce animations (movie files) and text output.

14.2.2 Example of Vissim application

An example study by Rouphail *et al.* (2005), using Vissim to address vehicle–pedestrian interaction, specifically pedestrians with vision impairments, was conducted, as follows: the objective of the research was to capture the gap acceptance behaviour for both sighted and blind people near roundabouts (revealed from a parallel research effort) and integrate that behaviour into a microscopic simulation model of pedestrian and vehicular traffic at a roundabout. This process enabled quantification of the mutual impact of pedestrian crossing behaviour on vehicle operation and of motorised traffic impact on pedestrian access (or delay). In addition, it enabled the conduct of experiments to evaluate various pedestrian treatments at or near the roundabout.

The methodology used in the study consisted of three steps.

(a) Selection of an appropriate simulation tool.
(b) Selection and coding of a test roundabout, incorporating observational data of actual pedestrian gap perception behaviour.
(c) Conduct of modelling experiments related to the differential performance of sighted and blind pedestrians at roundabouts, as well as the evaluation of alternative signalisation schemes.

It is important to note the limitations of the study, including the necessary use of default values. The exploratory nature of the

Figure 14.1 Vissim rendition of pedestrian signal at a splitter island (Rouphail *et al.*, 2005). EB, eastbound; NB, northbound; SB, southbound; WB, westbound

Figure 14.2 Vissim rendition of pedestrian signal at a midblock location (Rouphail *et al.*, 2005). EB, eastbound; SB, southbound; WB, westbound

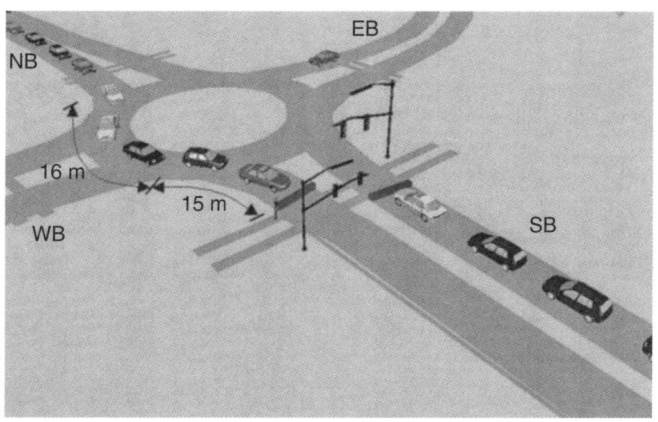

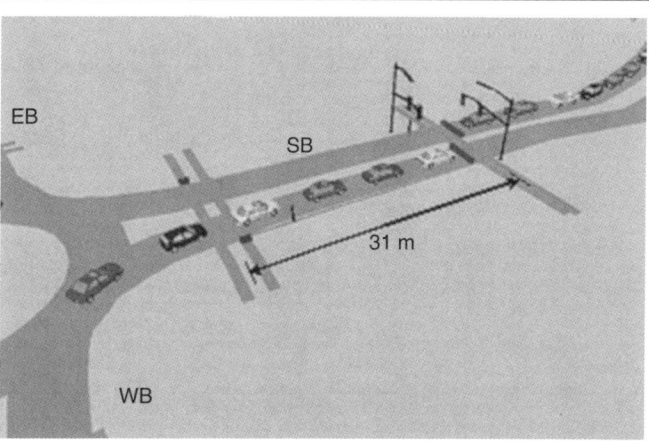

work should, therefore, be recognised, as well as the need for formal model calibration and validation.

14.2.2.1 Selection of simulation tool

Several roundabout analytical and simulation tools were reviewed and Vissim provided the most appropriate platform to achieve the study objectives. Other models either did not have the ability to explicitly model pedestrian movements, or required extensive coding to incorporate necessary pedestrian performance attributes.

14.2.2.2 Outputs

Examples of the outputs of the model included diagrams of the locations and tables and graphs of the numerical values. The locations of the experiments are shown in Figures 14.1 and 14.2; examples of the numerical results are shown in Table 14.2 and Figure 14.3.

The simulation results indicated that pedestrian delay increased in a non-linear fashion as vehicle volume increased. In addition, while there was a small difference in delays to sighted pedestrians at the entry and exit legs, the difference was more pronounced for blind pedestrians, who experienced higher delays on the exit side. Finally, an analysis of pedestrian crossing treatments indicated that placing a pedestrian-actuated, signalled crossing upstream or downstream of the roundabout resulted in delays to blind pedestrians that were comparable to those experienced by sighted pedestrians who crossed at the splitter island. The location of the crossing was assessed on the basis of queue spillback probabilities onto the roundabout exit leg.

The study concluded that the effective use of computer modelling in this case suggested that computer modelling might present a viable alternative to traditional field data-collection methods, in which subjects are placed at risk for the sake of

Table 14.2 Simulated sighted pedestrians' crossing delays at entry and exit legs (Rouphail *et al.*, 2005)

Item	Signal at splitter island with pedestrian actuation		Midblock signal with pedestrian actuation	
	6 pedestrian/h	40 pedestrian/h	6 pedestrian/h	40 pedestrian/h
Scenario	1	2	3	4
Delay: s	17.5 (2.69)[a]	24.2 (3.51)	17.2 (2.89)	26.8 (4.35)
Crossing time[b]: s	10.4	10.4	8.8	8.8
Access time[c]: s	0.0	0.0	45.5	45.5
Travel time: s	27.9	34.5	71.5	81.2

[a] Mean and standard deviation of delay based on ten Vissim replications.
[b] Single-stage crossing, kerb-to-kerb; crossing distance is shorter at the midblock location.
[c] Round trip 'sidewalk' travel time from roundabout crossing to signal location.

Figure 14.3 Simulated sighted pedestrians' crossing delays at entry and exit legs (Rouphail *et al.*, 2005)

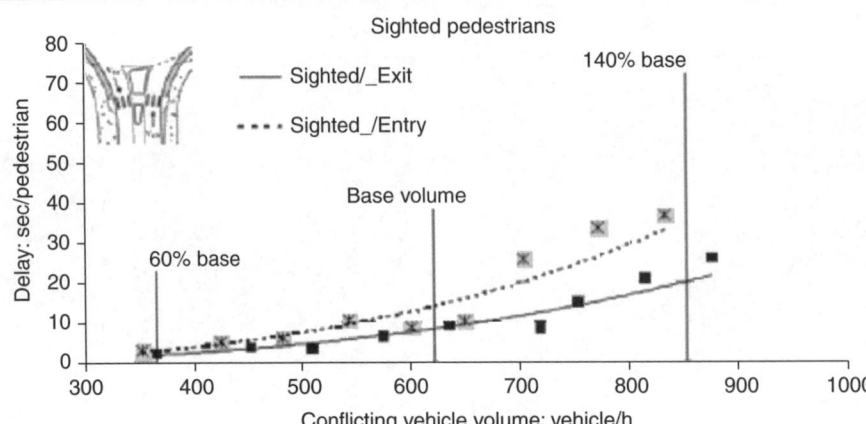

evaluation. While modelling does not rule out the need for eventual evaluation of effects 'in the field', it does permit an approach to operational field evaluations with the knowledge (obtained from the model) that the treatments being evaluated have been shown to have a high probability of success.

14.2.3 Crowd movement

As with other simulation models, Vissim is also used to portray dynamic visualisations of large crowd movements. In this regard, issues of access, safety, delay and related variables can be readily explored and, often, converted to numerical values to assist design and compare alternatives. Figure 14.4 shows an example of pedestrian crowd movements following a soccer game (Fellendorf and Vortisch, 2010). This model's outputs can be particularly useful in illustrating pedestrian movement for

Figure 14.4 Pedestrian bridge between stadium and bus stop 50 min after the finish of a soccer game (Fellendorf and Vortisch, 2010)

Reprinted with permission from Springer Nature: Jaume Barceló (ed.) (2010) *Fundamentals of Traffic Simulation*. Springer, New York, NY, USA.

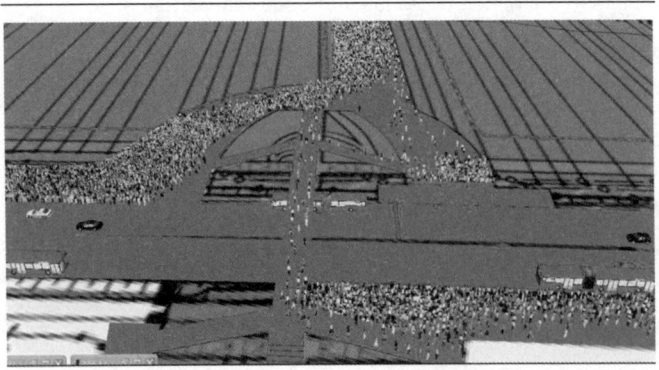

management as well as designers. This can reinforce decision-making about access control, emergency and related issues.

14.3. Legion SpaceWorks (Bentley, 2018)

These models, among other uses, can represent pedestrian movements in large transport terminals and other areas where people congregate (e.g. sports stadia, airports, major events, public and commercial buildings, retail environments and streetscapes). Such models simulate passenger arrivals, movement along corridors, queueing at processing locations, such as ticket counters, assembly on platforms, alighting and boarding of public transport vehicles and other activities where the extent and nature of people's activities requires design of the related facilities. Arriving and departing passengers are modelled simultaneously and the net number of passengers at any location in the system can therefore be accurately estimated. Other output includes use of space, pedestrian journey times, speeds and density. Density can be reported as a level of service (LOS), such as described in Chapter 13. Facilities design can therefore be conducted and tested in order to attain optimum space and layouts.

14.3.1 SpaceWorks – applications

The programme permits virtual experiments on the design or operation of a space and assessment of the impact of different physical designs or levels of pedestrian demand. The program comprises three applications

- Model Builder
- Simulator
- Analyser.

In combination, these applications simulate pedestrian movement within a defined space, such as a railway station, sports stadium, sports park, airport, tall building, piazza, transport hub, town centre or any place that people assemble in or move through.

14.3.1.1 Model Builder
This module is used to build a model of the environment, pedestrians and operational procedures. Required inputs include plans of the facility, preferably in CAD form, pedestrian demand as origin and destination (O–D) matrices and arrival profiles, supporting information, such as escalator speeds and service times, and operational information, such as signage.

Assembly of the inputs then results in a visual representation of the model geometry, with associated tabular information for further manipulation.

14.3.1.2 Simulator
As the essence of the model, Simulator runs and records foot-step simulations of how pedestrians move within the space defined in Model Builder. The Simulator operations are

- import model files
- playback, view and 'sense-check' the simulation
- record the simulation as a results file ('.res') to be analysed
- record all or appropriate parts of the simulation as a video file ('.avi') for presentations.

Legion's SpaceWorks Simulator is based on intelligent entities (individual pedestrians), which negotiate a spatial continuum navigated according to a least-effort algorithm based on the notion that an individual will minimise his or her dissatisfaction during the trip. Crowd behaviour is thus characterised based on probabilistic assignment of social, physical and behavioural characteristics from empirically established profiles.

The model works in continuous (vector) space, thereby simulating complex pedestrian movement, including: bi-directional flow; multi-directional flow and crossing; overtaking; obstacle detection and avoidance; and navigation through stationary crowds. The following three factors contribute to the entities' dissatisfaction

- *inconvenience* – physical effort needed to travel distances
- *discomfort* – the perceived lack of adequate personal space
- *frustration* – having to deviate from a preferred walking speed.

The navigation process comprises choice of destination, macro-navigation and micronavigation. Having chosen a destination and identified the preferred route to it (macronavigation), micronavigation takes into account

- an individual's physical characteristics

- an individual's personal space preferences and preferred walking speed in different contexts
- physical obstacles to a pedestrian's movement
- an individual's neighbours and their movement intentions
- the individual preferences and attributes of pedestrians, which vary over time and by context
- local congestion and resulting changes to route
- learning, adaptation and accumulation of memories to reduce future effort.

The Legion microsimulation model has been calibrated and validated using measurement data collected internationally. Values of flow, speed and density of real crowds, as well as qualitative issues, such as the size and shape of emergent queues at bottlenecks are important key variables included in the simulation.

14.3.1.3 Analyser
The Analyser enables user-defined analyses of the simulation. Much of the pedestrian movement and density patterns are based on many of the values described earlier. In addition to any user-defined thresholds, the software can output results using the Fruin (1990) density banding and related parameters and levels of service outlined in Chapter 13. As well as the graphical depiction of pedestrian movements, questions that can be asked and answered include the following.

- Will the venue cope with projected demand?
- What is the gateline or security check configuration that attains the best flow?
- What are the density levels at bottleneck points, such as the bottom of stairs, main entrances or stadium vomitories?
- What is the average waiting time at facilities during peak periods?
- Which operational scheme optimises individual customer experience?
- Where are the best places for retail units?
- Are existing stairs and escalators effective at clearing a platform before the next train arrives, or are more means of vertical circulation needed?
- What is the interchange time distribution between lines?
- Can the venue be evacuated safely in the case of an incident?
- Will queues in front of facilities, such as ticket windows or cash machines, impede regular circulation?

These outputs can then be used in the geometric design of facilities to check on capacity and, therefore, the required dimensions to accommodate pedestrians at key points in a facility, whether a restricted space, or public streets and assembly areas.

14.4. Legion for Aimsun (Legion, 2018)

Aimsun (a simulation programme integrated with Legion) addresses detailed pedestrian issues associated with urban areas, including uncontrolled and controlled crossings and public transport. Significant parts of the three elements described for Legion apply to the Aimsun procedures. Selected aspects of the simulation of pedestrian movement, geometric design and the necessary inputs are outlined next.

14.4.1 Pedestrian characteristics

The following definitions of behaviour, speed and size of pedestrian are used

- *entity profile* defines a pedestrian's behaviour model
- *speed profile* defines a pedestrian's speed model
- *luggage size* determines a pedestrian's width.

Settings will be applied to all pedestrians within the group. Entity and speed parameter values are taken from the Legion database.

14.4.2 Entities and entity profiles

The simulated people who use the space modelled are called entities. Entities can be displayed in different shapes, the areas of which approximate the size of actual pedestrians. The sizes of entities vary according to their distribution and the size of their luggage. In tight spaces they can temporarily 'flux' in order to squeeze past one another (although this is not represented visually). Entity profiles define a set of parameters that supply the general walking preferences of all entities of a particular type.

14.4.3 Pedestrian speed profiles

This distribution defines the speed at which entities of a particular type would move if unimpeded. Figure 14.5 shows the applicable 'window' of speed distribution and Figure 14.6 a display indicating the pedestrian types selected.

14.4.4 Pedestrian crossings – uncontrolled

An example of a pedestrian crossing location is shown in Figure 14.7, together with centroids defining the potential locations of pedestrians either waiting or crossing.

The situation of pedestrians giving way to vehicles at an unsignalised crossing is shown in Figure 14.8. The instructions to simulate this may be given by selecting the appropriate program option. If the difference between the passing time of the pedestrian and the time in which the priority vehicle reaches the conflict point is greater than the safety margin, the pedestrian can cross.

14.4.5 Signalised pedestrian crossings

Pedestrian crossings may also be included in signal-controlled junctions, using either fixed-time signals or actuated signals,

Figure 14.5 Pedestrian speed distribution (Legion, 2018)

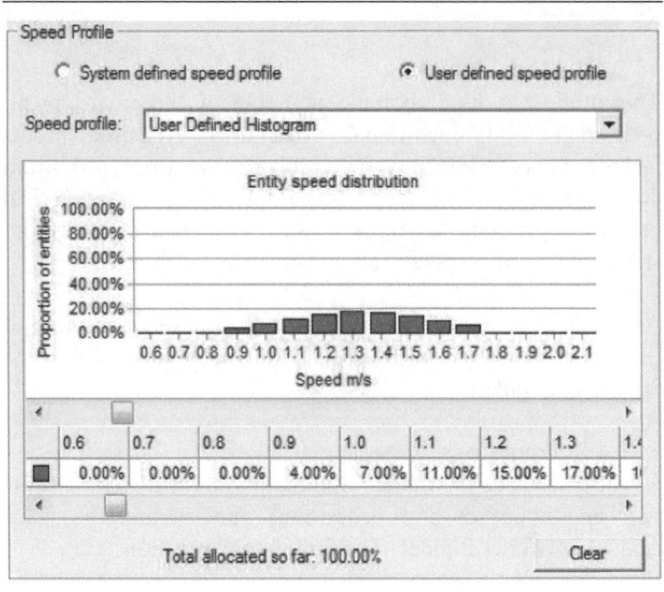

Figure 14.6 Pedestrian type selection (Legion, 2018)

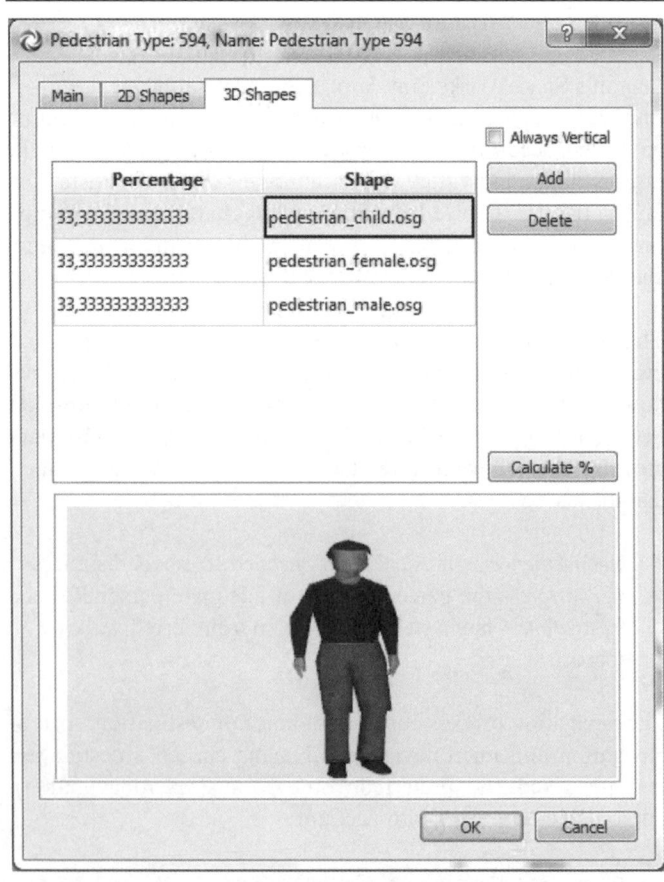

Figure 14.7 Pedestrian crossing showing pedestrian centroids (Legion, 2018)

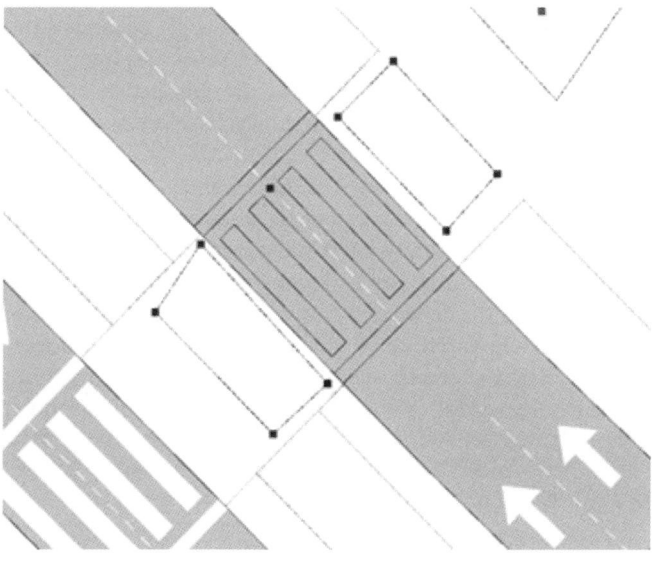

triggered by pedestrian presence. For this, the crossing includes a signal group for a node. If the controller is an actuated controller, push-buttons are automatically added.

Pedestrians will usually wait until their crossing is given a green signal by the signal stage attached to it and controlled by the signal control plan. Also included is a 'cross on red' option to allow pedestrians to cross, using the crossing against the normal signal. This simulates pedestrians who, on observing that it is safe to cross, will do so without waiting for the appropriate signals.

For each pedestrian type and the aggregated value for all pedestrian types, the following time series will be linked to the pedestrian crossings on simulated sections.

Figure 14.8 Unsignalled crossing movements (Legion, 2018)

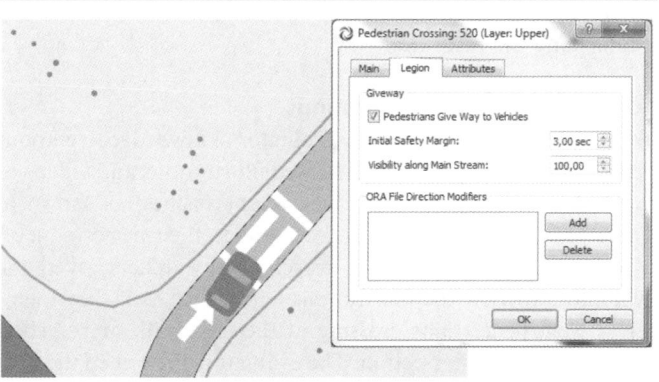

Figure 14.9 Mean calculated time in crossing influence area (Legion, 2018)

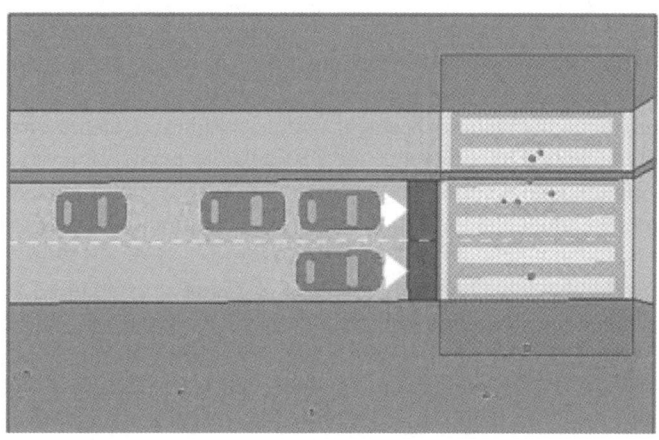

- *Number of pedestrians crossing*. Number of pedestrians in the network who cross the pedestrian crossing.
- *Pedestrian crossing time*. Mean time calculated to cross the crossing's zone of influence, highlighted in Figure 14.9.
- *Pedestrian effective crossing time*. Mean time calculated to cross the crossing alone, as shown by the highlighted area in Figure 14.10.

This brief outline of the key features of Legion for Aimsun has addressed a selection of the major applications of computerised simulation. A useful review of Legion for Aimsun is made in *Pedestrians in Microscopic Traffic Simulation* (Alexandersson and Johansson, 2013). As developments in human factors and software continue, it is likely that such methods will increasingly find application in the design of pedestrian facilities.

Figure 14.10 Mean calculated time on actual crossing (Legion, 2018)

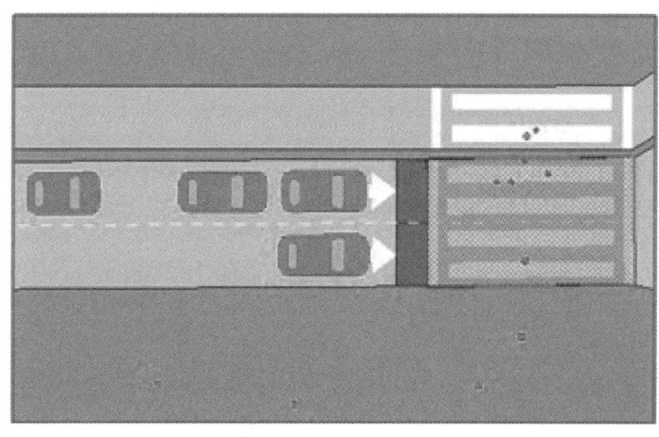

14.5. SafeNET (Software for Accident Frequency Estimation for Networks)

Originally developed by the Department for Transport, Safe-NET 2 (TRL, 2006) is an interactive software package to assist traffic engineers, transport planners and road safety officers in the design of safe road networks. It allows estimates to be made of the personal injury accident frequency (number of personal injury accidents per year) for particular elements of a road network (junctions of different types and road sections), including the trunk road network.

SafeNET 2 provides the user with a forecast of accident levels as a function of junction and road design, network management changes and the form of control and traffic assignment. It can be used as a stand-alone product to rapidly assess the safety of a scheme or in conjunction with traffic assignment models to provide a comprehensive overall assessment of delay, journey time and accident frequency.

To estimate the frequency of personal injury accidents on a road network, the vehicle flow, pedestrian flow and geometric data for each junction and road within the road network are entered. SafeNET uses this information to calculate how many personal injury accidents would be expected to occur within a one-year period. These values are then summed to provide the number of accidents that would occur across the network as a whole. Various levels of modelling may be employed, meaning that predictions can be made using only limited input data, although the more information that is available, the more accurate the predictions are likely to be.

SafeNET 2 is capable of modelling road networks which can include

- roundabouts and mini-roundabouts
- traffic signal junctions
- urban and rural priority T-junctions
- urban and rural crossroads and staggered junctions
- urban single-carriageway roads and urban roads including minor junctions
- traffic calming measures
- modern rural single- and dual-carriageway road links
- motorway links
- COBA models for urban and rural road links
- rural and urban speed models.

Using the number of accidents estimated by application of SafeNET, individual junction designs may be modified and, through a series of iterative steps, the physical geometry or other features of the network may be modified until a desired level of safety has been determined.

14.6. Ped ISI and Bike ISI

14.6.1 Overview

The Pedestrian and Bicyclist Intersection Safety Indices (Ped ISI and Bike ISI, FHWA, 2007) comprise a set of models that enable users to identify intersection crossings and intersection approach legs that should be given the greatest priority for undergoing detailed pedestrian and bicycle safety assessment.

Using observable characteristics of an intersection crossing or approach leg, such as number of lanes and traffic volume, the tool produces a safety index score, with higher scores indicating greater priority for a detailed safety assessment. The score is used as the dependent variable in a regression equation, the independent variables of which are the numerical values of the intersection characteristics. The equation is then used in a predictive sense to estimate potential scores of proposed designs.

14.6.2 Pedestrian and bicycle intersection safety indices

Each leg of an intersection may have different characteristics affecting pedestrian or bicyclist safety; therefore, the tool is intended to provide a rating of the safety of an individual crossing (Ped ISI) or approach leg (Bike ISI) rather than evaluating the intersection as a whole. A practitioner can use the tool to develop a prioritisation scheme for a group of pedestrian crossings or bicyclist approaches. This method enables the practitioner to prioritise and proactively address sites that are most likely to be a safety concern for pedestrians or bicyclists, without having to wait for crashes to occur.

14.6.3 Development

Researchers developed the Ped ISI and Bike ISI based on safety ratings, or expert opinion of the safety of a site, and observed behaviours, or observed interactions between pedestrians and motorists or bicyclists and motorists. These measures formed the basis for a method of estimating the relative safety of a pedestrian crossing or bicycle approach leg. Observations from 68 unsignalised and signalised intersections throughout the USA provided the data on which regression equations were developed. These equations provide the quantitative basis for establishing the priorities.

14.6.4 Safety measure: ratings

To develop the safety ratings, evaluators knowledgeable about pedestrian and bicyclist issues viewed illustrations and videos of the pedestrian crossings and bicycle approaches, as shown in Table 14.3, and rated the sites according to their perceived level of safety for a pedestrian or bicyclist. Researchers asked the evaluators to view the illustrations and videos as if they themselves were pedestrians crossing at the crosswalk or bicyclists approaching the intersection. The evaluators then rated the sites

Table 14.3 Ped ISI variable descriptions (Federal Highway Administration, 2007)

Ped ISI variable	Safety index value (pedestrian)	
SIGNAL	Signal-controlled crossing	0 = no 1 = yes
STOP	Stop-sign controlled crossing	0 = no 1 = yes
THRULNS	Number of through lanes on street being crossed (both directions)	1, 2, 3, …
SPEED	85th percentile speed of street being crossed	Speed in miles per hour
MAINADT	Main street traffic volume	Number of vehicles per day divided by 1000
COMM	Predominant land use on surrounding area is commercial development (e.g. retail, restaurants)	0 = not predominantly commercial area 1 = predominantly commercial area

on a scale of one (most safe) to six (least safe), according to their sense of safety and comfort.

14.6.5 Model application

Indices were developed, the purpose of which was to assign safety index values to crossings and bicycle approaches, with the goal of providing the practitioner with the means of prioritising these sites for the purpose of further safety evaluation. For example, a practitioner might have 30 crossings in his or her jurisdiction to evaluate with respect to pedestrian safety. The practitioner would use Ped ISI to assign safety index values to each of the 30 crossings. The crossings with the highest values would be the first crossings for which the practitioner should conduct in-depth evaluations to determine whether pedestrian safety problems exist at those sites.

It should be noted that the model does not dictate specific 'warrants' for safety treatments or thresholds at, or beyond which, treatment may be justified. Also, it should be noted that Ped ISI and Bike ISI were developed for signalised and stop-controlled intersections only, and are not suitable for roundabouts.

The evaluation may include an investigation into the crash history of the site, which may lead to countermeasure recommendations from other resources, such as PEDSAFE. Other evaluations might include pedestrian counts and behaviour studies or pedestrian conflict analysis.

Regarding safety indices for the whole intersection, if the user wishes to produce a safety index value for an entire intersection, the suggested method is to average the index values from all legs. However, caution should be used in this approach. Some intersections might have one leg that is high priority for safety evaluation (high index value) and three legs that are low priority (low index values). If the leg safety index values are averaged at

this intersection, the result will be a low intersection safety index, and the high priority of the one leg may go unnoticed.

Ped ISI and Bike ISI were developed using three- and four-leg intersections. Since the models produce safety index values for individual legs instead of entire intersections, it is possible to use Ped ISI and Bike ISI at intersections with five or six legs. Many of the factors that affect pedestrian or bicyclist safety at four-leg intersections would similarly affect safety at five- or six-leg intersections. However, safety index values produced for five- or six-leg intersections should be used only with the understanding that the models were not developed using intersections of that type.

14.6.6 Example

A pedestrian crossing location has characteristics, and corresponding Ped ISI variables (see Table 14.3) as follows (Federal Highway Administration, 2006).

- Signalised intersection (SIGNAL = 1).
- Stop-controlled (STOP = 1).
- Four through lanes on the main road, two in each direction (THRULNS = 4).
- 85th percentile speed on the main road of 67.6 km/h (42 mph) (SPEED = 42).
- Main road average daily traffic flow is 22 000 vehicles per day (MAINADT = 22).
- Surrounding area is commercial (COMM = 1).

Figure 14.11 illustrates the crosswalk of interest at the intersection; Table 14.3 summarises the model and its input values.

The Ped ISI regression equation is

$$\textbf{Ped ISI} = 2.372 - 1.867\textbf{SIGNAL} - 1.807\textbf{STOP} \\ + 0.335\textbf{THRULNS} + 0.018\textbf{SPEED} \\ + 0.006(\textbf{MAINADT} \times \textbf{SIGNAL}) + 0.238\textbf{COMM}$$

Figure 14.11 Crosswalk view of the example intersection (Federal Highway Administration, 2006)

Applying the regression equation by inserting the values (independent variables) for this example, the Ped ISI value is calculated as

$$\textbf{Ped ISI} = 2.372 - 1.867 \times \textbf{1} - 1.807 \times \textbf{1} + 0.335 \times \textbf{4} \\ + 0.018 \times \textbf{42} + 0.006(\textbf{22} \times \textbf{1}) + 0.238 \times \textbf{1} = \textbf{2.7}$$

The average daily traffic flow must be entered in thousands (i.e. 22 instead of 22 000) for the equation. A spreadsheet calculator allows the user to enter the average daily traffic flow as a whole value.

Another way to determine the Ped ISI value is to go directly to the quick reference tables of Ped ISI values. These tables are provided for various combinations of traffic control and area

type. For the example above, for a signalised crossing in a commercial area and four through lanes, an 85th percentile speed of 67.6 km/h (42 mph) (use the column of the nearest value, i.e. 40 miles per hour (64.4 km/h)), and 22 000 average daily traffic flow (use the row with 20 000 average daily traffic flow), the Ped ISI value is 2.7, as shown in Table 14.4 and as calculated by means of the regression equation.

14.6.7 Countermeasure selection: PEDSAFE

Once pedestrian crossings and bicycle approaches to intersections have been prioritised for in-depth safety evaluation using Ped ISI and Bike ISI, the practitioner will have many options for evaluation, analysis and treatment. It is recommended that PEDSAFE be used to assist in the selection of appropriate countermeasures for pedestrian facilities. PEDSAFE is available from the Federal Highway Administration (Zegeer *et al.*, 2013).

PEDSAFE is designed to recommend treatments for specific safety problems. To make full use of the information provided in these tools, knowledge of the most common safety problems at each site will be required. Examining the types of crash that occur at the site or analysing behaviour of pedestrians, bicyclists and motorists at the site can provide knowledge of these safety problems.

The PEDSAFE guide provides details for 49 different types of safety treatment, which can be used to improve pedestrian safety or mobility. It also includes information on the specific types of countermeasure that might be appropriate for addressing such objectives as

- reducing the speed of motor vehicles
- improving sight distance and visibility for motorists and pedestrians
- reducing the volume of motor vehicles

Table 14.4 Quick reference table – pedestrian Ped ISI values (signalled and commercial area) (Federal Highway Administration, 2006)

Main Rd Thru Lns	1 Through Lane					2 Through Lanes					3 Through Lanes					4 Through Lanes				
Main Rd Speed	25	30	35	40	45	25	30	35	40	45	25	30	35	40	45	25	30	35	40	45
ADT 1000	1.3	1.4	1.5	1.6	1.7	1.6	1.7	1.8	1.9	2.0	2.0	2.1	2.1	2.2	2.3	2.3	2.4	2.5	2.6	2.7
5000	1.3	1.4	1.5	1.6	1.7	1.7	1.7	1.8	1.9	2.0	2.0	2.1	2.2	2.3	2.4	2.3	2.4	2.5	2.6	2.7
10000	1.4	1.4	1.5	1.6	1.7	1.7	1.8	1.9	2.0	2.0	2.0	2.1	2.2	2.3	2.4	2.4	2.4	2.5	2.6	2.7
15000	1.4	1.5	1.6	1.7	1.7	1.7	1.8	1.9	2.0	2.1	2.1	2.1	2.2	2.3	2.4	2.4	2.5	2.6	2.7	2.7
20000	1.4	1.5	1.6	1.7	1.8	1.7	1.8	1.9	2.0	2.1	2.1	2.2	2.3	2.4	2.4	2.4	2.5	2.6	2.7	2.8
25000	1.4	1.5	1.6	1.7	1.8	1.8	1.9	2.0	2.0	2.1	2.1	2.2	2.3	2.4	2.4	2.4	2.5	2.6	2.7	2.8
30000	1.5	1.6	1.7	1.7	1.8	1.8	1.9	2.0	2.1	2.2	2.1	2.2	2.3	2.4	2.5	2.5	2.6	2.7	2.7	2.8
35000	1.5	1.6	1.7	1.8	1.9	1.8	1.9	2.0	2.1	2.2	2.2	2.3	2.4	2.4	2.5	2.5	2.6	2.7	2.8	2.9
40000	1.5	1.6	1.7	1.8	1.9	1.9	2.0	2.0	2.1	2.2	2.2	2.3	2.4	2.5	2.6	2.5	2.6	2.7	2.8	2.9
45000	1.6	1.7	1.7	1.8	1.9	1.9	2.0	2.1	2.2	2.3	2.2	2.3	2.4	2.5	2.6	2.6	2.7	2.7	2.8	2.9
50000	1.6	1.7	1.8	1.9	2.0	1.9	2.0	2.1	2.2	2.3	2.3	2.4	2.4	2.5	2.6	2.6	2.7	2.8	2.9	3.0

- reducing pedestrian exposure to traffic (e.g. reducing crossing distance)
- improving compliance with traffic laws
- eliminating behaviours that lead to crashes.

A listing of pedestrian-related treatments for each of these performance objectives is given by 'categories' of treatments, including pedestrian facility design, roadway design, intersection design, traffic calming, traffic management and signals and signage. For example, to reduce the speed of motor vehicles, possible roadway design treatments include adding a bike lane or shoulder, road narrowing, reducing the number of lanes, driveway improvements, kerb radius reduction and adding a right-turn slip lane.

The PEDSAFE guide also gives a description of 12 specific pedestrian crash types (e.g. dart or dash, walking along roadway, turning vehicle, multiple-threat), with corresponding countermeasure options for each crash type. PEDSAFE also contains write-ups of 71 case studies of pedestrian improvements that have been implemented in the USA.

The expert system software is provided to allow a user to input the type of pedestrian safety problem, along with the location or roadway section characteristics, such as intersection or lanes and traffic volume. The software will then generate a 'shortlist' of countermeasure options based on the type of pedestrian safety problem and site characteristics.

The PEDSAFE system also enables community participation in the improvement process – often a major element in the success of a project. Key features of such an example (Zegeer *et al.*, 2013, Case study no 69) are as follows.

- *Problem.* Speeding by cut-through traffic was increasing the number of vehicle–pedestrian conflicts for residents of a local community.
- *Results.* The community worked with the US Department of Transportation to install new sidewalks, a crosswalk, caution signs and school zone signs in the same area. A 200 ft sidewalk, crosswalks and signs were installed. Although no formal evaluation of the changes was made, residents were pleased with the changes and county officials feel that a viable alternative was worked out through the negotiation process.

14.7. Summary

This brief description of several examples of simulation, prediction, computer applications and combined methods has provided an overview of the demand for, and performance of, specific pedestrian facilities, ranging from individual junctions to entire terminals and other places where pedestrians move and assemble. A variety of models has been illustrated. This modelling can provide valuable information with which to guide and inform the analysis and geometric design configuration and dimensions for pedestrian facilities and provide insights into pedestrians' needs.

Together with imaginative and sound application of design principles, as well as caution in realising that all models are, by definition, only interpretations of reality, this can assist in formulating and implementing pedestrian facilities, which encourage walking and the use of public transport. Yet much remains to be done to encourage further modelling of pedestrian capabilities and facilities. It is hoped that this brief introduction will stimulate further applications, research and development of the relevant techniques.

REFERENCES

Alexandersson S and Johansson E (2013) *Pedestrians in Microscopic Traffic Simulation: Comparison Between Software Viswalk and Legion for Aimsun.* MSc thesis, Chalmers University of Technology, Gothenburg, Sweden.

Bentley (2018) Legion Simulation and Modeling Software. https://www.bentley.com/en/products/brands/legion-skid = CT_WRO_LGN_REDIRECT_18 (accessed 17/10/2018).

Federal Highway Administration (2006) *Pedestrian and Bicycle Crash Analysis Tool.* No. FHWA-HRT-06-13. Federal Highway Administration, Washington, DC, USA.

Federal Highway Administration (2007) *Pedestrian and Bicyclist Intersection Safety Indices.* Federal Highway Administration, McLean, VA, USA.

Fellendorf M and Vortisch P (2010) Microscopic traffic flow simulator Vissim. In *Fundamentals of Traffic Simulation* (Barceló J (ed.)). Springer, New York, NY, USA, pp. 63–93.

Fruin JJ (1990) *Pedestrian Planning and Design, Revised Edition.* Elevator World, Mobile, AL, USA.

Legion (2018) *Legion for Aimsun: Users' Manual.* Legion, London, UK.

PTV AG (2018) PTV Vissim. https://www.ptvgroup.com/en/solutions/products/ptv-vissim/ (accessed 17/10/2018).

Rouphail N, Hughes R and Chae K (2005) Exploratory simulation of pedestrian crossings at roundabouts. *Journal of Transportation Engineering* **131(3)**: 211–218.

TRL (Transport Research Laboratory) (2006) *SafeNET (Software for Accident Frequency Estimation for Networks).* TRL, Crowthorne, UK.

TRL (2010) Ratemystreet is Launched: How Walking Friendly is Your Street? https://trl.co.uk/news/prev/4057 (accessed 18/10/2018).

TRL (2018a) ARCADY: Assessment of Roundabout Capacity and Delay. https://trlsoftware.com/products/junction-signal-design/junctions/arcady/ (accessed 18/10/2018).

TRL (2018b) Transit: A Traffic Network Study Tool. https://trl.co.uk/reports/LR253 (accessed 18/10/2018).

TRL (2018c) *OSCADY: Optimised Signal Capacity and Delay.* https://trlsoftware.com/products/junction-signal-design/ junctions/oscady/ (accessed 18/10/2018).

TRL (2018d) Streetaudit – PERS: Pedestrian Environment Review Software, version 1.1.10.211. https://trlsoftware. com/products/road-safety/street-auditing/streetaudit-pers/ (accessed 18/10/2018).

TRL (2018e) Streetaudit – CERS: Cycling Environment Review Software, version 1.1.10.211. https://trlsoftware. com/products/road-safety/street-auditing/streetaudit-cers/ (accessed 18/10/2018).

Zegeer CV, Nabors D, Lagerway P *et al.* (2013) *PEDSAFE 2013: Pedestrian Safety Guide and Countermeasure Selection System.* Federal Highway Administration, McLean, VA, USA. http://www.pedbikesafe.org/PEDSAFE/ (accessed 18/10/2018).

Pedestrian Facilities, Second edition

Schoon, John G
ISBN 978-0-7277-6309-9
https://doi.org/10.1680/pfse.63099.229

Chapter 15
Examples and case studies

Numerous new and updated pedestrian facilities are being implemented. The examples and case studies included here combine a variety of sources – some actual projects, some hypothetical and some a combination of features and analysis. Each attempts to illustrate a selection of the key points presented in the earlier chapters.

It is emphasised that individual organisations and designers may approach similar cases differently and that the examples shown illustrate one possible approach only. Therefore, considerably more information and analysis will be required in practice.

15.1. Background

The format for several of the examples outlines the original conditions, the configuration of the proposals and comments on the rationale for the design. Along with each example is a commentary on relevant design features, likely information sources and other comments. Several sources of examples of actual installations are available, such as those cited in *Manual for Streets* (DfT, 2007) and in *Traffic Calming Techniques* (IHCSS, 2005).

An important aspect of examining any proposed or implemented project is the documentation associated with it. Not only is this important as a checklist of design procedures but it also provides a basis for making informed changes at local and regional levels when future development is planned. For this reason, where appropriate, summaries of existing conditions have been tabulated and these are related to the features of the schemes. The tabular format associated with identifying potential project conditions and options shown in Tables 6.6 and 6.7 is a typical method of compiling this documentation.

15.2. *Example 1.* Crossing at unsignalled trunk road and entrance to residential street of a medium-sized town

15.2.1 Existing conditions

The junction of Principle Street and South Lane is shown in Figure 15.1. Vehicle speeds, ease of crossing for a variety of pedestrians, including disabled and schoolchildren, and parking, have been considered. Applicable guidance is the *DMRB* (DfT, 2018) and the *Manual for Streets* (DfT, 2007), owing to the

junction's suburban residential location on a radial arterial route. Table 15.1 summarises the key features.

15.2.2 Proposed design approach

Refer to TA 23/81 (DfT, 1981), TD 42/95 (DfT, 1995b, Sections 5.1 and 6.2), *Manual for Streets* (DfT, 2007, Section 7), *Inclusive Mobility* (DfT, 2005a) and related documents.

Design features that reduce the distance walked and make drivers more aware of pedestrians in the crossing would be beneficial. Potential design devices for consideration therefore include

- a pedestrian refuge
- build-out of kerbing at entrance to South Lane
- raised table at entrance to South Lane
- reducing corner radii
- repositioning of pedestrian access ramps.

15.2.3 Selected design features (see also Figure 15.2)

- Reduce kerb radii – to reduce vehicle turning speeds and enable the pedestrian crossing ramps to be placed closer to the pedestrian route lines.
- Construct a build-out to reduce the distance crossed by pedestrians and provide better sight lines for pedestrians and drivers.
- Relocate ramps to pedestrian route lines in order to
  - improve pedestrians' sight line to approaching vehicles, thereby improving pedestrian safety
  - improve pedestrians' travel time by reducing their current diversion into South Lane.
- Widen pedestrian ramps to better accommodate passing wheelchairs and mobility scooters.
- Delineate crossing width to emphasise the crossing's presence.
- Recommend that the speed limit of Principle Street be reduced to 20 miles per hour, possibly by use of traffic calming measures.

15.2.4 Comments

- Raised table was not proposed because similar junctions at other locations along Principle Street do not employ

Figure 15.1 Example 1: existing conditions

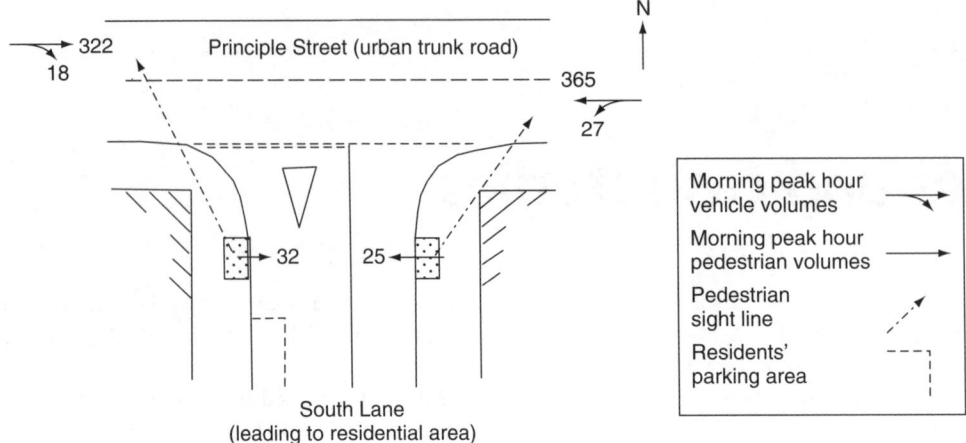

South Lane
(leading to residential area)

Table 15.1 Example 1: summary of existing conditions

Item	Physical and operational conditions, and comments
Vehicle speed	30 miles per hour limit; 85th percentile, 32.7 miles per hour
Vehicle traffic flow, peak hour	As shown in Figure 15.1
Pedestrian traffic flow, peak hour	Volumes as shown in Figure 15.1; approximately 12% of total pedestrian traffic comprises older people, of whom 3% are wheelchair or mobility scooter users
Accidents	5 injuries to pedestrians on and in vicinity of crossing, one serious, in 2008
Signal details	No signals at present
Crossing width	Informal crossing, approximately 2.4 m wide, with no markings
Kerb radii	Approximately 10 m and considered to result in excessive speed of vehicles turning into and out of Principle Street
Nearby junctions	None
Other crossings nearby	No formal or informal crossings
Entrances within 100 m	None
Disabled facilities	Dropped kerbs, damaged surface and not in accordance with applicable standards
Visibility	Satisfactory, based on LTN 2/95 (DfT, 1995a) and TD 50/04 (DfT, 2004a)

these devices, it being considered that some standardisation of devices is preferable.

- Central refuge was not proposed because crossing distance with the build-outs would reduce the available lane width to less than necessary to enable safe passage of motor vehicles and bicycles.
- Some delay to vehicles is expected on Principle Street, owing to vehicle drivers' greater need to slow down at the reduced radii. This is not expected to be severe in peak hours, for other reasons.

15.3. *Example 2*. Signalised pedestrian crossing of an inner city road adjacent to a pedestrianised area

15.3.1 Background

The objectives of this scheme were to substantially reduce the dominance of motor vehicles and improve public transport and pedestrian facilities in the city centre. The existing crossing did not cater for pedestrian desire lines and pedestrian safety was seriously compromised. The design approach, therefore, focused on two key considerations

Figure 15.2 Example 1: proposed design

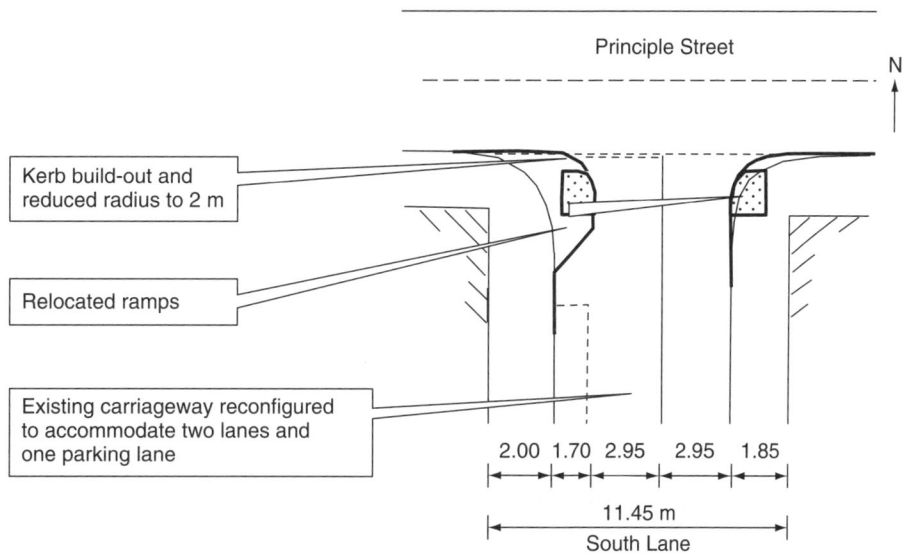

- pedestrian desire lines
- safety of pedestrian movements across the highway.

15.3.2 Existing conditions
Key physical and operational features of the site and its immediate environs are shown in Table 15.2; Figure 15.3 depicts the existing layout and proposed scheme.

15.3.3 Future demand and environs
Key determinants of the future design are as follows.

- Vehicle traffic volumes and composition will remain essentially unchanged.
- Pedestrian traffic volume and desire lines are expected to remain unchanged.

Table 15.2 Example 2: summary of existing conditions

Item	Physical and operational conditions, and comments
Vehicle speed	20 miles per hour limit; 85th percentile, 23.4 miles per hour
Vehicle traffic flow, peak hour	625 vehicles per hour in peak afternoon hour, approximately evenly distributed between two lanes, one-way southbound
Pedestrian traffic flow, peak hour	585 pedestrians per hour approximately, evenly distributed between north and southbound in morning and afternoon peak hours; approximately 3% using wheelchairs, 4% using other mobility aids and 3% using mobility scooters, mostly generated by Shopmobility scheme
Accidents	6 injuries to pedestrians on and in vicinity of crossing, one serious
Signal details	No signals at present
Crossing width	3 m, involves up to 2 min waiting time for pedestrians; this feature, together with observed crossings to adhere to desire lines, results in considerable premature, unsafe crossing by a significant number of pedestrians; considerable conflict between pedestrians in opposing streams on crossing
Nearby junctions	None
Other crossings nearby	No formal or informal crossings
Entrances within 100 m	None
Disabled facilities	Dropped kerbs and tactile paving in accordance with DETR *Guidance on the Use of Tactile Paving Surfaces* (DETR, 1998)
Visibility	Satisfactory, based on LTN 2/95 (DfT, 1995a) and TD 50/04 (DfT, 2004a)

Figure 15.3 Example 2: layout and design

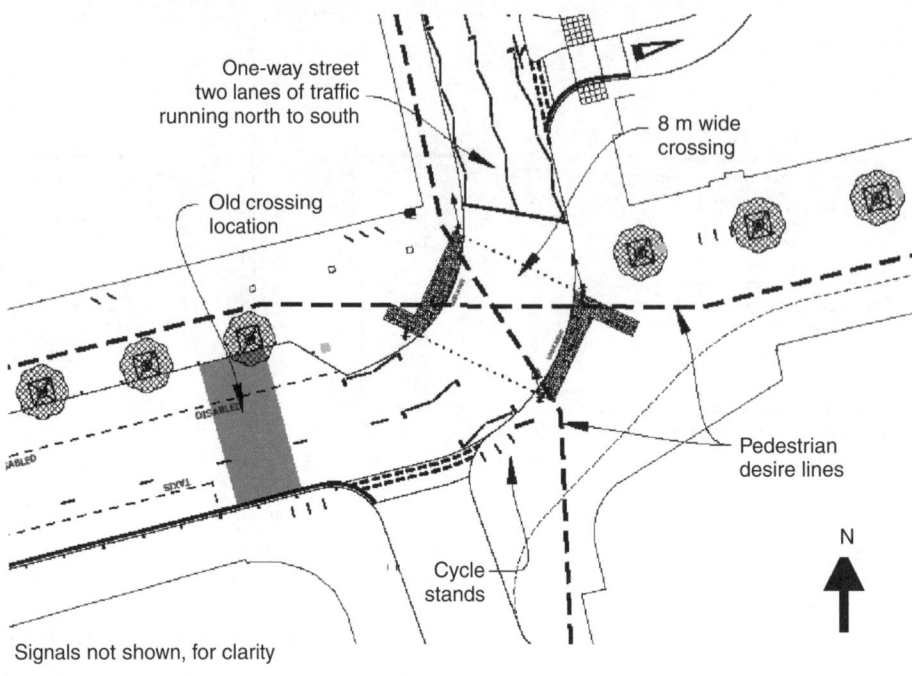

Signals not shown, for clarity

- Expansion of the current Shopmobility scheme, however, may be expected to generate more mobility scooter crossing, of the order of 1–2%.
- No significant land use changes, new junctions or new entrance ways are scheduled in current development plans.
- Streetscape and furniture items will not adversely affect intervisibility between pedestrians and vehicles.

15.3.4 Objectives and design responses
- Better serve the desire lines of pedestrians by relocating the crossing to the location shown in Figure 15.3, and widen the crossing.
- Provide a crossing that accommodates pedestrian and vehicular traffic flow and avoid frequent stops by vehicular traffic by considering the following.
  - A formal crossing would be required, owing to the vehicle and pedestrian volumes.
  - A zebra crossing would be unsuitable, owing to high volumes of pedestrians, who would require frequent stops for vehicular traffic.
  - A puffin crossing was selected in order to hold pedestrians during vehicle traffic flow and then allow pedestrians to cross in one go.
  - The crossing is made wide enough to accommodate high pedestrian flows with relatively short crossing times.
  - Accommodate pedestrian conflicts on crossing caused by intersecting desire lines.

15.3.5 Check on crossing dimensions and user requirements
The initially assumed cycle length is 68 s, as

Vehicle green time	42 s
Amber	3 s
Pedestrian green walking figure	9 s
Pedestrian clearance	14 s
Total cycle time	68 s

On this basis, the effective vehicle green time per hour is

$$3600 \times \frac{42}{3.60} \times 68 = 0.62 \text{ h}$$

The effective vehicle volume is

$$\frac{625}{2} \text{ lanes} \times 0.62 = 504 \text{ vehicle(h lane)}$$

This is an acceptable vehicle flow and should not cause significant delays or congestion.

The average number of pedestrians crossing per signal cycle (within the demand flow) is 11.

The pedestrian crossing time is calculated as 9 s at 1.2 m/s, and a 3 s comfort time should be allowed. This is well within the green-figure and clearance times and therefore acceptable.

Observations of the number of pedestrians crossing the present crossing indicate that its width of 3.5 m sometimes causes noticeable conflicts and delays. The increase to 8 m should quite adequately accommodate the volume anticipated. The employment of a puffin crossing will accommodate slower pedestrians, owing to its ability to extend the pedestrian clearance time, if set to do so. As with all signal installations, final adjustments should be made on site, along with appropriate monitoring of the scheme.

15.3.6 Justification and proposed design details

- A more traditional approach of pedestrian crossing design, which minimises crossing width, would have required two smaller crossings either side of the bend. This was ruled out as the crossing locations would not be on the pedestrian desire line and would compromise vehicle movements.
- Strategically located street furniture will be installed to guide pedestrians to the crossing points.
- Safety will be increased, as pedestrians have a crossing facility that caters for their desired crossing movements, thereby minimising random crossing.
- One crossing requiring one set of lights reduces the visual impact on the streetscape.
- The distance of the stop line from the crossing line is 2 m at the west kerb line of the crossing and 4.5 m from the east kerb line of the junction. Guidance from the *Traffic Signs Manual* (DfT, 2003, Chapter 5) notes that the set-back from the crossing should typically be between 1.7 m and 3 m. However, in this instance, in agreement with the local authorities it was increased to 4.5 m at the eastern kerb line.
- Typically, red coloured tactile paving would be used at a controlled crossing to provide a contrast with the surrounding paving. However, in this instance, to conform to the authority's streetscape design manual, granite tactile paving was used.
- The tactile paving at the crossing was designed in accordance with *Guidance on the Use of Tactile Paving Surfaces* (DETR, 1998). While the tactile stem should typically extend to the back of the footway or building line, in this instance it was felt that this would be excessive and, in agreement with the local highway authority, the north side of the junction was restricted to a 2.8 m stem and the southern side to a 5.2 m stem.
- The depth of tactile paving from the dropped kerb, where the carriageway meets the footpath, is 1.2 m. Where the direction of pedestrian flow is directly in line with the crossing (as opposed to having to turn to use the crossing), a tactile depth of 1.2 m from the dropped kerb is advised by *Guidance on the Use of Tactile Paving Surfaces* (DETR, 1998).

15.4. *Example 3.* Replacement of staggered signalised crossing with straight-across crossing

15.4.1 Overview

Two staggered, signalled crossings accommodated the high pedestrian volumes crossing the two arms of Upper Thames Street, as shown in Figure 15.4. Upper Thames Street provides the major link between East and West London and the Docklands, and is a part of Transport for London's improvement projects (TfL, 2014).

Because many pedestrians were crossing the second half of the street behind the guardrail around the central reservation, a large number of pedestrian–vehicle collisions occurred. To

Figure 15.4 Example 3: original layout showing staggered crossings (simplified; not to scale) (TfL, 2014)

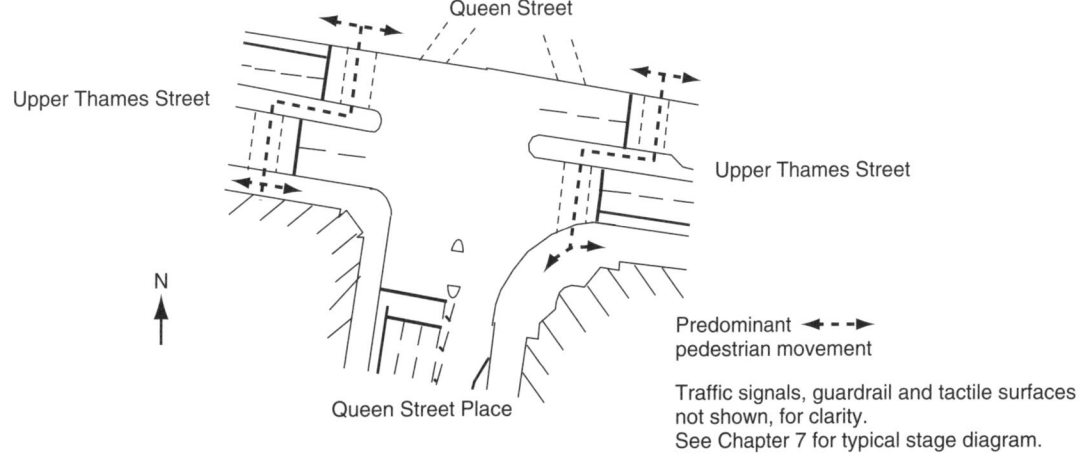

Table 15.3 Example 3: summary of existing conditions

Item	Physical and operational conditions, and comments
Vehicle speed	30 miles per hour limit; 85th percentile, 27.6 miles per hour
Vehicle traffic flow, peak hour	1260 vehicles per hour in afternoon peak hour, approximately evenly distributed between two eastbound and two westbound lanes
Pedestrian traffic flow, peak hour	1800 pedestrians per hour (900 per crossing), approximately evenly distributed between north and southbound in morning and afternoon peak hours; approximately 1% wheelchair users, 3% using other mobility aids and 1% using mobility scooters
Accidents	Above average number of accidents and numerous potential pedestrian–vehicle collisions, based on site observations
Signal details	Toucan pedestrian-activated signals on each half of the staggered crossings
Crossing width (each crossing)	3.5 m
Visibility	Acceptable, based on LTN 2/95 (DfT, 1995a) and TD 50/04 (DfT, 2004a)
Nearby junctions	Junction with Queen Street Place – see Figure 15.4; a relatively low volume of pedestrians cross this mouth of this arm of the junction; this is an uncontrolled crossing and pedestrians cross with the vehicle traffic signals
Other crossings nearby	Informal crossings at mouth of Queen Street Place
Entrances within 100 m	Queen Street has no vehicular connection to this junction, except for emergency vehicles; however, a significant number of cyclists cross Upper Thames Street at the subject crossings to and from Queen Street
Disabled facilities	Dropped kerbs and tactile paving in accordance with applicable standards at the crossing (DETR, 1998)

reduce the number of accidents and to make pedestrian crossings more attractive for walking (a policy of Transport for London), the staggered crossings were redesigned and implemented as 'straight-across' configurations. Key physical and operational features of the site and its immediate environs are summarised in Table 15.3.

15.4.2 Revised situation

The signalisation of the staggered crossings was redesigned for the 'straight-across' geometric layout shown in Figure 15.5. The physical features are

- straight-across crossings of Upper Thames Street arms of junction
- guardrail retained along central median
- crossing width between studs, 3.5 m (unchanged)
- dropped kerbs relocated
- no scope for pedestrians to short-cut the system.

15.4.3 Operational features

A selection of key features of the revised scheme is summarised in Table 15.4.

15.4.4 Summary of results

- There is a net decrease of over 25% in person travel time at this junction. This saving may also depend on other

junctions along the east–west corridor yet should contribute significantly to pedestrian amenity.
- There are increased delays to vehicular traffic (cycle time of junction increased to help counter delays).
- A longer time is required for the pedestrian stage.
- The waiting time for pedestrians to cross the road is reduced, since there is only one crossing to wait for.
- Pedestrians still cross against the green figure but in far less volume.
- Bicycle traffic from Queen Street runs at the same time as the pedestrian stage.

The scheme appears to function satisfactorily, although some increased local delay to vehicle traffic occurs, the impacts along the entire east–west corridor appear to be minimal. Monitoring of the scheme's performance is continuing.

15.5. *Example 4.* Context report for non-motorised user audit of preliminary design stage

This example is an excerpt from Annex B of HD 42/05 (DfT, 2005b). It provides a hypothetical example of an NMU context report, which could result from a preliminary design stage non-motorised user audit. Major features of this example are shown below.

Figure 15.5 Example 3: revised layout showing straight-across crossings (simplified, not to scale) (TfL, 2014)

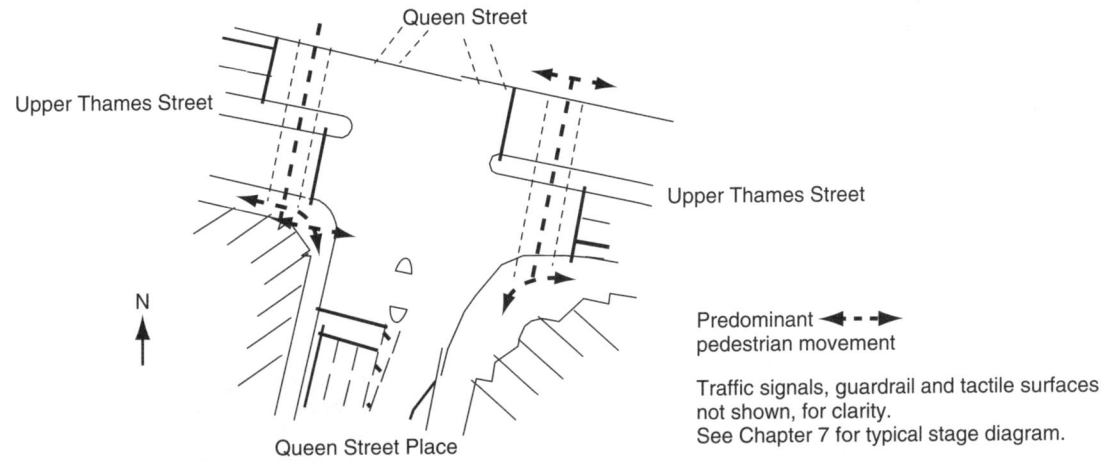

Queen Street

Upper Thames Street

Upper Thames Street

N

Predominant ◄- -► pedestrian movement

Traffic signals, guardrail and tactile surfaces not shown, for clarity.
See Chapter 7 for typical stage diagram.

Queen Street Place

Table 15.4 Example 3: Crossing operation analysis

Performance indicator (peak hour)	Original situation	Revised situation	Comments
Signal cycle time: s	88	96	Increased to cover longer pedestrian stage time
Green-figure time: s	6	9	Increased to improve pedestrian volume and allow for greater crossing distance
Average pedestrian crossing time (including waiting for green figure): s	$\left(\dfrac{(88-6)}{2}\right) + (88-6) + \left(\dfrac{7}{1.2}\right) \times 2$ $= 41 + 82 + 12 = 135$	$\left(\dfrac{(96-9)}{2}\right) + \left(\dfrac{18}{1.2}\right) = 58.5$	Originally waiting at two crossing points
Average green-figure time per hour: s	$\dfrac{6 \times 3600}{88} = 245.5$	$\dfrac{9 \times 3600}{96} = 337.5$	Average green-figure time given to pedestrians in the revised situation is used to cross the whole road width
Average number of cycles per hour	$\dfrac{3600}{88} = 41$	$\dfrac{3600}{96} = 38$	Reduced, as longer period
Average vehicle green time/cycle ratio	$\dfrac{65}{88} = 0.738$	$\dfrac{50}{96} = 0.521$	Vehicle traffic has increased delay because of the long intergreen period required following the pedestrian phase
Delay analysis for peak hour: Pedestrian: h Vehicle: h Average vehicle occupancy Total delay: h	34 8 1.4 $34 + 8 \times 1.4 = 45$	15 13 1.4 $14 + 13 \times 1.4 = 32$	Although vehicular delays increase, the savings in pedestrians' time in crossing more than offsets this, resulting in a net reduction of more than 25% in person travel time in the peak hour
Pedestrian–vehicle collisions: number/year	Unknown	Unknown	Too early for reliable figures

15.5.1   Scheme description – key points

- The proposed scheme is shown in Figure 15.6.
- A new dual carriageway, running from Newbridge, to the north of Anytown, to Oldcross Roundabout, bypassing the existing A999, running through the centre of Anytown. The dual carriageway will be approximately 1.75 km long and will pass 0.1 km east from Anytown at the closest point. A new roundabout will be constructed at Newbridge.
- The village of Eastfield, with its business park, lies approximately 1 km to the northeast of the proposed alignment of the scheme. The scheme will cross the Eastfield Road northeast of Anytown.

Figure 15.6 Example 4: schematic map showing proposed scheme and contextual information (DfT, 2005b)
© Crown Copyright, 2005
This information is licensed under the Open Government Licence v3.0. To view this licence, visit http://www.nationalarchives.gov.uk/doc/open-government-licence/ **OGL**

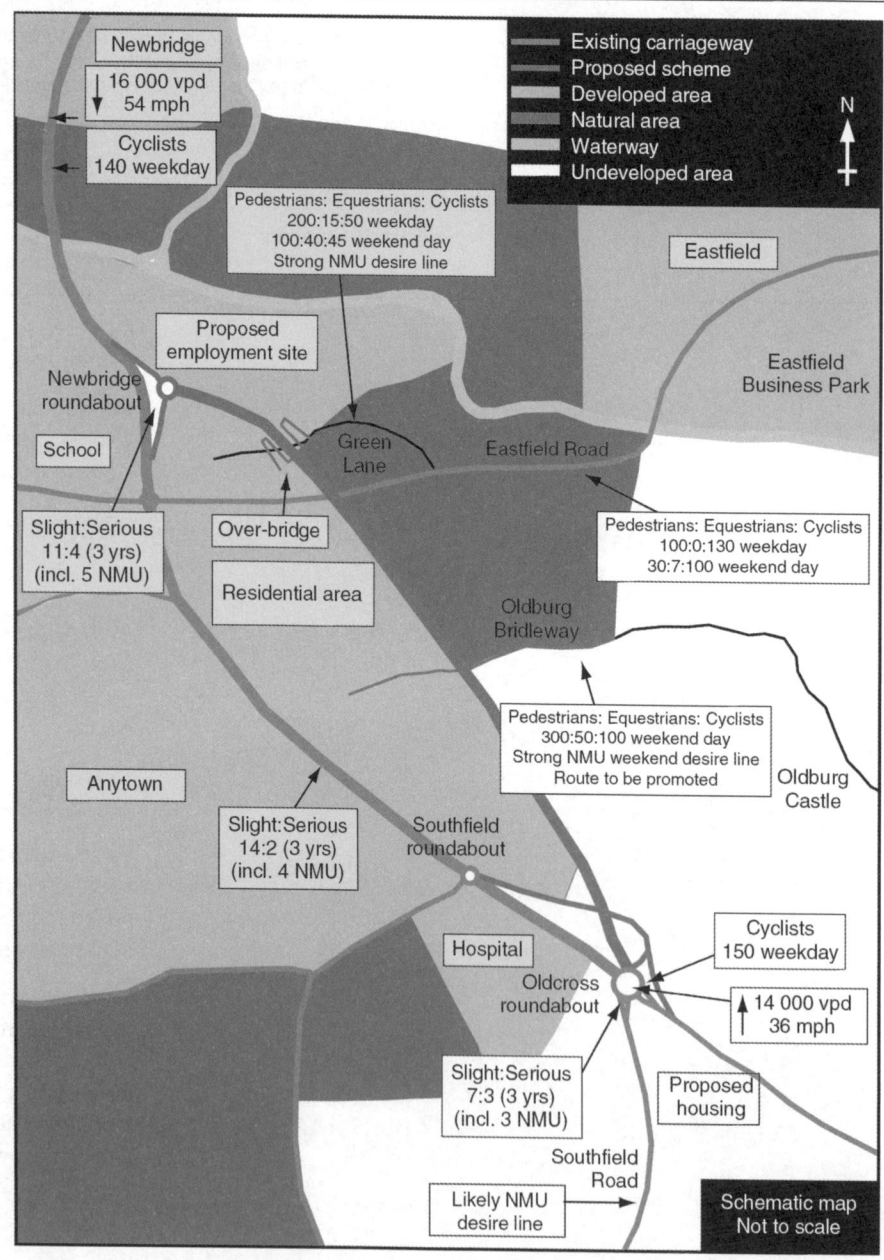

- Oldburg Castle, a popular tourist attraction, lies 1.5 km east of Anytown. It is linked to Anytown by a bridleway.
- An environmental impact assessment has been completed; figures for vehicle and NMU flows are taken from this source.

This report results from a preliminary design stage non-motorised user audit carried out on the A999 Anytown bypass. The audit was carried out by the design team from May to August 2005, in accordance with HD 42/05 (DfT, 2005b).

An NMU context report was prepared in accordance with HD 42/05 (DfT, 2005b) by the design team in July 2004.

The design team comprised: P Smith (NMU audit leader) BSc, CEng, MICE; R Jones (design team leader) BSc, CEng, MICE; M Lewis MSc.

The audit consisted of the following.

(a) An examination of the context report prepared at the feasibility stage. It was considered that this was still valid; no material changes had taken place since the compilation of the report.

(b) A continuous assessment of designs against the needs of NMUs.

(c) Meetings with the Anyshire County Council Access Officer, the Anytown Cyclists Touring Club, the Anytown Access Group and the Anytown Horse Club. Correspondence was also received from the Anyshire County Council 'Safe Routes to School' Officer and the Anytown Pedestrian Society.

(d) M Lewis, a principal engineer and member of the design team, visited the scheme location on three occasions between 1st and 30th June. Inspections were carried out during the hours of darkness and in wet conditions, and on cycle as well as on foot.

15.5.2 Objectives and design features

Seven key objectives were agreed for NMUs in the A999 Anytown Bypass scheme. These, and the design features that were incorporated to satisfy them, were included in the preliminary design.

The items raised in the report are summarised in Box 15.1; Figure 15.7 shows a scheme layout plan with references to the locations of the issues identified in the report.

Box 15.1 Example 4: items raised in the audit and the actions taken (DfT, 2005b)

Items raised in this audit

3.1 Issue. It was not clear how equestrians would get access to the proposed equestrian route near the Southfield Roundabout.

Action taken. A break in fencing and a pathway has been included in the design to provide an access route.

3.2 Issue. Pedestrians and cyclists may have been tempted to cross the dual carriageway near Oldcross Roundabout to access the route that runs along the bypass, rather than use the subway, because it is a shorter route.

Action taken. The subway has been aligned as closely as possible to the pedestrian desire line in order to reduce the advantage to pedestrians of circumventing it. A central reserve barrier has been specified and a pedestrian prohibition will need to be made.

3.3 Issue. Access to the Green Lane overbridge for northbound NMUs from the pedestrian/cycle route was considered circuitous.

Action taken. A link for northbound NMUs via the service road to the bridge ramps has been added to the scheme.

3.4 Issue. There were no footway/cycleway links to Anytown on the south side of the proposed Newbridge Roundabout. Some pedestrians and cyclists were considered likely to cross the bypass and use the verge/road rather than the longer route provided across the other three arms of the roundabout.

Action taken. A shared footway/cycleway has been added to the scheme on the south side of the roundabout. Features to segregate users on this route should be specified at detailed design stage.

3.5 Issue. The width of the equestrian route at chainage 2300 was initially shown as 2.7 m but should be a minimum of 3 m.

Action taken. The route has been amended to allow 3 m to be maintained along the entire route.

3.6 Issue. The gradients for the crossing at the Green Lane overbridge are approaching the limit of acceptability for wheelchair users.

Action taken. A level rest point has been added to the ramp.

3.7 Issue. The initial design of the subway would lead to concerns over personal security due to its alignment, which could lead to NMUs crossing the dual carriageway at grade.

Action taken. The approaches and location of the subway have been designed to allow good through visibility. A high-quality lighting system has been allowed for. Anti-vandal surfaces should be specified at detailed design stage. The design should provide for adequate drainage of the subway to minimise flooding.

3.8 Issue. Pedestrians from the houses adjacent to the bypass may be tempted to cross the dual carriageway between the Green Lane bridge and subway, where they would be in conflict with fast-moving vehicles.

Action taken. The two crossing points have been selected to satisfy the key desire lines identified by user groups.

3.9 Issue. Equestrians using the route between chainages 2200–2600 and 2800–3200 will not be visible to other road users. This leads to concerns over personal security.

Action taken. Anytown Horse Club was consulted about this and expressed no concern; therefore no action was taken.

3.10 Issue. The possibility of NMUs not being segregated on the Oldburg Bridleway bridge led to concern among users about potential conflicts. The bridge was not sufficiently wide for all users.

Action taken. The bridge has been increased in width by 0.75 m in order to allow pedestrians to be segregated from cyclists and equestrians by a kerb level difference while still providing a route to standard width for all users.

3.11 Issue. The Southfield, Oldcross and proposed Newbridge Roundabouts continue to represent a source of hazard to NMUs.

Action taken. The Newbridge and Oldcross Roundabouts are to be signalised. The geometry of the Southfield Roundabout will be changed to increase vehicle deflection and decrease flare on entry arms. Speed reduction features will be included on all approach arms to all three roundabouts.

3.12 Issue. NMUs and other traffic using Eastfield Road will be in conflict with motorised traffic using the bypass since the crossing is at-grade.

Action taken. A signalised junction is to be provided. A recommendation has been made to Anyshire County Council that Eastfield Road be subject to a 40 miles per hour speed limit (it is currently derestricted).

15.6. *Example 5*. St Neots North pedestrian/cycle river crossing (Skanska, 2017)

This example illustrates a scheme to improve access for pedestrians and cyclists between Eaton Ford and St Neots across the River Ouse. Considerations included provision of a crossing by widening an existing road bridge or providing a new bridge exclusively for pedestrians and cyclists.

For the existing road bridge, it was determined that the kerbs could not be moved out because of the effect this would have on the structural integrity of the bridge, thus limiting the available space for new lanes. It was decided, therefore, to proceed with investigating the options for a new bridge; key elements of the options and the geometric characteristics are summarised in Figure 15.8.

The options considered are

- *Location 1* – a new bridge landing in the field at the end of St Anselm Place, north of St Neots Rowing Club

- *Location 2* – a new bridge landing to the south of the Priory Centre

- *Location 3* – a new structure to be cantilevered from the north side of St Neots River Bridge

- *Location 4* – a new bridge to the south of the St Neots River Bridge, landing in line with River Terrace and Market Place.

(Skanska, 2017)

15.6.1 Design criteria – bridge geometry

There are various geometrical requirements for foot/cycle bridges in current design standards.

- In terms of the required width, a number of published documents give recommendations. Among these are LTN 1/12 [(DfT, 2012)], LTN 2/08 [(DfT, 2008)], BD 29/04 [(DfT, 2004b)], TA 90/05 [(DfT, 2005c)], and more recently the draft version of the Sustrans *Design Manual* on bridges and other structures [(Sustrans, 2015)]. Both the draft Sustrans Design Manual and

Figure 15.7 Example 4: layout plan with reference to issues/actions taken. Numerical references refer to numbering in Section 3 of report HD 42/05 (DfT, 2005b)
© Crown Copyright, 2005
This information is licensed under the Open Government Licence v3.0. To view this licence, visit http://www.nationalarchives.gov.uk/doc/open-government-licence/ **OGL**

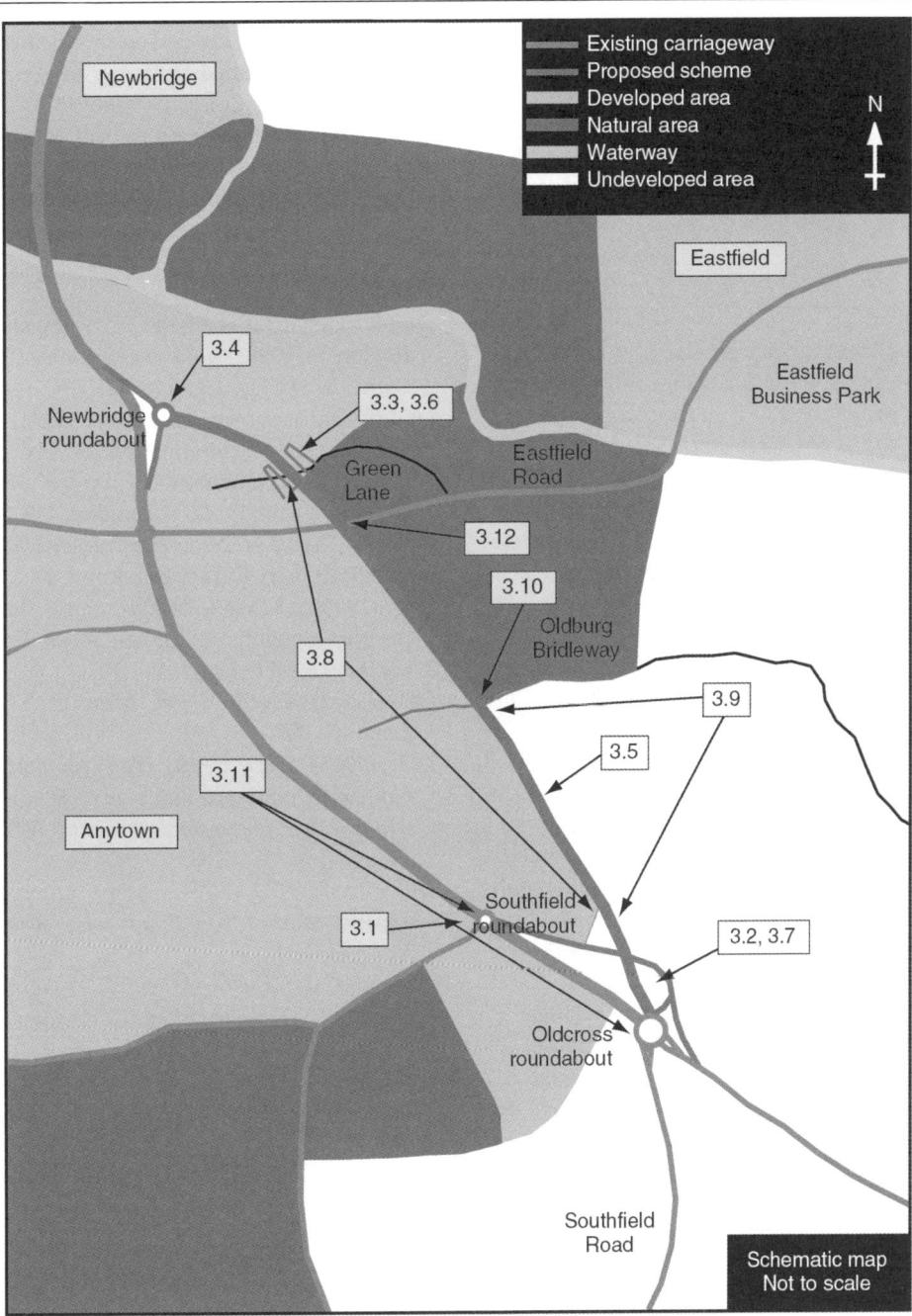

LTN 1/12 are found to be more onerous than the requirements given in BD 29/04. LTN 1/12 states that the preferred minimum effective width for an unsegregated cycle track is 3.0 m. Additional 500 mm horizontal clearances to vertical features above 600 mm high, such as parapets, are also required. This leads to an overall minimum clear width of 4.0 m for an unsegregated foot/cycle path on the bridge. The

Figure 15.8 Example 5: potential location of new pedestrian/cycle bridge (Locations 1–4) (Skanska, 2017)

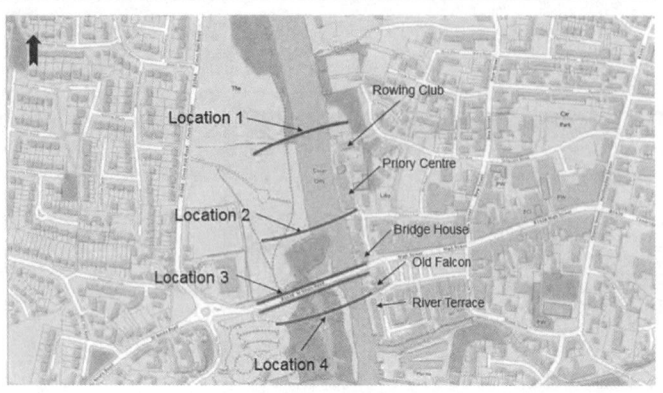

Sustrans *Design Manual* gives a similar recommendation, particularly for main routes, although it does permit a relaxation to 3.5 m for secondary routes or where 4 m is not achievable. Segregation would increase the minimum width further.

■ Generally, gradients on any footbridge ramp should be no steeper than 1 in 20. In addition, 2 m long landings with a maximum rise of 2.5 m between them shall be provided on any straight or spiral ramps. In some cases, it can be beneficial to relax the gradient to 1 in 21. The length of inclined ramp is increased but the omission of landings can make its overall length shorter and the line of the elevation can remain continuous rather than being broken by the landing.

■ It is possible to increase the slope to a maximum of 1 in 12 (8%) if additional landings are introduced, in accordance with BD 29/04, as long as the rise between landings does not exceed 650 mm. This is generally only done where space is limited and a longer ramp would not be feasible. The aforementioned Sustrans *Design Manual* recommends limiting the length where a gradient of up to 7% is used to a maximum of 30 m.

■ For spiral and curved ramps, the minimum inside radius of walkway, measured 900 mm from the edge of the walkway surface on the inside of the curve, shall be 5.5 m.

■ In addition to the approach ramps, stairs need to be provided at convenient locations if the ramped access leads to a significant detour for pedestrians.

(Skanska, 2017)

The report notes that there are numerous options and associated design requirements for segregation, but these can be considered at a later stage in the design process. Considerations need to be given to connections with existing or new paths. Headroom must be at least 2.4 m above paths or above any ramp returning upon itself – an important requirement.

15.7. Summary

The examples shown in this chapter have illustrated selected key points of actual and hypothetical pedestrian facilities, with an emphasis, where available or appropriate, on the numerical values underlying the layout and geometric features. In many instances, other factors besides the numerical will decide the nature of the features and even whether a new or improved installation is required. In many instances, also, the effectiveness of an installation is not subject to a uniform or rigorous evaluation. Continuing efforts will be needed to establish such an evaluation procedure in order to continually improve the safety, convenience and cost-effectiveness of the schemes proposed and, ultimately, encourage more walking.

REFERENCES

DETR (Department of the Environment, Transport and the Regions) (1998) *Guidance on the Use of Tactile Paving Surfaces*. DETR, London, UK.

DfT (Department for Transport) (1981) TA 23/81: Junctions and accesses: determination of size of roundabouts and major/minor junctions. In *Design Manual for Roads and Bridges (DMRB)*. DfT, London, UK.

DfT (1995a) LTN 2/95: *The Design of Pedestrians Crossings*. The Stationery Office, London, UK.

DfT (1995b) TD 42/95: Geometric design of major/minor priority junctions. In *Design Manual for Roads and Bridges (DMRB)*. DfT, London, UK.

DfT (2003) *Traffic Signs Manual*. The Stationery Office, London, UK.

DfT (2004a) TD 50/04: The geometric layout of signal-controlled junctions and signalised roundabouts. In *Design Manual for Roads and Bridges (DMRB)*. DfT, London, UK.

DfT (2004b) BD 29/04: Design criteria for footbridges. In *Design Manual for Roads and Bridges (DMRB)*. DfT, London, UK.

DfT (2005a) *Inclusive Mobility: A Guide to Best Practice on Access to Pedestrian and Transport Infrastructure*. DfT, London, UK.

DfT (2005b) HD 42/05: Non-motorised user audits. In *Design Manual for Roads and Bridges (DMRB)*. DfT, London, UK.

DfT (2005c) TA 90/05: Geometric design of pedestrian, cycle and equestrian routes. In *Design Manual for Roads and Bridges (DMRB)*. DfT, London, UK.

DfT (2007) *Manual for Streets*. Thomas Telford, London, UK.

DfT (2008) LTN 2/08: *Cycle Infrastructure Design*. The Stationery Office, London, UK.

DfT (2012) LTN 1/12: *Shared Use Routes for Pedestrians and Cyclists*. The Stationery Office, London, UK.

DfT (2018) *Design Manual for Roads and Bridges (DMRB)*. DfT, London, UK.

IHCSS (Institute of Highways and County Surveyors Society) (2005) *Traffic Calming Techniques*. IHCSS, London, UK.

Skanska (2017) *St Neots North Pedestrian/Cycle River Crossing: Feasibility Study Feasibility Report*. Skanska, Rickmansworth, UK.

Sustrans (2015) Bridges and other structures (draft). In *Sustrans Design Manual*. Sustrans, Bristol, UK, Chapter 8.

TfL (Transport for London) (2014) *Consultation on Proposed Safety Improvements on Upper Thames Street: Consultation Report*. Transport for London, London, UK.

Schoon, John G
ISBN 978-0-7277-6309-9
https://doi.org/10.1680/pfse.63099.243

Index